978-1-931504-27-0

The information presented in this book was developed by occupational hygiene professionals with backgrounds, training, and experience in occupational and environmental health and safety, working with information and conditions existing at the time of publication. The American Industrial Hygiene Association (AIHA), as publisher, and the authors have been diligent in ensuring that the materials and methods addressed in this book reflect prevailing occupational health and safety industrial hygiene practices. It is possible, however, that certain procedures discussed will require modification because of changing federal, state, and local regulations, or heretofore unknown developments in research. As the body of knowledge is expanded, improved solutions to workplace hazards will become available. Readers should consult a broad range of sources of information before developing workplace health and safety programs.

AIHA and the authors disclaim any liability, loss, or risk resulting directly from the use of the practices and/or theories discussed in this book. Moreover, it is the reader's responsibility to stay informed of any changing federal, state, or local regulations that might affect the material contained herein, and the policies adopted specifically in the reader's workplace.

Specific mention of manufacturers and products in this book does not represent an endorsement by AIHA.

ISBN 978-1-931504-27-0

AIHA Press
American Industrial Hygiene Association
2700 Prosperity Avenue, Suite 250
Fairfax, Virginia 22031
Tel.: (703) 849-8888 • Fax: (703) 207-3561
http://www.aiha.org • email: infonet@aiha.org

AIHA Stock No. EHWG01-440

CONTENTS

PREFACE

We are surrounded by chemicals. These vital substances make up the fabric of life, but, while chemicals, especially the synthetic vareity, provide great benefits to humans, they can also cause significant problems. Many of our chemical problems arise from careless handling and improper disposal that in turn creates situations where people are exposed to potentially harmful amounts of chemicals. Recognizing that these problems occur even under the best of circumstances, the Occupational Health and Safety Administration (OSHA) has promulgated regulations to reduce human exposures to hazardous chemicals. Although such exposures can occur anyplace, including the home, this book focuses on the federal regulations in 29 CFR 1910.120 that apply to hazardous waste sites.

The book arose out of the authors' experiences in working at hazardous waste sites and with hazardous chemicals. In conveying this experience to others, we have found that simply the recognition of a chemical hazard is probably the most difficult concept to teach. Many people are killed and injured every year because they are exposed to a chemical hazard without knowing it existed. Such recognition is difficult to teach because it is more of an attitude than a set of facts. In writing this book we have tried to instill the need to maintain an observant and cautious attitude when working with any chemical.

The first 14 chapters discuss the basics of chemical recognition, evaluation, and protection. However, the material presented is not intended to be an OSHA compliance guide, but is the practical application of the regulations. Only general information is presented; specific is a starting point for the recognition of and protection from chemical hazards; it is not a recipe for a specific course of action in a specific situation.

The last six chapters focus on the application of the basic principles. Preparing health and safety plans; handling and sampling chemical containers; entering confined spaces; excavating, trenching, and shoring; and transporting hazardous wastes are among the topics discussed.

Writing about chemical hazards is like writing about the weather; before you finish, the situation has changed. Chemical hazards change due to the production of new chemicals, updated means of detection, advances in protective clothing, and new regulations, just to mention a few. Therefore, this book should be used as a starting point for understanding and controlling chemical hazards and not as a definitive source of information.

ABOUT THE AUTHORS

Richard C. Barth, PhD, holds environmental degrees from Michigan State University, Colorado State University, and the University of Arizona. He has spent 26 years in the environmental and safety areas, both in the United States and in developing countries. Twelve of these years were spent as the Senior Research Ecologist for the Colorado School of Mines Research Institute. For the past 13 years Dr. Barth has been professor and program director of the Environmental and Safety Technology Program at Front Range Community College in Westminster, Colorado. In this capacity he has taught numerous OSHA courses and has been involved in a wide variety of health and safety issues at the local and regional level. Dr Barth has published four books and ten technical articles.

Patricia D. George, BS, MA, AAS, has been an educator for over 30 years, specializing in the earth sciences and environmental education and training. She has taught and trained for community colleges and universities as well as for private medical facilities and industry. She has 12 years of experience as a petroleum geologist supervising drilling. She has degrees in education/earth science, physical geography/photo interpretation, and environmental technology. She is coauthor of a book on training for hazardous waste sites workers.

Ronald H. Hill, MSPH, CIH, CSP earned a MSPH in Environmental Sciences and Engineering from the University of North Carolina at Chapel Hill, and a BS with Honors from North Carolina State University. He is a Certified Industrial Hygienist and Certified Safety Professional. His activities include Board of Directors of the American Industrial Hygiene Association (AIHA), President of the Rocky Mountain AIHA Section, and Vice President of the Colorado Industrial Hygiene Council. His publications include chemical protective clothing, passive dosimetry, urine mutagenicity, and chemical skin contamination. Work experience includes manufacturing, corporate, academic, public health, and consulting positions; and he has implemented occupational health programs at petroleum, petrochemical, synthetic fuels, and biotechnology facilities. Teaching experience includes conducting hazardous waste operations courses in the U.S. and Mexico, and lecturing at several state universities. He has worked as an expert consultant for litigation involving hazardous wastes, noise, asbestos, and chemically induced lung disease. Over the last ten years, he has been involved in the writing and implementation of over 50 Health and Safety Plans for hazardous waste sites.

ONE

Brief History of Health and Safety Law

1 | CONTENTS

Over 2,400 years ago, the Greeks became aware of the harmful effects of an unhealthy workplace. Two thousand years later a German physician documented the adverse clinical effects of various toxic substances. In 1713, Bernardo Ramazzini, sometimes referred to as the father of occupational health and safety, wrote *Diseases of Workers,* documenting the dangers of a less-than-healthful work environment. Occupational cancer was diagnosed in 1775 in English chimney sweeps. However, it was not until the 19th century that positive steps were taken to ensure worker health and safety, and not until 1970 that comprehensive health and safety measures for American workers were passed into law. A brief history of events leading to passage of the U.S. Occupational Safety and Health Act (OSH Act) is presented below.

1.1 Early Provisions

Prior to and during the Industrial Revolution, the law did not try to prevent workplace accidents and exposures but did attempt to relieve the financial hardship of industrial accidents on individual workers and their families. This concept of employer responsibility for the welfare of its employees continued a centuries-old principle of English law. Such law required employers to provide a safe work environment, to employ workers of known skill to decrease the risk of injury to themselves and others, to furnish safe materials, and to avoid exposing workers to extraordinary risks that could not be anticipated.

Initially, worker safety laws in the United States mimicked English common law. However, American courts added that in a free market, a worker assumed all the ordinary risks of employment by entering into a contract with an employer. This brought credence to the idea that if workers found a dangerous situation in the workplace that was required as part of the work process, they had the freedom to decide for themselves whether or not they wanted to work there. The facts that jobs might be scarce, that most workers were coming into cities from farms with little or no training or knowledge of industrial procedures, or that wages were tempting enough to lure people to positions regardless of the risks was ignored by the courts.

Gradually the regulation of workplace conditions to protect the health and safety of workers was put into place despite protests from employers. However, while England had many existing government institutions to deal with economics and trade, the United States had to build its own regulatory agencies. Considering the close connections between the two nations, it is not surprising that American safety legislation tended to mirror existing English statutes.

1.2 The Age of Industrialization

Initially, there was little worker safety legislation at either the state or federal level during the 19th and early 20th centuries. Serious attempts to improve workers' health and safety conditions were questionable. The following is a brief chronology of actions, orders, regulations and reforms that were put in place through the first quarter of the 20th century.

1837 and 1841	First Congressional reports on occupational health hazards;
1840	Executive order restricting naval shipyard workers to a 10-hour day;
1865	First bill submitted to create the safety-oriented Federal Mining Bureau;
1869	Massachusetts established the first State Bureau of Labor Statistics (BLS);
1881	American Public Health Association founded;
1888	First Federal Bureau of Labor established;
by 1890	21 states had passed occupational safety and health laws;
1893	Safety equipment specified for railroad cars and engines;
1902	Public Health Service was established; Congress passed embryonic safety act that regulated the sale and control of viruses, serums, and toxins;
1911	The United Society of Causality Inspectors (now the American Society of Safety Engineers) was founded;
1912	First important American occupational health legislation passed—The Esch Act—that placed a very large tax on the sale of white phosphorous matches, nearly eliminating a facial disease, phosphonecrosis, among match factory workers;
by 1921	46 States had some form of individual health law and workers' compensation.

State legislative concern over health and safety intensified during the period of 1890–1910 with the passage of laws that now seem archaic or simply commonsense. One law mandated meal breaks for any woman or young person working 16 hours or more. Five states passed legislation specifying that the place where workers ate be separated from noxious fumes or poisonous substances. Other legislation on a state level required shirt factories to sprinkle water on the floors to keep down cotton dust and set up industrial safety boards to propose occupational safety standards.

Such regulations did little to stem the tide of industrial accidents. For example, in 1906, one Pennsylvania county recorded "45 one-legged men, 100 hopeless cripples with crutches or canes, 45 men with a twisted, useless arm, and 30 men with an empty sleeve" from industrial accidents. Additionally, 500 men in the county died that year in work-related accidents.

In 1911, the Triangle Shirtwaist Company, housed in a 10-story building in New York City, caught fire and burned. A total of 146 employees, mostly young women, burned, suffocated, or jumped to their deaths. Most deaths were due to locked doors, inadequate fire escapes, poor ventilation, and overcrowding. Unfortunately, this tragedy had little positive impact on promoting safer working conditions.

1.3 Into the 20th Century

During the first half of the 20th century, industry in the U.S. flourished. Companies adopted mass production methods that made it possible to lower unit costs, while mass distribution helped guarantee larger markets. In this transition period, many types of industries and crafts were eliminated, while new ones, such as chemicals and assembly lines, became more important to the economy.

This rapid change in the nature of production had a direct and adverse effect on worker health and safety. Mechanization led to an increase in the pace of work.

Machines continued to replace manpower at an accelerating rate in U.S. industry. In the 60 years after 1859, the output of goods increased 33 times with only a seven times increase in manpower. This, coupled with the lack of worker health and safety training and safety controls in the working environment, contributed to a sharp increase in industrial accidents.

While accidents were the most visible threat on the job, work-related diseases were also taking their toll. Stonecutters died of silicosis. Tobacco industry workers suffered heart and respiratory illnesses. Hat makers contracted nervous system disorders from the mercury used to treat felt and fur. Women who painted watch and clock faces with radium shaped the delicate brushes by wetting them on their tongues and succumbed to a high incidence of radiation poisoning. Coal miners suffered from black lung disease, while bakers suffered the equivalent of white or gray lung from breathing flour.

Throughout the first half of this century, little attention was paid to the problem of occupational disease, and little protection was offered to workers who might be at risk. Some reforms were instituted, such as requiring employers to pay compensation for on-the-job injuries (workers' compensation). There was a backlash from this, however. Employers found the cost of compensation (dollar payout for a leg lost, an eye lost, etc.) so low that interest in spending money and time on safety education was quickly lost. Workers who protested against harsh working conditions were sometimes violently attacked or killed by thugs hired by industrialists.

A notable pioneer in American occupational health emerged during the early 20th century. Dr. Alice Hamilton conducted industrial health surveys in mines, lead works, mercury works, and munitions plants, publishing extensively on her findings. She was posthumously recognized in 1995 on a U.S. postage stamp for her work as a social reformer.

1.4 The Depression

The Depression was an economic setback to occupational health and safety concerns. Employers were unable to afford safe workplaces, and workers had little option but to take the positions they were able to find. In a report presented at the 1936 United Auto Workers Convention, it was noted that there had been 13,000 cases of lead poisoning in auto factories since 1929. A plant worker's description of 1932 conditions indicated there was no attempt to ventilate the work area or to remove pollutants from the air. It was an accepted fact that thousands of metal finishers suffered from lead poisoning. Other dangerous trades such as manufacturing, mining, and construction had similar unsafe conditions.

In the midst of deep economic crisis, workers in the Depression were willing to face many known and unknown dangers in order to receive paychecks. At the same time, industry felt little pressure to create hazard-free work environments. A monumental tragedy occurred in 1930 and 1931, as a water diversion tunnel was being drilled through a mountain at Gauley Bridge, West Virginia. Workers in the tunnel inhaled airborne dust containing a high concentration of silica and soon complained of shortness of breath. According to testimony at a subsequent hearing, the employers knew that the rock had high silica content. Over a two-year period, some 2,000 workers were employed; 476 eventually died and 1,500 were disabled from the occupational lung disease called silicosis. In 1936, a subcommittee of the U.S. House of Representatives held hearings on the Gauley Bridge Disaster and concluded that "the whole driving of the tunnel was begun, continued, and completed with grave and inhuman disregard of all consideration for the health, lives, and future of the employees" and that "such negligence was either willful or the result of inexcusable and indefensible ignorance." It determined that "silicosis is one of the greatest menaces among occupational diseases and that state laws governing prevention and compensation are totally inadequate." Nevertheless, a motion by the subcommittee for the Congress to fund a study of silicosis was turned down. Punitive action was not taken against any of the companies or contractors and the hearings and publicity had little effect on curtailing silicosis.

By the late 1930s, the effects of President Franklin D. Roosevelt's "New Deal" laws were being felt in the workplace. In 1938, the Fair Labor Standards Act implemented the first effective national child labor regulations. Opponents of previous bills claimed that a great burden would be placed on industry if child labor was eliminated. Many families were also against these regulations since they eliminated desperately needed income. However, the law passed and provided a major strand in today's web of nationwide worker protection laws.

1.5 World War II

During WW II a new wave of industrial accidents occurred with increased demands on production for war-related materials and the employment of millions of workers new to industry (primarily women). During the first three years of the war, more Americans were killed and injured in work-related accidents than on the battlefield! This finally became unacceptable when the country found itself short of available labor. Through the Division of Labor Standards, the federal government took a more active role in plant inspections, training, education, and industrial safety. This by no means meant that workers were protected from harm. Ship builders, for example, still sprayed asbestos into the hulls of Liberty Ships without benefit of respiratory protection.

Following the war, reaction against the New Deal labor laws and the return of male workers to the labor force resulted in the effective dismantling of many existing programs. Americans were once again left with little protection for occupational health and safety.

1.6 The New Era of Industry

Some stop-gap safety measures were implemented during the postwar boom. The Labor Management Relations Act was passed in 1947 providing workers with a provision to walk off the job if the work was "abnormally dangerous." In 1948, President Truman tried to decrease industrial accidents by calling the first Presidential Conference on Industrial Safety, but little was accomplished for the workforce as a whole. In the area of mine safety, some action took place following a disaster in 1951 when 119 miners were killed in an accident in Illinois. This led to the passage of the Coal Mine Safety Act in 1952.

In addition to a higher technological level of mechanized production, post-World War II industry saw an enormous increase in the use of new petrochemical-based materials in both industrial and consumer products. Prior to World War II, U.S. production of synthetic organic chemicals totaled less than one billion pounds per year. By 1976 production was nearly 163 billion pounds per year. In 1958 there were 17,000 commercially available synthetic compounds. By the mid-1980s this number had increased to over 90,000 with several thousand new materials coming onto the market yearly.

This proliferation of chemicals and chemical-based products such as polyesters, nylons, detergents, plastics, lubricants, pesticides, and herbicides introduced toxic chemicals into nearly every aspect of life in the United States. Most of these new synthetic chemicals and their byproducts were unregulated and untested for environmental and health impacts. At the same time nitrates, phosphates, toxic residues, smog, carcinogenic exhaust, as well as mounds of undegradable plastic containers, wrappings, and other materials from the petrochemical industry were added to the workplace and the environment. Additionally, new information technologies were created that globalized production by increasing communication and reducing costs. This electronics revolution created yet another set of occupational health hazards.

1.7 The Health and Safety Crisis of the 1960s

In the 1960s, industry-specific safety statutes were still being enacted, such as those that protected employees who conducted maintenance service for federal agency contractors. In 1965, the National Foundation of the Arts and Humanities Act made receipt of federal grants conditional on providing safe and healthful working conditions for performers, laborers, and mechanics. Another step in job and health safety was the passage by Congress of the Metal and Nonmetallic Mine Safety Act of 1966.

These feeble efforts failed to protect the majority of workers. Health and safety issues were limited to a few unions representing workers in industries where hazardous work was common such as auto, mining, construction, and electrical work. Wide-reaching state and federal laws were nonexistent or unenforceable.

By the late 1960s, evidence of both a safety and health crisis could not be ignored. From 1961 to 1970, the incidence of industrial accidents rose by 29 percent. Though difficult to verify statistically, occupational illness emerged as an increasingly serious problem. An environment and health-conscious public was responding to reports of asbestos-associated lung disease, cancers associated with coke oven emissions, and mining's black lung disease. Adding urgency to the public's concern and outrage were a 1968 West Virginia mine explosion, which killed 78 people, and the 1968 Santa Barbara oil spill, which covered 13 miles of ocean beach. Owing to media coverage, both events touched the public nerve and added momentum to the demand for change. In addition, affected individuals began to demand compensation. In 1969 coal miners marched on the West Virginia capital. This action helped to promote the passage of the Coal Mine Health and Safety Act of 1969.

Many industries were familiar with existing government health and safety regulations, although inspection and enforcement had not been widespread. Long dissatisfied with state safety codes and enforcement, organized labor began to bring pressure on industry under the Walsh-Healey Public Controls Act. Passed in 1936, this legislation addressed minimum wage, work week hours, and jobsite safety regulations. The Act had been updated from time to time, with the last major revision in 1956. Since this was the most comprehensive law of its kind on the books, labor used it as often as possible. In 1960, realizing that the Act was greatly outdated, the U.S. Department of Labor (DOL) began to institute piecemeal inspections and guidelines, but lacked a comprehensive plan of enforcement. The years between 1963 and 1969 saw proposals for revisions, proposals for issuing new laws, and proposals for adopting Atomic Energy Commission and other organizational law in order to cover all sectors of the workforce. This development of a "cut-and-paste" regulation package was defeated by state, industry, public, and political opposition.

In 1965, a proposed OSH Act was introduced, but no Congressional action was taken. The increase in industrial accidents led to the introduction of another such bill in 1968, but again without positive results. In January 1968, President Johnson proposed the nation's first comprehensive occupational safety and health program. It never reached a vote.

1.8 The OSH Act

In January 1969, the OSH bill was introduced by Congressman James O'Hara. Senator Harrison Williams introduced the Senate version in May, and in August the Nixon Administration introduced its proposal. Congress was presented with statistics stating that each year 14,500 persons were killed in the workplace, 2.2 million were disabled on the job, and an estimated 390,000 incurred some sort of occupational disease. The impact to the economy was staggering. Over $1.5 billion was wasted annually in lost wages with an $8 billion loss each year to the nation's Gross National Product. Technological advances and new processes were being introduced at a faster rate than occupational health expertise could be developed. It was estimated that every 20 minutes a new and potentially toxic chemical was introduced into industry!

After more than a year of amendments, revisions, and party-line politics, the OSH Act of 1970 was signed into law on December 29, 1970, taking effect 120 days after the signing. As shown by Appendix A, the OSH Act is only part of many areas of law falling under the authority of the DOL, codified in section 29 of the Code of Federal Regulations (CFR). Likewise, occupational safety and health rules for hazardous waste sites in 29 CFR 1910.120 and 126 are only a part of the Occupational Safety and Health Administration's (OSHA's) body of regulations.

1.9 Implementation of the OSH Act

As this history indicates, the OSH Act was the result of numerous compromises among the stakeholders. Although it is one of the more controversial pieces of legislation ever enacted, Congress has not changed it in any substantial way. The OSH Act remains largely intact since its passage in 1970; however, a policy change occurred in 1990 when many of its penalties were increased sevenfold.

One result of the passage of the OSH Act was the significant advance in protecting the health and safety of workers in all fields, not just a few select industries. Yet the enactment of a law does not make a problem disappear. Serious problems still persist, new problems have been identified, and other concerns have been created as new technologies emerge. Some of the leading work-related diseases and injuries as reported in 1999 included:

- occupational lung disease (asbestosis, silicosis, etc.)
- musculoskeletal injuries (neck, back, arms, etc.)

- occupational cancers (leukemia, mesothelioma, bladder)
- amputations, fractures, eye loss, lacerations, and traumatic death
- cardiovascular disease (hypertension, coronary disease)
- reproductive disorders (infertility, spontaneous abortion, etc.)
- neurotoxic disorders
- noise-induced hearing loss (NIHL)
- dermatologic conditions (dermatoses, burns, etc.)
- psychological disorders (drug dependency, personality disorders, alcoholism)
- violence in the workplace

In industry, the risk of illness and death on the job is not evenly distributed. Certain industries are far more dangerous than others, with construction and farm work being among the most dangerous. Hazardous waste site work often involves construction-type activities, thus adding a greater chance for dangerous working conditions. For 1999 (the last year for available statistics), the BLS (Bureau of Labor Statistics) listed the incidence rate of occupational injuries, including those not covered by the OSH Act, and illness for all full-time workers as 6.3 per 100 full-time workers as compared to manufacturing at 9.2 cases per 100 full-time workers, construction at 8.6, agricultural production at 7.3, transportation and public utilities at 7.3, retail trade at 6.1, and mining at 4.4. Total time lost from work in 1999 due to work injuries was 17 million worker days, with a total cost to industry of $25.7 billion.

While new technologies eliminate some dangerous work, they often present new unanticipated dangers. Emerging technologies in manufacturing and office work are generating a set of conditions that are only beginning to be evaluated. Some areas of concern emerging from the explosion of computer and microelectronics-based industries are:

- chemical hazards (such as solvents);
- musculoskeletal problems and cumulative trauma disorders (such as carpal tunnel syndrome);
- keyboarding (such as arm, neck, and upper back muscular and nerve disorders);
- stress.

OSHA has been the target of more criticism than nearly any other federal regulatory agency. Critics, however, have not questioned the objective of the agency, which is to "assure so far as possible every working man and woman in the nation safe and healthful working conditions and to preserve human resources." Despite OSHA's efforts, every year nearly 6,000 Americans die from workplace injuries (5915 in 2000), an estimated 50,000 people die from illnesses caused by workplace chemical exposures, and 6 million people suffer nonfatal workplace injuries.

Some of the more publicized criticisms have focused on OSHA's confrontational character. Administrations during recent years have unsuccessfully tried to temper this approach, and to encourage a more cooperative business/government attitude to promoting workplace safety by striking a balance between the competing objectives of regulation and production. In response to criticism that inspections are few and far between, inspection policies and the number of inspections are being changed. Monetary fines are increasing in response to charges that low fines do little to improve working conditions.

1.10 Hazardous Waste Operations

Hazardous waste operations provide a unique set of chemical and physical hazards that can impact worker health and safety. OSHA has responded to this threat by promulgating regulations specifically for hazardous waste sites. The following examples show why such concern is valid.

Example 1: More than 50,000 drums of chemical waste were packed into a 13-acre disposal site near Seymour, Indiana, a community of 15,000. This dumping ground for nearly 400 companies included cyanide, arsenic, PCBs, toxic metals, solvents, naphthalene, and 200 pounds of explosive materials.

Example 2: The 37-acre, Publicker Industries site in Philadelphia County, Pennsylvania housed a liquor and industrial alcohol distillation process plant from 1912 until 1985. The facility was abandoned in 1986. It contained nearly 440 tanks, plus storage drums, product stock, warehouses, a power plant, and several hundred miles of above-ground process lines, some of which were covered with asbestos. One hundred eighty cylinders contained toxic, flammable, and reactive gases while electrical equipment contained PCBs. Two million gallons of hazardous materials were on-site when the facility was abandoned, with many vessels and transfer lines in disrepair. In 1987, a portion of the plant was destroyed in a multi-alarm fire with explosions and fire flares reported. Shallow groundwater was contaminated with toluene and xylene, as was the aquifer supplying drinking water to 185,000 people. Air monitoring found elevated levels of volatile organic compounds (VOCs).

Example 3: The five-acre Woodbury Chemical Company site in Colorado began operation in the 1950s as a pesticide production facility. The plant burned down in 1965. Debris and rubble, including contaminated soil and over 1,500 pounds of water-soaked pesticides, were moved to an adjacent lot. The plant was then rebuilt and continued its operations until 1971. A variety of pesticides and VOCs were produced or used on the site, which was surrounded by other industries. Nearly 3,000 people live within 0.5 miles of the site.

Approaching, evaluating, and working on such sites requires special regulations, equipment, and training. The object of this book is to present the standards of safety required during hazardous waste operations. The materials discussed in this manual are not all specific compliance standards, but include guidelines for recognizing and dealing with chemical and physical hazards during hazardous waste operations. Detailed site-specific hazards assessment according to 29 CFR 1910.120 must be performed for every working situation and for every hazardous waste and emergency response operation.

1.11 Sources of Information

Arbuckle, J. Gordon, et al: *Environmental Law Handbook.* Rockville, MD: Government Institutes, Inc., 1991. pp. 370–395.

Ashford, Nicholas A. and C.C. Caldart: *Technology, Law, and the Working Environment.* New York: Island Press, Van Nostrand Reinhold, 1995.

Bureau of Labor Statistics: *Annual Occupational Injuries and Illnesses.* Washington, DC: U.S. Government Printing Office, 2000.

Corn, Jacqualine K.: *Response to Occupational Health Hazards, A Historical Perspective.* New York: Industrial Health and Safety Service/Van Nostrand Reinhold, Wiley, 1992.

Hamilton, Alice, *Exploring the Dangerous Trades.* Boston, MA: Northeastern University Press, 1985. p. 433.

Miller, Ken: Workers Sacrificed for Poorly Run Firms. *Denver Post.* (12 September 1993).

National Safety Council: *Accident Facts, 1997 Edition.* Itasca, IL: National Safety Council, 1997.

"Occupational Safety and Health Act," Public Law 91–596, 84 Stat. 1590, 29 U.S. Code, Section 651 *et seq.* 1970, updated in 1999.

Rothstein, Mark A.: *Occupational Safety and Health Law.* 3rd ed. St. Paul, MN: West Publishing Co., 1990. pp. 1–43.

Section of Labor and Employment Law, ABA. Stephen A. Bobat and H.A. Thompson III, Editors-in-Chief. Washington, DC: The Bureau of National Affairs, Inc., 1988. pp. 3–56.

Viscusi, W. Kip: Reforming OSHA Regulation of Workplace Risks. In *Regulatory Reform, What Actually Happened.* Boston, MA: Little Brown & Co., 1986. pp. 234–268.

Worobec, Mary D. and C. Hogue: *Toxic Substances Controls Guide.* 2nd ed. Washington, DC: BNA Books, 1992.

APPENDIX

The Title 29 Labor Law is the only one of 50 titles involving federal regulations. These regulations are available at the Government Printing Offices, libraries or accessible on the World Wide Web.

Elements of 29 CFR—Labor Law

Title 29 of the Code of Federal Regulations, parts 0 to 2800, covers many areas of labor law in addition to 1910.120 (Occupational Health and Safety at Hazardous Waste Sites). For those individuals with questions relating to other areas of labor law/safety, the following outline may act as a basic reference. It includes all general titles under 29 CFR, both specific and general. Those sections and parts which appear to have been omitted or overlooked have been set aside pending the future inclusion of additional rules and regulations.

SUBTITLE A:

Parts 0–99 Office of the Secretary of Labor: Includes ethics and conduct of DOL employees; general regulations and procedures; public works contractors; labor standards and provisions for federal service contracts; any contracts covered by the Contract Work Hours and Safety Standards Act; general and specific rules for DOL employees and contracts with Federal money as to nondiscrimination, equal opportunity, arbitration, fraud, Privacy Acts, lobbying requirements, audit requirements, grants, etc.

SUBTITLE B:

Parts 100–199 National Labor Relations Board

Parts 200–299 Office of Labor-Management Relations and Cooperative Programs under the Department of Labor

Parts 300–399 National Railroad Adjustment Board

Parts 400–499 Office of Labor-Management Standards, Department of Labor

Parts 500–899 Wage and Hour Division of the Department of Labor: Includes Fair Labor Standards Act; migrant and seasonal agricultural worker protection/investigations/inspections/record keeping; employment of students, apprentices, messengers, workers with disabilities, seasonal workers; child labor; over-time; specific jobs such as seamen, agriculture, aquatic products, forestry, motor carriers, radio/television station employees and bulk petroleum distributors among others; Employee Polygraph Protection Act; Family and Medical Leave act; Garnishment and restriction.

Parts 900–999 Construction Industry Collective Bargaining Commission

Parts 1200–1299 National Mediation Board .

Parts 1400–1499 Federal Mediation and Conciliation Service

Parts 1600–1699 Equal Employment Opportunity Commission

Parts 1900–1999 Occupational Safety and Health Administration

Parts 1900–1909 Procedures for State agreements; enforcement of state plans; inspections, citations, and proposed penalties; recording and reporting occupational illnesses and injuries; Rules of Practice; consultation standards

Part 1910 Occupational Safety and Health Standards

Subpart A (1910.1–1910.8): purpose, definitions, issuance/amendment/repeal of standards, definition and requirements for a nationally recognized testing laboratory; OMB numbers

Subpart B (1910.11–1910.19) Adoption and Extension of Established Federal Standards: Includes construction work, shipyard employment, longshoring and marine terminals; changes in established Federal standards, special provisions for air contaminants.

Subpart C Reserved

Subpart D (1910.21–1910.30) Walking–Working Surfaces: Includes guarding floor and wall openings and holes; industrial stairs; ladders; scaffolding, and others

Subpart E (1910.35–1910.38) Means of Egress: Includes employee emergency plans and fire prevention plans.

Subpart F (1910.66–1910.68) Powered Platforms, Manlifts, and Vehicle-Mounted Work Platforms

Subpart G (1910.94–1910.98) Occupational Health and Environmental Control: Includes ventilation, occupational noise exposure, ionizing radiation, non-ionizing radiation.

Subpart H (1910.101–1910.126) Hazardous Materials: Includes specific compressed gases, flammable and combustible liquids, spray finishes, dip tanks, explosives and blasting agents, LPG, anhydrous ammonia; process safety management of highly hazardous chemicals; hazardous waste operations and emergency response.

Subpart I (1910.132–1910.139) Personal Protective Equipment

Subpart J (1910.141–1910.147) General Environmental Controls: Includes sanitation, temporary labor camps, accident prevention color codes, signs, and tags, permit-required confined spaces, control of hazardous energy (lockout/tagout).

Subpart K (1910.151–1910.152) Medical and First Aid

Subpart L (1910.155–1910.165) Fire Protection: Includes portable fire suppression equipment, fixed fire suppression equipment, other fire protection systems, and five technical appendices.

Subpart M (1910.166–1910.169) Compressed Gas and Compressed air Equipment

Subpart N (1910.176–1910.184) Materials Handling and Storage: Includes powered industrial trucks, overhead and gantry cranes, crawler locomotive and truck cranes, derricks, helicopters, and slings.

Subpart O (1910.211–1910.219) Machinery and Machine Guarding

Subpart P (1910.241–1910.244) Hand and Portable Powered Tools and Other Hand-Held Equipment

Subpart Q (1910.251–1910.255) Welding, Cutting and Brazing

Subpart R (1910.261–1910.272) Special Industries: Including pulp, paper, and paperboard mills, textiles, bakery equipment, laundry operations, sawmills, pulpwood logging, agricultural operations, telecommunications, grain handling facilities.

Subpart S (1910.301–1910.399) Electrical: Includes design safety standards for electrical systems, safety-related work practices, safety-related maintenance requirements, safety requirements for special equipment, plus three technical appendices.

Subpart T (1910.401–1910.441) Commercial Diving operations

Subparts U–Y (Reserved)

Subpart Z (1910.1000–1910.1450) Toxic and Hazardous Substances: Includes nearly 30 specific materials such as asbestos, benzidine, lead, bloodborne pathogens, etc.; Hazard Communication Standard; occupational exposure to hazardous chemicals in laboratories.

Parts 1911–1925 Occupational Safety and Health Administration: National Advisory Committee on OSH and its Standards; Shipyard and Marine Terminal OSH standards; longshoreman and Harbor Worker's Compensation Act; safety and health standards for Federal service contracts.

Part 1926 Safety and Health Regulations for Construction: Includes 26 subparts with many of the same items as 1910.120.; also includes signs, signals, and barricades, materials handling, storage, use, and disposal, excavations, concrete and masonry construction, steel erection, underground construction, caissons, cofferdams, and compressed air, demolition, blasting and use of explosives, power transmission and distribution, rollover protective structures/overhead protection, diving, and toxic and hazardous substances.

Part 1927–Reserved

Parts 1928–1948 Agriculture

Parts 1949–1999 OSH Administration: Includes the Office of Training and Education, Federal employees and OSH programs, coverage of employers; discrimination of employees exercising their rights under the Williams-Steiger Occupational Safety and Health Act of 1970; Surface Transportation Assistance Act of 1982; the identification classification, and regulation of potential occupational carcinogens.

Parts 2200–2499 Occupational Safety and Health Review Commission: Regulations on the Freedom of Information Act; the Sunshine act; Equal Access to Justice Act; regulations implementing the Privacy Act.

Parts 2500–2599 Pension and Welfare Benefits Administration

Parts 2600–2699 Reserved

Parts 2700–2799 Federal Mine Safety and Health Review Commission

Parts 4000–4999 Pension Benefit Guaranty Corporation

TWO

Regulations

2 | CONTENTS

Health and safety regulations at hazardous waste sites must be understood in context with many other regulations. This unit gives an overview of various federal laws that govern the handling, use, transportation, and disposal of hazardous materials. While some of these are expanded upon in the appendices at the end of this unit, all may be accessed in the *Federal Register* and in the *U.S. Code of Federal Regulations*.

2.1 Regulations Overview

Federal legislation that impacts the use and disposal of chemicals can be organized into six major groups:

1. **Chemical Use Laws.** Such laws include the Toxic Substance Control Act (TSCA), the Federal Insecticide, Fungicide, and Rodenticide Act (FIFRA), the Federal Food, Drug, and Cosmetic Act (FFDCA), and the OSH Act.
2. **Chemical Discharge Laws.** The three laws included in this group are the Clean Air Act (CAA), the Clean Water Act (CWA), and the Safe Drinking Water Act (SDWA).
3. **Chemical Waste and Disposal Laws.** These laws include the Resource Conservation and Recovery Act (RCRA) and its "cradle to grave" control of hazardous waste, the Comprehensive Environmental Response, Compensation, and Liability Act (CERCLA) or "Superfund," and the Superfund Amendment and Reauthorization Act (SARA).
4. **Chemical Transportation Laws.** Primary regulatory authority for these regulations comes under the Hazardous Materials Transportation Act (HMTA), with additional requirements imposed by RCRA.
5. **Chemical Communication Laws.** Federal regulations require that potential chemical hazards be communicated to workers and the public. These laws include the hazard communication standard (HAZCOM) (29 CFR 1910.1200), National Contingency Plan (NCP), and Emergency Planning and Community Right-to-Know (EPCRA). These laws are summarized in Appendices A through C.
6. **Other Laws Affecting Chemicals.** A variety of other regulations can affect chemical use. Some of these laws include the Consumer Product Safety Act (CPSA), the Federal Hazardous Substances Act (FHSA), the Flammable Fabrics Act (FFA), the Poison Prevention Pack-

aging Act (PPPA), the Ports and Waterways Safety Act (PWSA), the Pipeline Safety Act (PSA), and the National Environmental Policy Act (NEPA).

Chemical control laws interweave, creating a complex web of regulations that, in some way, influence nearly everything manufactured in or imported into the United States. For example, a common cleaner, trichloroethane, became a controlled substance as of May 1, 1993. At that time companies using it in their processes had to suspend its use and/or label their materials as being made with this substance. In addition, disposal costs for spent trichloroethane increased considerably. Many companies were forced to find substitute products that would meet their needs while still being cost-effective. In this case, the CAA's efforts to control the emissions of evaporating trichloroethane were affected by the RCRA's regulations on the disposal of the product; meanwhile worker safety standards for substitute products were being set by OSHA.

2.2 The Scope of OSHA—Federal and State

The OSH Act covers most private sector employees in every state, the District of Columbia, and U.S. territories. It does not apply to federal employees (except for the United States Postal Service), workplaces protected by other federal agencies, and state and local government employees, although such workers are generally covered by similar measures. Also excluded are workers covered under other existing occupational safety and health laws such as nuclear energy employees. However, a 1992 Memorandum of Understanding between the DOL and the U.S. Department of Energy (DOE) formalized an agreement between these two agencies. It stipulated an exchange of technical data, on-site visits by OSHA to DOE offices and facilities, consultation, personnel exchange and, most importantly, the stipulation that DOE employees are covered under Executive Order 12196. DOE must furnish its employees with a safe and healthful workplace, comply with all relevant OSHA standards, including 29 CFR 1960, and must allow OSHA to make announced and unannounced inspections and to investigate complaints arising from, or related to, federal employers at DOE facilities. Limitations to OSHA enforcement have been declared for certain small businesses, the Outer Continental Shelf, and certain farms.

In 1980, Executive Order 12196, the Occupational Safety and Health Program for Federal Employees, went into effect. It stated that "all Federal Agencies are required to maintain safe and healthful working conditions for their employees," and that "agencies must comply with existing OSHA standards." However, even following several amendments to this law, some federal agencies are not yet covered by OSHA, including large segments of the legislative, judicial, and executive branches. Even those agencies on the exclusion list must provide a safety and health staff to whom incidents and/or worker concerns can be brought. For a discussion and clarification of health and safety policies unique to federal agencies, see notices from the *Federal Register* (www.cpm.gov/fedregis/index.htm) and in OSHA Directives pertinent to Executive Order 12196, 29 CFR 1960. OSHA does not currently have the authority to levy fines on federal agencies. However, there is recent legislation putting the U.S. Post Office under OSHA's jurisdiction. This would allow OSHA to issue citations and penalties for violations.

There are certain activities where other agency regulations preempt OSHA law in part or in whole. One specific example is the Coast Guard, whose regulations preempt OSHA coverage so long as it is "seamen" who are concerned and not "longshoremen" who come under OSHA regulations. Other areas include airlines, atomic energy, customs, explosives, meat packing, migrant farm workers, mines and milling, motor carriers (over-the-road truckers), offshore oil platforms, pipelines, and railroads. Although not directly under OSHA regulations, many of these groups must still adhere to OSHA recordkeeping procedures for labor statistics and accident information.

States are encouraged to assume responsibility for the development and enforcement of occupational safety and health programs. OSHA, which monitors these plans and provides for up to 50 percent of the plan's operating cost, must approve such state enforcement plans. These plans must include:

- a main enforcement agency
- standards at least as stringent as, or more stringent than, comparable federal standards

Table 2-1 **States and Territories Having Safety and Health Plans as Approved by the Secretary of Labor** (as of March 2000)

Alaska	New Mexico
Arizona	New York*
California	North Carolina
Connecticut*	Oregon
Hawaii	Puerto Rico
Indiana	South Carolina
Iowa	Tennessee
Kentucky	Utah
Maryland	Vermont
Michigan	Virgin Islands
Minnesota	Virginia
Nevada	Washington
New Jersey*	Wyoming

*Covers public employees only. (All others cover both private sector and state and local government employees.)

Source: *Directory of States with OSHA Occupational Safety and Health State Plans,* www.osha.gov/, June 2000.

- conduction of inspections to enforce its standards
- covering of public (state and local government) employees
- operation of occupational safety and health training and educational programs

After the approval of such a plan, when independent enforcement capability has been proven, the state plan becomes operational and federal OSHA suspends its enforcement in those activities covered by the state plan. When final approval or accreditation of the plan occurs, OSHA relinquishes its authority to matters covered by the state. Twenty-six states and territories currently have approved programs as listed in Table 2-1.

In instances such as a major disaster, a state's plan will undergo detailed reevaluation, may be revoked, may be amended sufficiently to be reapproved, or may undergo minimal change. North Carolina's plan came under such a review, following the Imperial Food Products fire of September 3, 1991, in which 25 employees at a chicken processing plant died. An investigation found that 6 of the 9 exit doors were locked or blocked, there was no sprinkler system or fire alarms, and the 11-year-old plant had never been inspected. While federal guidelines recommended 116 OSHA inspectors for the state, only 27 were hired. Although the owner was sentenced to nearly 20 years in prison for manslaughter, no fines have been collected, and charges against supervisors and managers have been dropped.

This type of incident points out that the mere presence of a safety and health program does not mean it is effective in protecting workers. An agency's internal organization, personnel, enforcement, monetary allotment, and desire to protect the worker must also be in place to make a plan successful.

2.3 Federal Rules Governing Health and Safety for Hazardous Waste Operations

Of primary concern in this book is the body of OSHA law governing workers on hazardous waste sites (see HAZWOPER, 29 CFR 1910.120). Workplace law is found in 29 CFR, a codification of the general and permanent rules published in the *Federal Register* by the executive departments and agencies of the federal government. The general industry regulations for workers on hazardous waste sites are found in 29 CFR Part 1910.120. Other hazardous waste site regulations that are essentially identical to 29 CFR 1910.120 include 29 CFR 1926.65 for the construction industry, and 40 CFR part 311, which regulates hazardous waste site work that is not controlled by the previous title 29 regulations. Other important chemical safety sections include 29 CFR 1910.1200, containing employees' "right-to-know" regulations concerning the hazards of chemicals with which they work (see summary in Appendix A), 29 CFR 1910.1020, toxic and hazardous substances, and 29 CFR 1910.1032, the bloodborne pathogen regulations.

2.4 The Occupational Safety and Health Adminstration and the Environmental Protection Agency

OSHA and the Environmental Protection Agency (EPA) are not interchangeable. The OSH Act is administered by OSHA, created by the Secretary of Labor within the DOL. However, the EPA's jurisdiction is much broader than OSHA's issues. The mission of the EPA is to protect human health and to safeguard the natural environment—air, water, and land—upon which life depends. Issues range from air pollution control to waste management. As a result, the EPA now administers 10 comprehensive environmental protection laws. In contrast, the mission of OSHA is to save lives, prevent injuries and protect the health of America's workers by stressing risk management and process safety. Environmental protection and human health and safety issues often overlap, so both agencies might regulate the same substances but in different ways. These regulations do not necessarily conflict since they generally apply to different uses of a material under different circumstances. For example, trichlorothene is regulated as an air pollutant under EPA and as a respiratory hazard to workers under OSHA.

2.5 Other Significant Occupational Health and Safety Groups

2.5.1 NIOSH

Although OSHA and NIOSH were created by the same Act of Congress in 1970, they are two distinct agencies with separate responsibilities. OSHA is within the DOL and is responsible for creating and enforcing workplace safety and health regulations. NIOSH is in the Department of Health and Human Services, specifically authorized to conduct research on safety and health issues and to recommend criteria for handling and using toxic or dangerous materials in the workplace; in other words, to prevent workplace hazards. NIOSH identifies the causes of work-related diseases and injuries and the potential hazards of new work technologies and practices. It may require employers to measure, record, and make reports on the exposure of their employees to potentially hazardous or known hazardous substances for its own studies. With this information, NIOSH determines new and effective ways to protect workers from chemicals, machinery, and hazardous working conditions.

Additionally, NIOSH responsibilities include investigating potentially hazardous working conditions as requested by employers or employees and providing education and training to individuals preparing for, or currently working in, the field of occupational safety and health. NIOSH publishes a yearly list of all known toxic substances as well as industry-wide studies on chronic or low-level exposure to a variety of industrial materials, processes, and stresses and their potential for illness, disease, or loss of functional capacity in adults.

2.5.2 OCCUPATIONAL SAFETY AND HEALTH REVIEW COMMISSION (OSHRC)

OSHA created OSHRC as an independent agency with only one function—to hear and decide contested OSHA citations and penalty proposals. It consists of a number of administrative law judges who hear cases and issue rulings. These rulings may then be reviewed and changed by a three-member, President-appointed commission whose decisions, in turn, can be appealed to the United States Court of Appeals. Its mandate is to provide fair and impartial hearings and decisions on contested matters. All decisions are put into writing with detailed discussion of the results. These provide valuable guidelines for businesses evaluating their own operations or the potential outcome of a case if taken to the OSHRC judges.

2.5.3 HAZCOM

When it first went into effect in 1985, HAZCOM was considered one of the most significant job safety and health regulatory actions ever adopted. Its purpose was to give workers information concerning the existence of potentially dangerous substances in their workplaces as well as the proper steps they could take to protect themselves against them. In addition to these hazard assessments and associated worker training, this "right-to-know" law required the labeling of chemical containers and ensured worker access to safety data and training.

HAZCOM requires employers to adopt a written hazard communication program, keep Material Safety Data Sheets (MSDS) for each material on the premises containing a hazardous chemical, provide their employees with education and training concerning those hazards, and make certain that proper warning labels are in place. MSDS are key to the program since they are technical bulletins containing information about a hazardous chemical on which employee training and information are based. These documents are the primary vehicles for transmitting detailed hazard information to all concerned parties. A MSDS must be prepared by each manufacturer, distributor, or importer of every product containing a hazardous chemical (see Appendix D). A copy must also be provided to each purchaser of the product.

The required written HAZCOM program must be readily available to all employees regardless of work assignment and must be provided to OSHA representatives upon request. Surveys of OSHA citations have shown that this requirement is the one most often neglected, and programs are often unavailable at many businesses.

2.5.4 OFFICE OF FEDERAL AGENCY PROGRAMS

The mission of the Office of Federal Agency Programs (OFAP) is to ensure that each federal agency has the guidance needed to produce and implement an effective occupational safety and health program within the agency. OFAP also tracks the progress of such agencies, making reports and inspections of the H&S programs. In reality, OFAP functions as a "mini-OSHA" by carrying out basically the same activities for approximately three million federal employees as OSHA does for private sector employees. To this end, the Office carries out inspections, data assessment, abatement issues, allegations of reprisal, and employee complaints. It also encourages communication among federal agencies through support of local federal councils and an annual federal safety and health conference.

2.6 OSHA Requirements for Hazardous Waste Operations

Following are the specific OSHA requirements for work on hazardous waste sites. These regulations are often collectively referred to as HAZWOPER or Hazardous Waste Operations and Emergency Response Regulations. The following materials contain only highlights of the law. A current issue of CFR 29 Part 1910.120 must be consulted for detailed information. These are available in many public libraries as well as being accessible on the Internet at www.cfr.gov, and at federal offices.

2.6.1 TO WHOM DO OSHA HAZWOPER REGULATIONS APPLY?

OSHA's HAZWOPER regulations (1910.120(a)(1)) apply to nearly everyone working with hazardous substances or at hazardous waste sites. Covered work sites include those on the National Priority List (NPL), State Priority List, sites recommended for NPL, and others identified for inspection. Corrective actions at RCRA sites, voluntary actions at government-identified uncontrolled hazardous waste sites; operations at treatment, storage, and disposal (TSD) facilities regulated by RCRA, and emergency response operations are also included.

2.6.2 SAFETY AND HEALTH PROGRAM

Employers must develop and implement a site-specific, written safety and health program for their employees who are involved in hazardous waste operations (1910.120(b)). This program must identify, evaluate, and control safety and health hazards and provide the proper response procedures for the workplace. The basic plan may contain some generic information, but must address the specific site hazards present during all phases of the operation. This plan must be readily available to all employees.

2.6.3 SITE CHARACTERIZATION AND ANALYSIS AND CONTROL

All hazardous waste sites must be evaluated to identify specific hazards and to determine appropriate safety and health procedures. Site characterization (1910.120(c)) requires identification of chemical and physical hazards, location and accessibility of the site, the pathways contaminants might take to disperse, available emergency response personnel, and hazards that are known or might be expected at the site. Personal protective equipment (PPE) and monitoring equipment must be provided based upon these findings. Finally, once the presence and concentrations of specific hazardous substances and health and safety hazards have been established, the risks associated with these substances must be identified and made available to workers on the waste site.

2.6.4 SITE CONTROL

Site control refers to the processes and procedures used to prevent access to the site by unauthorized or unprotected people (1910.120(d)). It is implemented by establishing work zones, specifying appropriate activities in each zone, and physically marking the boundaries of each zone.

2.6.5 TRAINING

All employees working on a hazardous waste site that has the potential for exposing them to safety and health hazards must receive OSHA-mandated training before they are permitted to work on site (1910.120(e)). For general site workers, these standards include a minimum of 40 hours of off-site instruction, a pre-entry briefing, and a minimum of three days (24 hours) of actual field experience under the direct supervision of a trained, experienced supervisor. Workers who are on site only occasionally and not required to wear respiratory protection must receive a minimum of 24 hours of off-site instruction site and a minimum of eight hours (one day) actual field experience under the direct supervision of a trained supervisor. Supervisors must receive an additional 8 hours of specialized training. In addition, all OSHA-certified employees must receive 8 hours of refresher training annually to acquaint them with new training examples, incidents, and relevant topics.

2.6.6 MEDICAL SURVEILLANCE

Any employee working on-site who has the potential for chemical exposure at or above OSHA exposure limits for more than 30 days per year must be on a medical monitoring program (1910.120(f)). Physicals should be conducted on a set schedule by a physician, with the cost paid by the employer. Medical surveillance includes examinations prior to employment, and periodic examinations during employment and at the termination of employment, with special examinations as required to complete the record. All medical records must be made available to the employee and kept as part of the company records for a period of 30 years following termination.

2.6.7 ENGINEERING CONTROLS, WORK PRACTICES, AND PERSONAL PROTECTIVE EQUIPMENT

Engineering controls, work practices, PPE, or a combination of these must be used whenever the threat of chemical exposure exists (1910.120(g)). Engineering controls include techniques such as ventilation, machinery design, and enclosure of processes (Appendix E). The employer is responsible for providing and maintaining equipment appropriate to the jobs that will be performed.

2.6.8 MONITORING

Monitoring (1910.120(h)) refers to that part of the site characterization in which instruments determine the presence and concentrations of hazardous materials, especially in the air. Since concentrations of chemicals can change with time, weather, or work procedures, continuous or frequent monitoring is necessary.

2.6.9 INFORMATIONAL PROGRAMS

As part of the in-house health and safety program, employers are required to develop and implement a plan that provides hazard information to employees, contractors, and subcontractors (1910.120(i)). Information as to the actual hazards and level and degree of likely exposure in doing everyday work must be disclosed. This section does *not* cover employees, contractors, and subcontractors who work outside of the actual operations site. However, the OSHA Hazard Communications standard applies to both on-site and off-site personnel.

2.6.10 HANDLING DRUMS AND CONTAINERS

29 CFR 1910.120(j) deals with the handling, transporting, labeling, and disposing of hazardous or contaminated substances in drums and containers. The manner in which containers are moved, opened, closed, inspected, organized, and sampled are part of the OSHA training program. Special handling is required for certain wastes such as radioactive, shock sensitive, and laboratory wastes.

2.6.11 DECONTAMINATION

All equipment and people exiting a hazardous waste site exclusion zone (unit 13) must be decontaminated (1910.120(k)). Therefore, site operations must have specific decontamination procedures. In addition, the disposal of all waste products generated during decontamination operations must be specified.

2.6.12 EMERGENCY RESPONSE

For each hazardous waste site, the employer must prepare and make available to all employees a written emergency response plan (1910.120(l)). This plan should cover such items as what constitutes an emergency, who to call in an emergency, responsibilities of each person, how to evacuate and decontaminate, and the emergency first-aid available.

2.6.13 ILLUMINATION AND SANITATION

People working at hazardous waste cleanups and at other associated locations require adequate illumination and appropriate eating, drinking, washing, and toilet facilities that are compatible with decontamination, site control, and safety and health (1910.120(m)(n)). Location, condition, and disposal of all resulting waste materials must also be addressed.

2.6.14 NEW TECHNOLOGY PROGRAMS

The employer must inform and train workers in new technologies and equipment that have been developed for the improved protection of employees working with hazardous waste cleanup operations (1910.120(o)). These procedures are part of the site health and safety program, and help ensure that employee protection is an ongoing process.

2.6.15 CERTAIN OPERATIONS CONDUCTED UNDER RCRA

Employers at TSD facilities are regulated under the RCRA of 1976. These facilities must have site-specific safety and health programs, as well as hazard communication programs, medical surveillance, decontamination, new technology, material handling, training, and emergency response programs (1910.120(p)).

2.7 Other Pertinent Regulations

The following OSHA regulations and guidelines may be pertinent to a hazardous waste site operation.

2.7.1 EMERGENCY RESPONSE TO HAZARDOUS SUBSTANCE RELEASE

An emergency response program is required for any employer whose employees are engaged in emergency response operations that involve hazardous substances (29 CFR 1910.120(q)). This regulation covers many manufacturing, processing, and transportation facilities not covered by the preceding regulations. A site that uses chemicals must be prepared to respond to a chemical release.

This regulation requires that a comprehensive emergency response plan be prepared for each site. The plan must include coordination with outside parties, lines of communication, emergency recognition, evacuation routes, decontamination, medical treatment, PPE equipment, etc. All responders receive training based on their duties during an emergency. Training levels are as follows:

First Responder Awareness Level—training includes hazard recognition and notification responsibilities;

First Responder Operations Level—initial response training limited to defensive actions without trying to stop the release;

Hazardous Materials Technician Level—training includes techniques to stop releases and other aggressive actions to control releases;

Hazardous Materials Specialist Level—training is similar to that described above, but responders must have more specific knowledge of chemical hazards and response options; and

On-Scene Incident Commander—this training is required for those assuming control of the incident.

2.7.2 HEARING CONSERVATION STANDARD

In 1983, the Hearing Conservation Standard went into effect. This standard, 29 CFR Part 1910.95, was designed to protect employees from the effects of noise at its source in the workplace. Employees exposed to greater than 85 decibels per eight-hour workday require training, hearing protection, and noise dosimeters for a continuous record of daily noise exposure, and must be provided with an initial baseline hearing test. On a yearly basis, such employees must be tested to determine if loss of hearing has taken place. If hearing loss is detected, the employer must provide further protective measures.

2.7.3 OCCUPATIONAL EXPOSURE TO BLOODBORNE PATHOGENS

Workers in medicine, laboratories, and emergency rescue or trauma can be exposed to bloodborne diseases. Two such diseases include Hepatitis B (HBV), which attacks and destroys healthy liver cells, and Human Immunodeficiency Virus (HIV), which generally leads to Acquired Immunodeficiency Syndrome (AIDS). The Bloodborne Pathogen Standard, 29 CFR Part 1910.1030, limits occupational exposure to such hazards.

Since it is possible to become infected from only one exposure, a major part of this standard deals with exposure control, workplace protection procedures, vaccinations for HBV, postexposure evaluation and follow-up, hazard communication and training, and accurate record keeping for long-term followup.

Emphasis is placed on HBV and HIV, although other bloodborne pathogens such as Hepatitis C, syphilis, malaria, babesiosis, brucellosis, and leptospirosis, among others, are included. Techniques to prevent exposures are stressed in addition to techniques for remediating existing or potentially hazardous situations.

2.7.4 PERMIT-REQUIRED CONFINED SPACES

One of the most recent OSHA regulations, 29 CFR Part 1910.146 (Permit-Required Confined Space), became effective April 15, 1993, with a major revision effective February 1, 1999. Confined spaces are one of the most dangerous American workplaces. Common hazards associated with confined spaces include toxic, explosive, and asphyxiating atmospheres. Examples of spaces of concern include process vessels (brewing and food processing industries), water towers, bulk material hoppers, silos, auger-type conveyors, mixers, tank cars, pipelines, and utility conduits. The regulations emphasize ensuring acceptable entry conditions, isolating the permit space, purging and ventilating, testing and monitoring the air, training workers, knowing the duties of the authorized entrants, their attendants, and supervisors, and rescue.

2.7.5 ERGONOMIC GUIDELINES

With many industries changing technologies from hand-written to computer storage systems in the 1980s and 1990s, a type of workplace hazard proliferated. Cumulative Trauma Disorders (CTDs), such as carpal tunnel syndrome, are now some of the fastest growing complaints registered by workers in a variety of jobs.

Guidelines for helping solve such ergonomic (or human factors engineering) problems were issued by OSHA in late 1990, but these have not been adopted under required rulemaking procedures and are, therefore, not enforceable standards as of 2001. They are suggestions, recommendations, and guidelines made by OSHA that employers can accept in full or in part or even totally reject. Since so many types of jobs such as office and retail recordkeeping, grocery checkout, assembly lines, construction (nailing/sawing), copier operations, processors, and so on require repetitive motion, OSHA hoped that employers would incorporate some of the guideline suggestions to prevent worker disability. In November 1999, OSHA proposed an Ergonomics Programs rule. A final rule, however, is unlikely in the near future because of changes in politics and administrations in Washington, DC.

2.7.6 TRENCHING AND SHORING

Excavation cave-ins cause serious and often fatal injuries to workers. The BLS and OSHA data suggest that nearly 1000 such worker-related injuries occur each year in the United States. Statutes are in effect for the express purpose of protecting those who work in trenching and excavation situations, especially in the construction industry where the majority of these incidents occur.

Federal regulations 29 CFR 1926.650-653 address the practice of using shoring and shields for trenches as well as presenting information for access and egress from excavations, employee exposure to vehicular traffic, falling loads, hazardous atmospheres, water accumulation, unstable structures in and adjacent to excavations, and daily inspections by a qualified individual.

2.7.7 ELECTRICAL SAFETY

Electrical protective equipment is in constant use during electrical power generation, transmission, and distribution work. In spite of this, over seven percent of OSHA-documented yearly occupational fatalities come from worker contact with electrical current or from contact with overhead power lines. In 1971, the National Electrical Code was established to set safety criteria for work around electrical systems as well as to define necessary specifications for electrical equipment. In 1990, the Electrical Safety-Related Work Practices standard became effective. This standard (29 CFR 1910.331-335) addresses work performed on or near exposed energized and de-energized parts of electric equipment and the safe use of electric equipment.

In 1994, a new OSHA standard (29 CFR 1910.269) addressed the work practices to be used during the operation and maintenance of electrical power generation, transmission, and distribution facilities. The standard includes requirements relating to enclosed spaces, hazardous energy control, working near energized parts, grounding for employee protection, work on underground and overhead installations, line-clearance tree trimming, work in substations and generating plants, and other special conditions. At the same time, OSHA revised the electrical protective requirements found in the General Industry Standards (29 CFR 1910.301–303) by including performance-oriented requirements for the design of electrical protective equipment. Finally, OSHA issued new requirements for the safe use and care of electrical protective equipment to backup these design provisions.

2.7.8 PPE HAZARD EVALUATION

In 1998, OSHA revised several of the PPE regulations (29 CFR 1910.132-139) with the standards applying to almost all workplaces under OSHA's jurisdiction. The general requirements still make employers responsible for providing PPE wherever it is necessary, to lessen the risk to workers. The new sections add additional responsibilities for the employer. First, they must select PPE for the workers based upon an assessment of the hazards in the workplace. Such an assessment should incorporate all data possible to evaluate potential hazards. These include MSDSs, operating manuals for machines, warnings from the companies making tools, machines and chemicals used in the workplace, as well as any applicable safety and health standards, among others. Proof of this assessment must be made by written verification. A variety of forms have been compiled for this purpose with industries composing their own to meet their requirements; the forms range from very simple to more site specific. Employers must also ensure that the PPE is properly used and maintained, even when it is employee owned.

The second addition is a requirement for training the employees in the proper use of the assigned PPE. This helps to ensure that the employee is protected and that the PPE does not create new hazards in itself. Even with these changes, it should be remembered that PPE is not an acceptable substitute for more permanent measures, such as engineering controls or administrative controls, which provide more reliable protection. However, some employers are facing difficulties in implementing their PPE standard for very cold environments, for fire brigades, and for electrical distribution work.

2.7.9 STATE STANDARDS

As covered in section 2.2, 25 states and territories administer their own OSHA programs. Each entity faces a variety of specialized situations for public and private workers within its boundaries. Some states have implemented laws far more stringent and more widely reaching than federal standards. One such state is California, where Cal/OSHA has promulgated regulations that are used as models for other states, and for the federal government itself.

One of the more recent Cal/OSHA rulings has been the California Ergonomics Rule which went into effect July 3, 1997. This became the first ergonomics standard in the nation. All employers are subjected to the regulations, even when the rule could impose "unreasonable costs." Opponents feel it is premature with questionable scientific support for repetitive motion syndrome.

Cal/OSHA also promotes a "Special Emphasis Program (SEP)" whose plan is to identify industries, employers, processors and behaviors that contribute to high incidence of workplace injuries, illnesses or fatalities. Once identified, Cal/OSHA Consultation Services provide interfacing with these facilities to help lower injury and illness losses. Most of this is done through employer and employee educational outreach in the areas of greatest workplace safety and health need.

2.8 Definitions

Certain terms will recur in this manual. While they may seem similar, they may have very specific legal meanings. For example, hazardous waste, hazardous substance, and hazardous material have very distinct definitions. A brief definition of some important regulatory terms are presented below.

2.8.1 HAZARDOUS SUBSTANCE (AS DEFINED BY CERCLA)

A hazardous substance is any substance designated as hazardous under any existing federal regulations. Therefore, CERCLA's hazardous substances include the hazardous waste designation by RCRA, designated substances under the CWA, characteristics identified under or listed pursuant to the Solid Waste Disposal Act, any hazardous pollutants listed in the CAA, and hazardous chemicals pursuant to the Toxic Substances Control Act. Hazardous substances may also refer to chemicals identified through OSHA's hazard communication regulation.

2.8.2 HAZARDOUS WASTE (AS DEFINED BY RCRA)

Waste is classified as hazardous under RCRA if it meets one or more of the following conditions:

- it has been named and listed as a hazardous waste by RCRA;
- it has not been excluded as a hazardous waste by RCRA;
- it is a mixture containing a listed hazardous waste and a nonhazardous waste; or
- it exhibits, on analysis, any characteristic of a hazardous waste. These characteristics include ignitability, corrosivity, or reactivity, or the waste meets the detection limits for substances determined by the toxicity characteristic leaching procedure.

2.8.3 HAZARDOUS MATERIAL (AS DEFINED BY DOT)

A hazardous material is a substance in a quantity or form posing an unreasonable risk to health, safety, and/or property when transported in commerce.

2.8.4 HAZARDOUS WASTE (NON-REGULATORY DEFINITION)

A generic, nonspecific term applied to RCRA hazardous wastes, CERCLA hazardous substances, and sometimes U.S. Department of Transportation (DOT) hazardous materials.

2.8.5 HAZARDOUS WASTE OPERATIONS (HAZWOPER)

This refers to the health and safety requirements for employers and employees working at hazardous substance or hazardous waste sites. It usually refers to regulations found in 29 CFR 1910.120 (a) through (p). The training for these regulations is often called "40-Hour OSHA." However, some include the emergency response regulations and training found in 29 CFR 1910.120 (q).

2.8.6 NONHAZARDOUS WASTE (AS DEFINED BY RCRA)

This definition includes solid waste, including municipal wastes, household hazardous waste, municipal sludge, and industrial and commercial wastes that are not hazardous.

2.8.7 SOLID WASTE (AS DEFINED BY RCRA)

Any garbage, refuse, sludge, waste, or other discarded material is a solid waste. All solid waste is not solid; it can be liquid, semisolid, or can contain gaseous material. It does not include solid or dissolved material in domestic sewage, certain nuclear material, or certain agricultural wastes since these are controlled by other agencies.

2.9 Sources of Information

Balden-Anslyn, Roxanna: Hazard Assessment and PPE Selection. *Occupational Health and Safety, Vol. 65, Number 10*: 154–158 (1996).

"Basic Program Elements for Federal Employee Occupational Safety and Health," *Code of Federal Regulations* Title 29, Part 1960. 1998.

Bureau of Labor Statistics: *Census of Fatal Occupational Injuries, 1992–1999—Fatal Occupational Injuries by Event or Exposure, 1992–1999*. Washington, DC: U.S. Government Printing Office, 2001.

Bureau of Labor Statistics: *Fatal Occupational Injuries by Event or Exposure, 1992–1999*. In www.osha.gov/oshstats/fataltbl.html. Washington, DC: Bureau of Labor Statistics, 2001.

Bureau of Labor Statistics in cooperation with State and Federal agencies: *Census of Fatal Occupational Industries—Fatal Occupational Injuries by Event or Exposure and Major Private Industry*. Washington, DC: U.S. Government Printing Office, 2001.

California Occupational Safety and Health Administration: *Emergency Provisions for the Bloodborne Pathogens Standard*. In www.dir.ca.gov/DOSH/DOSHBloodBorne.html. Sacramento, CA: August 1999.

California Occupational Safety and Health Administration: *Special Emphasis Program (SEP)*. In http://www.dir.ca.gov/sep/sep/html Sacramento, CA: 1999.

"Hazard Communication Standard," U.S. Code 29, Section 1910.1200 *et seq.* Washington, DC: U.S. Government Printing Office, August 1999.

Jones, Gary A.: Environmental Alert: California Ergonomics Rule Goes Into Effect. *Printers' Environmental Assistance Center*. In www.pneac.org. 21 April 1998.

Kerr, Mary Lee and B. Hall: Chickens Come Home to Roost. *The Progressive, 45*:29 (January 1992).

Kincaid, William: Tips and Techniques for Coping With OSHA's

1910.132. In *Occupational Health and Safety, Vol. 65, Number 10*:114–116 (1966).

Lacayo, Richard: Death on the Shop Floor. *Time, 138*:28–29 (16 September 1991).

"Occupational Exposure to Bloodborne Pathogens," *Code of Federal Regulations* Title 29, Part 1910.1030. 2001.

"Occupational Safety and Health Act," Public Law 91–596, 84 Stat. 1590, U.S. Code 29, Section 651 *et seq.* Washington, DC: U.S. Government Printing Office, as of August 1999.

Occupational Safety and Health Administration: *Directory of States with OSHA Occupational Safety and Health State Plans.* In www.osha.gov/. Washington, DC: Department of Labor, August 2001.

Occupational Safety and Health Administration: *Federal Agency Safety and Health Programs* (OSHA Directive Number FAP 1.3). Washington, DC: In, http://www.osha-slc.gov/OshDoc/ 17 May 1996.

Occupational Safety and Health Administration: "Memorandum of Understanding between the U.S. Department of Labor and the U.S. Department of Energy." Washington, DC: U.S. Government Printing Office, August 28, 1992. [Memo.]

"Permit-Required Confined Spaces," *Code of Federal Regulations* Title 29, Part 1910.146. 1998, effective 1 February 1999.

Poultry Plant Owner Gets 20 Years in Deadly Blaze. *JET 82*:29 (5 October 1992).

Rothstein, Mark A.: *West's Handbook Series, Occupational Safety and Health Law.* 3rd ed. St. Paul, MN: West Publishing Co., 1990. pp. 11–37.

Sullivan, Thomas S. (ed.): *Environmental Law Handbook.* 14th ed. Rockville, MD: Government Institutes, Inc., 1997.

Waks, J.W. and C.R. Brewster: OSHA Reform Efforts are Heating Up. *The National Law Journal*:16–18 (7 June 1993).

Worklink: *Legal Victory for Labor on California Ergonomics Standard.* In www.worklink.net. September 1997.

Worobec, Mary D. and G. Ordway: *Toxic Substances Controls Guide.* Washington, DC: BNA Books, 1992.

APPENDIX A

Hazard Communication Standard, 29 CFR 1910.1200

One of the most significant job safety and health regulatory actions ever adopted, the HAZCOM Standard, alerts workers to the existence of potentially hazardous substances in the workplace and the proper way to protect themselves against them. Popularly known as the "employee law," it requires employers to provide information in the form of hazard warning labels, material safety data sheets, and employee training programs.

Although the standard went into effect for the manufacturing industry in 1985, by 1987 it had been applied to most other employers. It requires each employer to develop its own written hazard communication program, to keep MSDS for each product containing a hazardous chemical on the premises, to educate the workers on those chemical hazards, and to make certain that proper warning labels are in place.

The material safety data sheet must contain a listing of health hazards from exposure, sign(s) of exposure, primary routes of exposure, OSHA and other exposure limits, whether the substance is a carcinogen, precautions for safe handling, and information on emergency first aid procedures. The name and address of the manufacturer must be included. These sheets must be available to employees at all times. The MSDS is important since it forms the basis of all labeling, as well as the training programs that must be implemented.

Finally, employers must establish a training and information program in hazardous substances handling. Such training must include information and the hazards of all chemicals used in work areas and of the measures employees can take to protect themselves form the hazards of the substances. Training must also take place whenever employees are shifted to another work area where new or different substances are used.

An example of a standard MSDS form recommended by OSHA is presented in Appendix D.

APPENDIX B

National Contingency Plan (NCP), 40 CFR 300

The National Contingency Plan is a federally mandated blueprint for government cleanup under the Superfund response program. It sets forth standards for cleanup activities and describes relationships between and among various federal and state agencies responsible for Superfund implementations. The NCP was first established in 1968 under the Federal Water Pollution Control Act (later the Clean Water Act), and was originally developed to deal with spills (chemical and oil) into navigable waters. It was amended in 1982 to cover Superfund releases. Another function was that of describing governmental roles and procedures for evaluating remedial options.

On March 8, 1990, the EPA issued major NCP revisions required by the 1986 SARA amendments (Appendix 2C). These amendments presented the EPA's principal interpretations of various CERCLA requirements and gave detailed plans for the EPA's cleanup policies. It is the only set of federal regulations which has been issued under the Superfund law. All CERCLA response actions must comply with the NCP cleanup standards.

In conjunction with the EPA, the NCP has set forth a structures program for evaluating sites and for placing some of them on the National Priorities List (NPL). The NPL is set out as Appendix B of the NCP and sites on this list are eligible for cleanup using Superfund dollars. The NCP set out nine criteria for remedial options at waste sites. These nine are:

- overall protection of human health and the environment;
- compliance with ARARs;
- long-term effectiveness and permanence;
- reduction of toxicity, mobility, or volume;
- short-term effectiveness;
- implementability;
- cost;
- state acceptance; and
- community acceptance.

The NCP also establishes Regional Response Teams to provide local assistance in an emergency, as well as providing for a system of coordinators. These on-scene coordinators are assigned to respond to each discharge into the environment and to assess what type of response is required.

APPENDIX C

SARA Title III—Emergency Planning and Community Right-to-Know Act, 42 USC 9601: Public Law 99-499

In 1984, the world's worst industrial accident to-date occurred at a Union Carbide pesticide plant in Bhopal, India. It was estimated that approximately 3,700 people were killed and 300,000 were injured when 40 tons of methyl isocyanate, a very toxic gas, leaked from a storage tank. The Bhopal disaster provided the impetus for establishing a regulation relating to catastrophe planning and preparedness. SARA, the Superfund Amendments and Reauthorization Act of 1986, included Title III—the Emergency Planning and Community Right-To-Know Act.

Through this Act, Congress has attempted to anticipate similar disasters by requiring state and local governments to develop emergency plans for responding to unanticipated environmental releases or explosions that could be caused by a number of acutely toxic materials known as "extremely hazardous substances." Industrial and commercial facilities are required to report annually the quantities of substances present in their places of business as well as the quantities that are released to the environment on a routine basis. The law also provides for emergency notification to the community if a hazardous chemical is released into the environment. Notice must be given immediately for releases of any materials above the EPA release thresholds on the Extremely Hazardous Substances (EHS) list as well as any chemicals on the CERCLA list.

Title III has four major sections: emergency planning, emergency notification/community right-to-know, reporting requirements, and a toxic-chemical-release reporting emissions inventory. As such, a number of public agencies and departments have roles in Title III activities. Concerns include issues related to the environment, natural resources, emergency management, public health, occupational safety and transportation.

To carry out this program, SARA mandated creation of State Emergency Response Commissions (SERCs) and Local Emergency Planning Committees (LEPCs). SERCs are appointed by state governors, and, in turn, designate local emergency planning districts, appointing a LEPC for each district. Over 4,000 districts and LEPCs have been formed. Districts are responsible for preparing a comprehensive emergency response plan that must:

- identify facilities and transportation routes of extremely hazardous substances;
- describe emergency response procedures and define facility and community coordinators to implement the plan;
- outline emergency response communication procedures;
- describe how to determine the occurrence of a release and the probable area and population affected;
- describe and identify community and industry emergency equipment and facilities;
- outline evacuation plans;
- describe training programs for emergency response personnel;
- annually review the plans for appropriateness; and
- industry must provide information about quantities and locations of chemicals on site as well as the hazards they pose.

APPENDIX D

CHEM CORP CHEM-WELD	MATERIAL SAFETY DATA SHEET	Date Revised: JUN 1994 Supersedes: APR 1992

Information on this form is furnished solely for the purpose of compliance with the Occupational Safety and Health Act and shall not be used for any other purpose. Chem Corp urges the customers receiving this Material Safety Data Sheet to study it carefully to become aware of the hazards, if any, of the product involved. In the interest of safety, you should notify your employees, agents, and contractors of the information on this sheet.

SECTION I

MANUFACTURER'S NAME Chem Corporation	Transportation Emergencies: CHEMTREC: (800) 424-9300 Medical Emergencies: (213) 222-3212 (L.A. Poison Center 24 Hour No.) Business: (310) 366-3300
ADDRESS 45 Main Street, San Francisco, CA 90327	
CHEMICAL NAME and FAMILY PVC/CPVC Primer Mixture of Organic Solvents	TRADE NAME WELD-ON P-70 Primer for PVC and CPVC Plastic Pipe FORMULA: Proprietary

SECTION II—HAZARDOUS INGREDIENTS

None of the ingredients below are listed as carcinogens by IARC, NTP or OSHA	CAS#	APPROX %	ACGIH-TLV	ACGIH-STEL	OSHA-PEL	OSHA-STEL
Tetrahydrofuran (THF)	109-99-99	45–55	200 PPM	250 PPM	200 PPM	
Methyl Ethyl Ketone (MEK)	78-93-3	47	200 PPM	300 PPM	200 PPM	300 PPM
Cyclohexanone	108-94-1	5–15	25 PPM Skin		25 PPM Skin	

SHIPPING INFORMATION

DOT Hazard Class: Flammable Liquid
DOT Shipping Name: Flammable Liquid, N.O.S. (Tetrahydrofuran, Methyl Ethyl Ketone)
Identification Number: UN 1993

SPECIAL HAZARD DESIGNATIONS

	HMIS	NFPA	HAZARD RATING
HEALTH	2	2	0-MINIMAL
FLAMMABILITY:	3	3	1-SLIGHT
REACTIVITY:	0	1	2-MODERATE
PROTECTIVE			3-SERIOUS
EQUIPMENT:	H		4-SEVERE

SECTION III—PHYSICAL DATA

APPEARANCE Purple or Clear, thin liquid	ODOR Ethereal	BOILING POINT (°F/°C) 151°F Based on first boiling component: THF
SPECIFIC GRAVITY @ 73 Typical 0.870	VAPOR PRESSURE (mm Hg.) 143 mm Hg. Based on first boiling component, THF @ 20°C	PERCENT VOLUME BY VOLUME (%) 100%
VAPOR DENSITY (Air = 1) 2.49	EVAPORATION RATE (BUAC = 1) >1.0	SOLUBILITY IN WATER Completely soluble in water

SECTION IV—FIRE AND EXPLOSION HAZARD DATA

FLASH POINT	FLAMMABLE LIMITS (Percent by Volume)	LEL	UEL
6°FT.C.C. based on THF		2.0	11.8

FIRE EXTINGUISHING MEDIA
Ansul "Purple K" potassium bicarbonate dry chemical, carbon dioxide, National Aero-O-Foam universal alcohol resistant foam, water spray.

SPECIAL FIRE FIGHTING PROCEDURES
Evacuate enclosed areas, stay upwind. Close or confined quarters require self-contained breathing apparatus, positive pressure hose masks or airline masks. Use water spray to cool containers, to flush spills from source of ignition and to disperse vapors.

UNUSUAL FIRE AND EXPLOSION HAZARDS
Fire hazard because of low flash point and high volatility. Vapors are heavier than air and may travel to source of ignition.

SECTION V—HEALTH HAZARD DATA

PRIMARY ROUTES OF ENTRY: X Inhalation X Skin Contact ___ Eye Contact ___ Ingestion

EFFECT OF OVEREXPOSURE

ACUTE: Inhalation: Severe overexposure may result in nausea, dizziness, headache. Can cause drowsiness, irritation of eyes and nasal passages.
Skin Contact: Skin irritant. Liquid contact may remove natural skin oils resulting in skin irritation. Dermatitis may occur with prolonged contact.
Skin Absorption: Prolonged or widespread exposure may result in the absorption of harmful amounts of material.
Eye Contact: Overexposure may result in severe eye injury with corneal or conjunctival inflammation on contact with the liquid. Vapors slightly uncomfortable.
Ingestion: Moderately toxic. May cause nausea, vomiting, diarrhea. May cause mental sluggishness.
CHRONIC: Symptoms of respiratory tract irritation and damage to respiratory epithelium were reported in rats exposed to 50000 ppm THF for 90 days. Elevation of SGPT suggests a disturbance in liver function. The NOEL was reported to be 200 ppm.

MEDICAL CONDITIONS AGGRAVATED BY EXPOSURE: Individuals with pre-existing diseases of the eyes, skin or respiratory system may have increased susceptibility to the toxicity of excessive exposures.

EMERGENCY AND FIRST AID PROCEDURES

Inhalation: If overcome by vapors, remove to fresh air and if breathing stopped, give artificial respiration. If breathing is difficult, give oxygen. Call physician.
Eye Contact: Flush eyes with plenty of water for 15 minutes and call a physician.
Skin Contact: Remove contaminated clothing and shoes. Wash skin with plenty of soap and water for at least 15 minutes. If irritation develops, get medical attention.
Ingestion: Give 1 or 2 glasses of water or milk. Do not induce vomiting. Call physician or poison control center immediately.

SECTION VI—REACTIVITY

STABILITY			CONDITIONS TO AVOID
	UNSTABLE		Keep away from heat, sparks, open flame and other sources of ignition.
	STABLE	X	

INCOMPATABILITY
(MATERIALS TO AVOID) Caustics, ammonia, inorganic acids, chlorinated compounds, strong oxidizers and isocynates.

HAZARDOUS DECOMPOSITION PRODUCTS
When forced to burn, this product gives out carbon monoxide, carbon dioxide, hydrogen chloride and smoke.

HAZARDOUS POLYMERIZATION			CONDITIONS TO AVOID
	MAY OCCUR		Keep away from heat, sparks, open flame and other sources of ignition
	WILL NOT OCCUR	X	

SECTION VII—SPILL OR LEAK PROCEDURES

STEPS TO BE TAKEN IN CASE MATERIAL IS RELEASED OR SPILLED
Eliminate all ignition sources. Avoid breathing of vapors. Keep liquid out of eyes. Flush with large amount of water. Contain liquid with sand or earth. Absorb with sand or nonflammable absorbent material and transfer into steel drums for recovery or disposal. Prevent liquid from entering drains.

WASTE DISPOSAL METHOD
Follow local, State and Federal regulations. Consult disposal expert. Can be disposed of by incineration. Excessive quantities should not be permitted to enter drains. Empty containers should be air dried before disposing. Hazardous Waste Code: 214

SECTION VIII—SPECIAL PROTECTION INFORMATION

RESPIRATORY PROTECTION (Specify type)
Atmospheric levels should be maintained below established exposure limits contained in Section II. If airborne concentrations exceed those limits, use of a NIOSH-approved organic vapor cartridge respirator with full face-piece is recommended. The effectiveness of an air purifying respirator is limited. Use it only for a single short-term exposure. For emergency and other conditions where short term exposure guidelines may be exceeded, use an approved positive pressure self-contained breathing apparatus.

VENTILATION
Use only with adequate ventilitation. Provide sufficient ventilation in volume and pattern to keep contaminants below applicable exposure limits set forth in Section II. Use only explosion proof ventilation equipment.

PROTECTIVE GLOVES	EYE PROTECTION
PVA coated	Splashproof chemical goggles

OTHER PROTECTIVE EQUIPMENT AND HYGIENIC PRACTICES
Impervious apron and a source of running water to flush or wash the eyes and skin in case of contact.

SECTION IX—SPECIAL PRECAUTIONS

PRECAUTIONS TO BE TAKEN IN HANDLING AND STORING
Store in the shade between 40°F–110°F. Keep away from heat, sparks, open flame and other sources of ignition. Avoid prolonged breathing of vapor. Use with adequate ventilation. Avoid contact with eyes, skin and clothing. Train employees on all special handling procedures before they work with this product.

OTHER PRECAUTIONS
Follow all precautionary information given on container label, product bulletins and our solvent cementing literature. All handling equipment should be electrically grounded.

APPENDIX E

Engineering Controls

Throughout this manual, emphasis is placed on proper personal protective equipment, especially in working at hazardous substance sites. Some references are also made to "engineering controls" which are more likely to be recognized in specific industries where they protect workers from specific mechanical, toxic or physical hazards. Many of these have been adapted to work with hazardous materials. Such controls are separate from PPE and are treated as such under OSHA regulations and policy.

Since its inception, OSHA has had policy relevant to the ways in which health hazards could be corrected. A hierarchy was established as an order of preference for generalized methods that could be used to control exposure to a range of harmful substances. The hierarchy, as established was:

1) Engineering Controls—the containment or elimination of a hazard through physical means, such as ventilation systems to control exposure to a volatile toxic solvent;
2) Administrative Controls—includes the manipulation of work schedules to reduce exposure, such as limiting the time employees may spend in a noisy area or in a radiation area; and
3) Personal Protective equipment—includes the use of protective devices such as earmuffs to attenuate noise, or respirators to filter out chemical vapors.

These three safety control methods have been utilized to enforce the General Duty Clause of Public Law 91-596, Section 5(a) (OSHA) that states each employer:

"shall furnish to each of his employees, employment and a place of employment which are free from recognized hazards that are causing or are likely to cause death or serious physical harm to his employees . . ."

In carrying out this edict, OSHA has made limited exceptions to this hierarchy if the implementation of engineering controls is not feasible for technical and/or economic reasons. Historically, when a hazardous situation was present in industry, PPE was generally the control of choice, in an effort to take care of the problem by the easiest means. More recently, however, fines, education, pollution prevention programs, and bottom-line numbers have made industry more inclined to "clean up" instead of living with the problem.

Engineering controls can be as simple as loading an electric drill trigger switch with a spring so that if a failure results on the part of the operator, the machine will shut off. At the other end of the engineering spectrum, ergonomically designed keyboards are used by computer operators to prevent repetitive motion syndrome.

The following list gives a sample of the scope of engineering controls which might be found under a variety of working conditions. Each industry or separate job situation may benefit from safety controls which have been developed specifically for them.

- machine guards
- dust collection system
- ventilation systems
- scaffolding designs
- hoist designs
- stairway design
- automatic sprinkler systems
- fall protection
- trench shoring
- two hand controls for operating some equipment (presses and cutters)

- using nonmetallic safety devices to prevent sparking
- rope specifications
- proper illumination
- explosion-proof monitors
- rollover protection for vehicles
- air cleaning/purifying devices
- proper size and insulation of wiring
- sound-deadening designs
- negative pressure ducts in ventilated areas
- fire detection and extinguisher design
- proper wiring/grounding receptacles for electrical equipment

Regardless of the planning that goes into eliminating or reducing hazards to workers, engineering controls will only be effective if allowed to perform their designated tasks. Individuals may defeat the purpose of engineering controls or safety devices if they perceive them as being cumbersome, a nuisance, or an impediment to production. The controls should not unduly encumber workers' activities.

Sources of Information

Asfahl, C. Ray: *Industrial Safety and Health Management.* 3rd ed. Englewood Cliffs, NJ: Prentice-Hall, Inc., 1995.

Berman, Daniel M.: *Death on the Job, Occupational Health and Safety Struggles in the United States.* New York: Monthly Review Press, 1978.

Lofgren, Don J.: *Dangerous Promises, An Insider's View of OSHA Enforcement.* Ithaca, NY: Cornell University/ILR Press, 1989.

Occupational Safety and Health Administration: *Occupational Safety and Health Standards for the Construction Industry (29 CFR Part 1926).* Washington, DC: U.S. Government Printing Office, July 1999.

Occupational Safety and Health Administration: *Occupational Safety and Health Standards for General Industry (29 CFR Part 1910).* Washington, DC: U.S. Government Printing Office, January 2001.

THREE

Chemistry of Hazardous Substances

3 | CONTENTS

Chemicals are an integral part of our daily lives. These include fat, antibiotics, fabrics, plastics, pesticides, chemotherapy drugs, and anti-ultraviolet (UV) coatings on sunglasses. Obviously, chemicals play a variety of critical roles in our world.

It is estimated that there are about 8 million known chemicals, of which about 70,000 are in production for domestic and industrial consumption in the United States. Of these, about 1,000 chemicals account for 99 percent (by weight) of all chemicals used. These are the focus of OSHA regulations.

Although most chemicals have some type of specific, intended use, all may be hazardous to some extent, even those considered to be "natural." A basic tenant of toxicology is that the dose differentiates a poison from a remedy. In addition, when chemicals are improperly used or disposed of, human health can be affected as well as environmental quality.

3.1 Nature of Chemical Substances

Following are definitions that describe the basic concepts of chemistry as they relate to hazardous substances:

Atom—Referred to as the building blocks of matter, atoms are the smallest units of an element that keep all of the chemical properties of that element. The basic components of an atom are a nucleus of positively charged protons and uncharged neutrons, and a cloud of electrons surrounding the nucleus.

Element—An element is matter composed of only one type of atom. Elements are listed in the periodic chart and include solids (silver, lead, gold), liquids (mercury and bromine), and gases (nitrogen, chlorine, radon). There are currently 115 known elements, 22 of which have been artificially produced and are not found naturally in the environment.

Periodic Table—This is a systematic arrangement of elements so that those with similarities in properties and electronic configurations are presented in columns and rows. The main function of the periodic table is to present a basic framework for the systematic organization of chemistry (Figure 3-1).

Molecule—A molecule is a chemical unit composed of one or more atoms held together by electronic bonds. Molecules may contain several different kinds of atoms, such as water (two hydrogen atoms bonded with one oxygen), and table salt (one sodium atom bonded with one chlorine atom).

Isotope—An isotope is an element having a specific number of protons in the nucleus, but differing in the number of neutrons in the nucleus.

1
H
1.008 ←Atomic Number ←Symbol ←Atomic Mass

1 IA	2 IIA	3 IIIB	4 IVB	5 VB	6 VIB	7 VIIB	8 ←	9 VIIIB	10 →	11 IB	12 IIB	13 IIIA	14 IVA	15 VA	16 VIA	17 VIIA	18 VIIIA
1 H 1.008																	2 He 4.003
3 Li 6.941	4 Be 9.012											5 B 10.81	6 C 12.01	7 N 14.01	8 O 16.00	9 F 19.00	10 Ne 20.18
11 Na 22.99	12 Mg 24.31											13 Al 26.98	14 Si 28.09	15 P 30.97	16 S 32.06	17 Cl 35.45	18 Ar 39.95
19 K 39.10	20 Ca 40.08	21 Sc 44.96	22 Ti 47.90	23 V 50.94	24 Cr 52.00	25 Mn 54.94	26 Fe 55.85	27 Co 58.93	28 Ni 58.69	29 Cu 63.55	30 Zn 65.39	31 Ga 69.72	32 Ge 72.59	33 As 74.92	34 Se 78.96	35 Br 79.90	36 Kr 83.80
37 Rb 85.47	38 Sr 87.62	39 Y 88.91	40 Zr 91.22	41 Nb 92.91	42 Mo 95.94	43 Tc 98.91	44 Ru 101.1	45 Rh 102.9	46 Pd 106.4	47 Ag 107.9	48 Cd 112.4	49 In 114.8	50 Sn 118.7	51 Sb 121.8	52 Te 127.6	53 I 126.9	54 Xe 131.3
55 Cs 132.9	56 Ba 137.3	57 La 138.9	72 Hf 178.5	73 Ta 180.9	74 W 183.9	75 Re 186.2	76 Os 190.2	77 Ir 192.2	78 Pt 195.1	79 Au 197.0	80 Hg 200.6	81 Tl 204.4	82 Pb 207.2	83 Bi 209.0	84 Po (210)	85 At (210)	86 Rn (222)
87 Fr (223)	88 Ra 226	89 Ac (227)	104 Rf (261)	105 Ha (262)	106 Sg (263)	107 Ns (264)	108 Hs (265)	109 Mt (266)	110 — (269)	111 — (272)	112 — (277)		114 — (298)		116 — (289)		118 — (293)

58 Ce 140.1	59 Pr 140.9	60 Nd 144.2	61 Pm (145)	62 Sm 150.4	63 Eu 152.0	64 Gd 157.3	65 Tb 158.9	66 Dy 162.5	67 No 164.9	68 Er 167.3	69 Tm 168.9	70 Yb 173.0	71 Lu 175.0
90 Th 232.0	91 Pa 231.0	92 U 238.0	93 Np (237)	94 Pu (244)	95 Am (243)	96 Cm (247)	97 Bk (247)	98 Cf (251)	99 Es (252)	100 Fm (257)	101 Md (258)	102 No (259)	103 Lr (262)

Figure 3-1 Periodic Table of the Elements

Matter—Anything that has weight and takes up space, from parts of an atom to skyscrapers is considered to be matter.

Compound—A compound is a type of matter consisting of atoms of two or more different elements chemically bonded together into a molecule that has a fixed composition. Examples are sodium hydroxide (NaOH), benzene (C_6H_6), and hydrogen cyanide (HCN). Although most compounds occur naturally, industrial and research chemists have identified over 8,000,000 compounds.

Inorganic Compounds—Substances that do not contain carbon atoms in their make-up are classified as inorganic. Examples are salt (NaCl), rust (Fe_2O_3), and sulfuric acid (H_2SO_4).

Organic Compounds—Substances that contain carbon, usually in a chain or ring structure, are considered organic. Organic compounds are more numerous than inorganic compounds and include sugar ($C_6H_{12}O_6$) and the pesticide endrin ($C_{12}H_8Cl_6O$).

Hydrocarbon—This is an organic compound containing only carbon and hydrogen atoms, such as benzene (C_6H_6), methane (CH_4) and propane ($CH_3CH_2CH_3$).

Halogenated Hydrocarbon—This is a compound whose molecular components are limited to hydrogen, carbon, and a halogen such as fluorine, chlorine, bromine, or iodine. Methylene chloride (CH_2Cl_2) is a halogenated hydrocarbon.

Mixture—A mixture is two or more substances that are physically, rather than molecularly, combined. It can be separated by physical means such as filtering. Examples are muddy water, tea, and milk.

Solution—This is a term usually applied to homogenous combinations of gases or solids dissolved in a liquid. Hazardous waste lecheate and salt water are examples.

Density—Density is defined as the amount of mass contained in a specific unit of volume at a given temperature. Lead has a high density while polystyrene foam has a low density.

Ionization—This is the change of a molecule or atom from an uncharged electronic state to a charged state. For example, if a molecule releases an electron, the molecule then has a positive charge. The ionization process is particularly important when vapors are to be detected electronically.

3.2 Physical States of Matter

All matter can exist in three different physical states—solid, liquid, or gas, depending on temperature and pressure. Ice, water, and steam reflect these states in water. For every element or compound, there is a "normal" physical state in which it commonly exists at normal temperature and pressure. At room temperature, iron and brass are solids, oxygen and carbon monoxide are gases, and mercury and water are liquids. Lowering the temperature can change a gas into a liquid or a liquid into a solid. Raising the temperature sufficiently will reverse the process. Some materials, such as carbon dioxide in its solid form (dry ice), can change directly from a solid to a gas, a process that is called sublimation.

Terms associated with the changes of states of matter are:

Freezing—Freezing is defined as the lowering of temperature (removing energy) and decreasing molecular motion, causing a change from liquid to solid; water to ice.

Melting—Adding energy and subsequently increasing molecular motion, causing a material to change from a solid to a liquid, is termed melting. Examples are iron to molten iron, or ice to water.

Boiling—Boiling is caused by adding energy, causing an increase of molecular motion and a change from liquid to gas as in liquid nitrogen changing to gaseous nitrogen, or water to steam.

Condensation—The removal of energy and lessening of molecular motion, causing a substance to change from a gaseous state to a liquid state is known as condensation. An example is water vapor changing to liquid water, a process that occurs when fog is formed.

Sublimation—Sublimation is the addition of energy such that the state changes directly from a solid to a gas; dry ice sublimates to carbon dioxide gas.

3.2.1 SOLIDS

Matter in the solid state is rigid and cannot flow. The molecules comprising a solid are packed closely together, so molecules in solids cannot be compressed to any appreciable degree. When compared to the liquid state, a solid has a greater density than a liquid of the same substance. Therefore, solid lead sinks in molten lead. One important exception to this rule is water. Solid water, or ice, floats on water in the liquid state.

Because of their lack of mobility, solids are generally less dangerous than liquids and gases, with some exceptions. Phosphorous is a nonmetallic solid that is highly reactive in air and must be stored in water so that it does not spontaneously ignite. Sodium is a reactive metal that will also ignite spontaneously in air at room temperature. It is generally shipped and stored in oil.

Definitions of the basic properties of solids are as follows:

Melting Point—Melting point is the temperature at which a solid changes to a liquid. Ice changes to water at a temperature greater than 32°F (0°C), while phosphorous melts at 111°F (44°C).

Specific Gravity—The ratio of the mass of a substance to the mass of the same volume of water is the definition of specific gravity. Solids having a specific gravity of less than 1.0 will float on water, whereas solids having a specific gravity of more than 1.0 will sink in water. This property of solids is important when dealing with hazardous substances. A solid with a specific gravity greater than 1.0, such as a block of cadmium, will be found at the bottom of a pond, stream, or water-filled container. In contrast, a block of polystyrene foam has a specific gravity of less than 1.0 and will be found floating on water.

Suspended solids form a special category because of their increased mobility and health hazard. As the particle size of a solid decreases, its mobility in air increases. Particulate matter with a diameter of 10 microns or less is especially hazardous because it can be readily transported in the air and easily inhaled. Asbestos, beryllium, lead, coal, silica, arsenic, and other finely-divided, airborne, particulate solids are regulated as air pollutants because of their health hazard.

3.2.2 LIQUIDS

Matter in the liquid state is not rigid and is able to flow and conform to the shape of its container. Molecular attraction in a liquid is not as great as in a solid. For most substances, the liquid state has a lower density than the solid state.

Liquids are relatively mobile, more difficult to avoid than solids, and can flow or splash onto workers. In addition, liquids that are separated into tiny droplets (mists) can become airborne and present a respiratory hazard. Furthermore, some liquids can readily release vapors into the air causing a respiratory hazard. Examples are gasoline, benzene, mercury, and halogenated hydrocarbons solvents such as methylene chloride and perchloroethylene.

Definitions of the basic characteristics of liquids are as follows:

Boiling Point—The boiling point is the temperature at which a liquid's vapor pressure equals the atmospheric pressure and the liquid changes into a gas. This property will help determine what physical state a substance will likely be at room temperatures. A substance with a boiling point of 2°F (−17°C) will likely be found as a gas at room temperatures, while a substance with a boiling point of 200°F (93°C) will likely be found as a liquid. Substances with high boiling points, such as 2000°F (1093°C) are usually found as solids at room temperatures. Boiling points are determined for sea level unless otherwise stated. As elevation increases above sea level, the boiling point decreases.

Solubility—Solubility is the amount of a solid, liquid, or gas that dissolves in a liquid. The liquid is usually water. In other words, solubility is the ability of a substance to mix with water. Salt is completely soluble in water, but carbon dioxide is not, so a carbonated beverage goes flat when exposed to air and the carbon dioxide is released. Solubility is important when recovering hazardous substances spilled in water; water-insoluble solids and liquids are easier to recover from ponds and streams than water-soluble chemicals.

Specific Gravity—Specific gravity is an expression of the density of a liquid as compared to water at the same temperature. Insoluble liquids such as tetrachloroethylene and other halogenated hydrocarbon solvents with a specific gravity greater than 1.0 will be found at the bottom of a pond, stream, or water-filled container. In contrast, gasoline has a specific gravity of 0.8 and will be found floating on water. This characteristic is important when deciding how to respond when insoluble liquid chemicals are released in water.

Volatile Liquid—This is a generic term given to liquids that readily vaporize, or evaporate, at room temperature. Alcohols and ether are considered volatile liquids. Volatilization is dependent primarily on temperature and pressure. For example, a liquid not generally considered volatile could readily volatilize when dropped onto a hot pavement.

3.2.3 GASES AND VAPORS

Compared to liquids and solids, gases and vapors have the lowest density and the greatest mobility. Due to their high energy, the molecules of a gas are far apart and move rapidly. Gases and vapors disperse rapidly to fill a container. They exert pressure as the molecules bounce off the walls of a container, and they are capable of being compressed. Unless gases and vapors are confined they escape into the atmosphere. When heated, gases and vapors can expand rapidly and sometimes violently.

Compared to solids and liquids, gases and vapors generally pose the greatest health and safety respiratory hazards. They can move quickly, some are difficult to detect and avoid, many can explode, and gases have easy access to the human body by way of the lungs.

Definitions for this state of matter are as follows:

Gas/Vapor—There is fundamentally no difference between these terms, but "gas" is commonly used to describe a substance which appears in the gaseous state under standard conditions of pressure and temperature, while "vapor" is used to describe the molecules or atoms released into air from a substance which appears ordinarily as a liquid (such as gasoline or mercury) or a solid. Common usage dictates whether "gas" or "vapor" is used (water vapor, chlorine gas). Note that aerosols are not vapors or gases, but are solid or liquid particles suspended in air (smoke, fog).

Vapor Density—Vapor density is the density of a gas as compared to an equal volume of dry air at the same temperature and pressure. If a substance has a vapor density less than 1.0, such as ammonia, it will rise in air and may be trapped near the ceiling in a room or in other confined spaces. If the vapor density is more than 1.0, such as hydrogen sulfide, the gas will sink in air and will likely be found in low areas such as ravines and ditches. These "heavier" gases may be dangerous since they often concentrate in the breathing zone.

The vapor density definition assumes that both the air and the gas are at the same temperature. If this is not the case, the above rule may not apply. For example, while ammonia has a vapor density of less than 1.0 and should rise in air, ammonia vapor at a temperature substantially lower than the surrounding air will likely stay at ground level until the vapor reaches surrounding air temperature.

Vapor Pressure—The pressure exerted by the motion of gas molecules against the sides of a closed container is its vapor pressure. Vapor pressure is dependent on temperature, for as heat energy is added, molecular action of a vapor increases and collisions of these molecules with the container sides increase in frequency and intensity. This pressure is usually expressed in pounds per square inch or atmospheres of mercury. In essence, vapor pressure is a measure of the rate of evaporation or volatilization from a liquid. As vapor pressure increases, the rate of evaporation also increases. Substances with high vapor pressures are usually dangerous because they readily evaporate or volatilize.

Boiling point and vapor pressure are related. As a general rule, liquids with a low boiling point have a high vapor pressure because they give off vapors at a low temperature. One of the hazards of low boiling point and high vapor pressure substances occurs when such substances are confined (as in a tank) and subsequently heated (as during a fire). This can create a "boiling liquid expanding vapor explosion," or BLEVE, a potentially deadly event.

3.3 Chemical Terms Relating to Fire and Explosion Hazards

Information concerning the fire or explosion hazard of a chemical is essential for workers using the substance, as well as for those confronted with emergencies. The use of any single term should be avoided when evaluating fire or explosion hazards. No single chemical property adequately describes a chemical's fire and explosion danger, and the use of a single property can lead to misunderstandings and possibly an improper response. When evaluating fire or explosion hazards, several properties must be considered:

Combustible—Any substance that will burn is considered to be combustible. This term includes all flammable materials. When used as a general definition, combustion refers to solids that are relatively difficult to ignite and that burn relatively slowly, or to liquids with a flash point greater than 100°F (37.8°C). The ease of combustion may depend upon the size, shape and/or chemical nature of the material. Example: metal powders can ignite and burn rapidly, but most are noncombustible as bulk solids.

Noncombustible—A solid, liquid or gas that will not ignite or burn is considered to be noncombustible. Examples: silicon dioxide, water, and carbon dioxide.

Flammable—A substance that is capable of being easily ignited or of burning quickly is said to be flammable. The definition according to 29 CFR 1910.106(a) reads: flammable liquid—any liquid with a flash point of less than 100°F (38°C) [except mixtures with components having flash points of 100°F or higher and comprising 99 percent or more of the total volume of the mixture]. Flammability also applies to the rate at which flames spread once ignition occurs.

The terms flammable, combustible, and noncombustible are not always easy to separate in application. Any material that will burn at any temperature is combustible by definition. Therefore, flammable applies to a special group of combustible materials that ignite easily and burn rapidly.

Flash Point—Flash point is the minimum temperature at which a substance gives off sufficient flammable vapors to momentarily ignite if a spark source is present. Example: a substance (usually a liquid) is evaporating fast enough to provide adequate vapors that will ignite near the surface of the liquid, but vapors are not produced fast enough to sustain combustion. The vapors flash and then go out. Regarding fire, the flash point is the danger point because liquids at this temperature produce enough vapors to momentarily burn.

Flash point is determined by either an open cup or closed cup method. Generally, the flash point for the open cup method is 5° to 10°F higher than when the closed cup method is used. The closed cup reading is generally preferred since the flash point is at a lower temperature and thus provides an extra measure of safety. A flash point of less than 140°F (60°C) closed cup, is the criterion used by the EPA to determine if a chemical is hazardous by ignitibility. DOT classifies materials with flash points less than 140°F (60°C) as flammable and above 141°F (60°C) and below 200°F (93°C) as combustible.

Fire Point—The temperature at which a liquid gives off enough vapor to continue to burn when ignited is its fire point. This temperature is higher than the flash point, but often only a few degrees higher. Therefore, a substance at the flash point may not require much additional heat to reach the fire point and burn freely.

Ignition (or Auto-Ignition) Temperature—The minimum temperature required to initiate or cause self-sustained combustion (burning) in the absence of any source of ignition is a material's ignition temperature. Example: a burning building may generate enough heat to increase the temperature of an adjacent fuel tank to the ignition temperature. At this temperature, the fuel will start to burn even though no spark has ignited the fuel. Ignition temperature is also called auto-ignition temperature and is higher than the flash point and the fire point. Acetone has a flash point of 0°F (–18°C), a fire point of 40°F (4°C), and an auto-ignition temperature of 869°F (465°C).

Many variables affect ignition temperature, such as the shape and size of the space in which the ignition occurs, the percentage of composition of the air-vapor mixture, and the rate and duration of heating. Therefore, listed ignition temperatures should be considered approximations.

Lower Explosive (or Flammable) Limit—The lower explosive limit (LEL) is the minimum concentration of vapors in the air necessary to have a fire or explosion, provided an ignition source is present. The LEL is usually expressed as a percentage. Therefore, at 100 percent LEL, there are sufficient vapors in the air for a fire or explosion, provided there is an ignition source. When the LEL is below 100 percent, vapor concentrations are too dilute or too lean to support combustion.

Upper Explosive (or Flammable) Limit—The upper explosive limit (UEL) is the maximum concentration of vapor in the air, above which a fire or explosion will not occur in the presence of an ignition source. Above the UEL there is too much fuel in the air and not enough oxygen for a fire or explosion to take place. This can be likened to "flooding" a carburetor in a car with gasoline, making ignition impossible.

Even though a fire or explosion is not possible above the UEL, it does not mean the situation is benign. In a UEL condition, there is potentially insufficient oxygen to sustain life as well as toxic concentrations of vapors. In addition, a puff of air can quickly decrease the fuel concentration to the point where a fire or explosion is now possible.

Flammable Range—The range from 100 percent LEL to the UEL is the flammable range. In this range there is sufficient, but not excessive, fuel in the air to support a fire or explosion. This range denotes an explosive or flammable environment that should not be entered, since any ignition source such as sparking from a motor, a light switch, or monitoring instruments could cause an explosion.

3.4 Other Chemical Terms

3.4.1 CORROSIVITY

Solids, liquids, and gases are considered corrosive if they cause visible destruction of human skin at the site of contact or if they have a severe deterioration rate on steel or aluminum. The EPA defines a corrosive as a substance having a pH of less than 2.0 or greater than 12.5.

The pH scale is used to define corrosivity. This scale expresses the concentration of hydrogen ions (H^+) in solution and is a logarithmic scale to the base 10. In other words, when the pH decreases from 3.5 to 2.5, the concentration of hydrogen ions increases ten times. The range of the pH scale is from 0 to 14. A pH of 7.0 is neutral. Substances above or below the neutral point are defined as follows.

Acid—An acid is a substance that has a pH less than 7.0. They typically have a sour taste, such as vinegar that contains acetic acid. Strong acids will blister and "burn" skin and can cause irreversible tissue damage.

Base—A base is a substance that has a pH greater than 7.0. A base is also called "caustic" or "alkaline." They typically have bitter tastes, such as baking soda, and make the skin slippery. Strong bases will emulsify skin and turn it into a thick liquid. As with acids, bases can cause irreversible tissue damage.

Every liquid has a pH; this value for some common substances follows.

Substance	*Approximate pH*
Battery acid (hydrochloric acid)	0.0–0.2
Stomach (gastric) juices	1.0–3.0
Lemon juice	2.2–2.4
Vinegar	2.4–3.4
Wine	2.8–3.8
Tomatoes	4.0
Black coffee	4.8–5.2
Milk	6.3–6.6
Pure Water	7.0
Blood	7.3–7.5
Sea water	7.9
Soap solutions	8.5–10.0
Aqueous Ammonia	11.6
Bleach	12.2
Sodium Hydroxide	14.0

When acids and bases combine, they can react violently. This neutralization process can create substantial heat and spattering.

3.4.2 REACTIVITY

Chemicals classified as reactive are capable of violent or explosive decomposition when exposed to air, water, shock, heat, or ignition sources. This decomposition may produce a shock wave, fire, toxic vapors, or other undesirable reactions. Such substances are often in common use and must be handled with extreme care. Reactivity can be described by the following terms:

Air Reactive—A substance that violently decomposes or burns vigorously when exposed to air is termed air reactive. White phosphorous will ignite and burn in air at temperatures above 85°F (29°C).

Water Reactive—A water reactive substance is one that violently decomposes or burns vigorously when in contact with water. Pure sodium violently reacts with water to form sodium hydroxide (a base) and hydrogen. Strong bases and acids often react violently with water to form heat and vapors.

Pyrophoric—Any liquid or solid that will ignite spontaneously in air is pyrophoric. Classes of pyrophoric chemicals include, but are not limited to: metal powders (aluminum, cobalt, iron, magnesium, zinc, etc.), metal hydrates (sodium hydrate), white phosphorus and alkali metals (potassium, sodium). In addition, some compounds are called pyrophoric because they spark when slight friction is applied (like a match).

Unstable—Unstable chemicals decompose, condense, or crystallize in a violent manner. Liquid picric acid will form crystals that are shock sensitive and can explode; nitroglycerin and old dynamite act in a similar manner.

Oxidation—Oxidation is a reaction that occurs when oxygen combines with another element or compound. Rust and fire are two forms of oxidation.

Oxidizer—An oxidizer is a substance that readily yields oxygen when involved in a reaction, particularly combustion. This oxygen can greatly accelerate and intensify burning.

Cryogenic—The term cryogenic relates to very low temperatures, often with liquefied gases that have a boiling point of –100°C (–148°F) or lower. For example, the boiling points of oxygen, hydrogen, and helium are –183°F (–119°C), –253°F (–158°C), and –269°F (–167°C), respectively. One major hazard of cryogenic liquids is their ability to expand rapidly when they revert to a vapor state. For example, liquified methane can expand approximately 630 times its original volume.

3.5 Sources of Information

Bassow, Herbert: *Air Pollution Chemistry, An Experimenter's Sourcebook.* Rochelle Park, NJ: Hayden Book Company, Inc., 1976.

Bryner, Gary C.: *Blue Skies, Green Politics—The Clean Air Act of 1990 and Its Implementation.* Revised ed. Washington, DC: Congressional Quarterly Press, Inc., 1995.

Cheremisinoff, Nicholas P., J.A. King, and R. Boyko: *Dangerous Properties of Industrial and Consumer Chemicals.* New York: Marcil Dekker, Inc., 1994.

Dickson, T.R.: *Introduction to Chemistry.* 7th ed. New York: John Wiley & Sons Inc., 1995.

Joesten, Melvin D., et al: *World of Chemistry.* 2nd ed. Philadelphia, PA: Saunders College Publishing, 1995.

Lappé, Marc: *Chemical Deception, The Toxic Threat to Health and the Environment.* San Francisco: Sierra Club Books, 1992.

Lewis, Richard J. Sr.: *Hawley's Condensed Chemical Dictionary.* 13th ed. New York: Van Nostrand Reinhold Company, 1997.

Lide, David R. (editor-in-chief): *CRC Handbook of Chemistry and Physics.* 78th ed. New York: CRC Press, 1997–1998.

Manahan, Stanley E.: *Fundamentals of Environmental Chemistry.* Boca Raton, FL: Lewis Publishers, 1993.

Meyer, Eugene: *Chemistry of Hazardous Materials.* 3rd ed. Englewood Cliffs, NJ: Prentice Hall, Inc., 1997.

Proctor, Nick H. and J.P. Hughes: *Chemical Hazards of the Workplace.* Philadelphia, PA: J.G. Lippincott Company, 1978.

Sax, N. Irving and R.J. Lewis, Sr.: *Hawley's Condensed Chemical Dictionary.* 11th ed. New York: Van Nostrand Reinhold Company, 1987.

Sullivan, Thomas F. (ed.): *Environmental Law Handbook.* 14th ed., Rockville, MD: Government Institutes, Inc., 1997.

U.S. Department of Health and Human Services: *NIOSH Pocket Guide to Chemical Hazards.* Washington, DC: U.S. Government Printing Office, 1997.

U.S. Department of Health and Human Services: Online supplements to *NIOSH Pocket Guide to Chemical Hazards.* In www.niosh.gov. Washington, DC: August, 2001.

U.S. Patent and Trademark Office: Patenting Trends in the United States—State Country Reports 1963–1996. In *Statistical Abstract of the United States, 1998: The National Data Book.* 118th ed. Washington, DC: U.S. Department of Commerce/Bureau of the Census, 1998.

FOUR

Basic Toxicology

4 | CONTENTS

Hazardous waste sites present chemical risks different from other occupational situations that involve chemical contact. Waste sites are generally characterized by uncontrolled or partially controlled chemical releases. These chemicals seldom occur alone, and are usually found in mixtures that have poorly understood health effects. Hazardous waste sites often have chemicals in more than one physical state (gaseous, liquid, or solid) and many sites contain unknown chemical hazards.

Such complex chemical conditions emphasize the need to understand chemicals and their effect on the human body. Toxicology is the science that studies the harmful effects of chemicals on living organisms, especially humans. The new sciences of behavioral toxicology, genetic toxicology, and neurotoxicology have been established over the past two decades to evaluate specific reactions to exposures to toxic substances. Presently it is difficult to say with certainty whether symptoms seen in large-scale human exposures to toxic substances (such as residents near toxic waste dumps) can be attributed to toxic exposure, psychosomatic illnesses, or other unknown combinations of previous illness or exposures. Some on-going studies of chemical exposures appear to show a connection between toxic material releases and health concerns, while other studies are not conclusive.

The term "toxin" is often used to describe harmful chemicals present as contaminants in air, water, or soil. Poison is a generic term that can include any substance having a negative effect (illness or death) when introduced into the body. It may be more correct to refer to poisons as toxic substances. However, poisons are usually limited to materials that are dangerous in small amounts and are often defined or controlled by specific government actions with reference to transportation or consumer products.

4.1 Dose/Response Relationships

Toxic substances that enter the human body may be characterized by a dose/response relationship. The response is dependent on the dose, or how much of a substance is taken over a period of time. A quart of alcohol gradually consumed over a year's time will likely have no measurable response. However, if the same quart of alcohol is consumed within one hour, the response could be fatal. Other factors contributing to response may include the chemical combinations and physical characteristics of the substance. For a substance to have a toxic effect, several conditions are necessary: a sufficient amount of a substance, a pathway to get the material into the body, absorption by the target organ of the body, and a limited period of dose time that overcomes the body's defense mechanisms (which vary considerably from substance to substance).

In the workplace, the dose/response relationship

(or the amount of a chemical to which a worker can be exposed over a set period of time) is expressed as an exposure limit. These limits have been established by various organizations including OSHA, NIOSH, the American Conference of Governmental Industrial Hygienists (ACGIH), and the American Industrial Hygiene Association (AIHA). All of these groups may study doses for acute and chronic responses and recommend or enact exposure limits.

4.2 Exposure Routes

For a chemical to exert a toxic effect on an organism, it must first gain access to that organism's cells and tissues. In humans, the major routes by which toxic chemicals enter the body are inhalation, ingestion, dermal (skin) absorption, and injection. Tissues of the gastrointestinal (GI) tract, mucous membranes, lungs, and skin all offer absorptive surfaces through which chemicals can enter the body.

4.2.1 INHALATION

The exposure pathway of greatest concern is inhalation, which brings chemicals into direct contact with the lungs. Most inhaled chemicals are gases, such as carbon dioxide, vapors of volatile liquids such as trichloroethene (TCE), or small particles such as smoke. Absorption in the lungs is high because the surface area in the alveoli (tiny air sacs located at the ends of the lung airway) is large and blood vessels are close to the exposed surface. The surface area of the lungs is 45 times greater than the skin. This is a difference of about 90 square meters for the lungs compared to about 2 square meters for the skin. Some vapors, especially organic or non-aqueous vapors, quickly reach deep interior regions of the lung and pass through the thin, permeable walls of the alveoli directly into the bloodstream. The rate of absorption depends on the solubility of the toxic agent in the blood, the rate of breathing and the concentration of toxic substances.

Some chemicals do not directly affect the lungs but pass through the lung tissue into the bloodstream where they are transported to other vulnerable areas of the body. For example, when a sufficient dose of benzene is inhaled, it is carried by the blood to the bone marrow where it may inhibit the production of blood cells.

While most inhaled chemicals are in a gaseous state, mist (tiny liquid droplets) and especially particles, can become lodged in the lungs. Particles smaller than 10 microns (aerodynamic diameter) can bypass the normal defense mechanisms of nose hair and sticky mucus and enter the lungs. Some particles can remain in the alveoli indefinitely and cause lung disease. Asbestos, silica dust, and coal dust are examples of particles that threaten the respiratory system.

4.2.2 INGESTION

Ingestion brings chemicals into contact with the tissues of the GI tract from the mouth to the large intestine. The normal function of the GI tract is the digestion and absorption of foods and fluids. The stomach and intestines are also effective in absorbing toxic chemicals that contaminate food and water, including cyanide and pesticides. Chemicals absorbed in the GI tract pass through the liver before entering the bloodstream, making this target organ particularly vulnerable.

While some large doses of toxins may be removed from the GI tract by physical means such as inducing vomiting, this cannot be done for small, chronic doses or for caustic materials. Accepted standard operating procedures forbid eating, drinking, smoking, chewing gum, or chewing tobacco to prevent the ingestion of toxic substances at a hazardous waste site.

Toxic chemicals can also be ingested from the air. For example, lead particles in roadside dust can be inhaled. Because of their usual large size, they become trapped in the mucous membrane. Most of this mucus is swallowed and once in the stomach, digestive acids dissolve the lead (or other heavy metals) and allow it to enter the bloodstream.

4.2.3 SKIN AND EYE ABSORPTION

Some chemicals may cause skin and/or eye injury following direct contact or exposure. Fat-soluble chemicals pass readily through the skin into the bloodstream where they are transported to target organs. For example, when phenol is absorbed in large quantities, it is carried to the nervous system where it may cause numbness, paralysis, or death, depending on the dose and susceptibility of the individual. It should be remembered that the blood supply to the skin is one of the richest in the body, so care should be taken to avoid exposure even to dry materials.

Skin absorption may be enhanced by abrasions,

cuts, heat, and moisture. Dry materials touching the skin on a warm day may mix with perspiration and be absorbed as a liquid. Moist skin is 10 times more permeable than dry skin, and oily solutions generally permit more absorption than water-based solutions. The eye is particularly vulnerable because airborne chemicals can dissolve on its moist surface and cause local damage, or can be carried to the rest of the body through the bloodstream.

Appropriate safety procedures can protect workers from dermal exposure to chemicals. Use of protective gloves and garments, not wearing contact lenses in contaminated atmospheres (since they may trap and hold chemicals against the eye surface), keeping hands away from the face, and minimizing contact with liquid and solid chemicals can help protect against skin and eye contact.

4.2.4 EXPOSURE BY PENETRATION OR INJECTION

The remaining primary route of chemical exposure is injection or penetration. Chemicals can be introduced into the body through puncture wounds, by stepping or falling onto contaminated sharp objects, or through injection from equipment such as spray guns or needles. This may happen quickly or without warning. Drums containing chemical waste can have sharp edges and slivers. Recently, a worker in a hospital was preparing a room for asbestos removal. While reaching behind the heat register, he punctured his finger with a needle that had been used on a patient with HBV. The worker caught the disease and survived, but with side effects that will persist throughout his life.

Workers must cover cuts or abrasions with waterproof bandages. Wearing safety shoes and appropriate hand wear, avoiding physical hazards (discussed in Unit 5), and taking common-sense precautions are important protective measures against chemical injection.

4.3 Exposure Limits

4.3.1 OSHA EXPOSURE LIMITS

OSHA, whose jurisdiction includes promulgation and enforcement responsibilities, has set chemical exposure limits that are legally enforceable. Exposure limits are specified in Tables Z-1-A, Z-2, and Z-3 of the OSHA General Industry Air Contaminants Standard (29 CFR 1910.1000). These values are based on a variety of considerations, including scientific data from NIOSH and ACGIH, as well as political and economic realities. The OSHA exposure limits given below are for airborne respiratory hazards.

Permissible Exposure Limit (PEL)—The PEL is a legally enforceable time-weighted average (TWA) concentration that must not be exceeded during any 8-hour shift of a 40-hour work-week. For example, the PEL for phenol is 5 parts per million (ppm). This means that during an 8-hour day, the average concentration of phenol in the air a worker breathes must not exceed 5 ppm, with some restrictions that are discussed below. The PEL definition implies that at or below the exposure limit no adverse effects should be experienced in a "normal" worker. In order to know that a PEL has not been exceeded, air monitoring must be done for the work environment through all shifts that utilize the product.

PEL-Ceiling (PEL-C)—The PEL-C is an airborne concentration of a substance that shall not be exceeded at any time. If instantaneous measurement is not possible, the ceiling is a 15-minute TWA not to be exceeded during any part of the workday; however, some exceptions to this time interval are specified in the regulation. Hydrogen chloride has a ceiling of 5 ppm for a 15-minute exposure.

Carcinogens (Ca)—Without establishing PELs, OSHA set standards in 1974 regulating the industrial use of chemicals identified as occupational carcinogens. Exposure of workers to these chemicals is regulated through the required use of engineering controls and work practices, reinforced by training, and stipulated with special PPE. Details of these specific chemicals and protection requirements are located in 29 CFR 1910.1003.

Short Term Exposure Limit (STEL)—This is the average exposure concentration for a 15-minute work period that must not be exceeded at any time during a work shift. The time period for averaging exposure may vary depending on the chemical, but is most often 15 minutes.

PEL-Peak (PEL-P)—These exposure limits are airborne concentrations that may not exceed the maximum duration and concentration specified by OSHA. For example, toluene has a PEL—8-hour TWA of 200 ppm; a PEL-C of 300 ppm; and a PEL-P of 500 ppm for a maximum duration of 10 minutes for an 8-hour shift.

Action Level—When this mandated or advisory level is detected, some type of protective action must be taken. Furthermore, employers at hazardous waste sites often take protective measures when exposure reaches 50 percent or even 10 percent of the applicable exposure limit (PEL). Such readings trigger certain actions.

Note that OSHA sets the above standards for *airborne* exposure limits. Skin contact exposure limits are generally approached through recommendations that all skin contact with chemicals be minimized by the use of engineering controls and/or PPE, and that any skin contaminated with chemicals be immediately cleaned.

OSHA's PELs are not often revised. When they are, a lengthy process may be required in order for changes to be approved. Public and private sectors may comment on such changes, causing delays. Court challenges are likely. In 1989, OSHA added PELs for 164 substances and made existing PELs more protective for 212 substances. On July 7, 1992, the 11th Circuit Court of Appeals refused these changes, and OSHA decided not to appeal the decision to the Supreme Court. OSHA formally revoked the 1989 standards and reinstated the limits which had previously been in effect since 1971. These 1971 PELs are based on research conducted primarily in the 1950s and early 1960s, and in many cases, do not adequately protect worker health. Industry is unsure how to respond to this PEL confusion. Many companies have already put in place procedures that comply with the 1989 standards, and some states have decided to continue enforcement of the 1989 standards. Some states have established their own list of allowable exposures, often relying on NIOSH and ACGIH limits for guidelines.

When a PEL is not available, the Recommended Exposure Limit (REL) set by NIOSH may be enforced as law under the OSHA General Duty Clause 29 CFR 1960.8 (a). The General Duty Clause mandates that each employee must be furnished with a workplace free from recognized hazards. OSHA sometimes uses this clause to enforce serious violations that are not addressed by a specific OSHA standard or program elements.

Some workplace hazards are controlled by OSHA through work performance standards rather than through specific airborne exposure limits. Work practice standards include, hazard communication, confined spaces, process safety, bloodborne pathogens, and chemical hygiene in laboratories.

4.3.2 NIOSH EXPOSURE LIMITS

NIOSH, a division of the U.S. Department of Health and Human Services, recommends exposure limits based on health effects. Unlike OSHA, NIOSH does not factor in the feasibility or practicality of enforcing the limits they set. For example, OSHA has established a PEL of 1.0 ppm for benzene, while NIOSH recommends an exposure limit of 0.1 ppm. Discussions by industry tried to show that equipment and protective devices for achieving the 0.1 ppm rate would be very costly, and at that level even people pumping their own gasoline would have to wear protective respiratory gear. In this case, economic practicality got in the way of adopting the NIOSH standard.

NIOSH, as an agency, has no enforcement authority. It releases exposure values as soon as its research has been completed. In contrast, OSHA must go through a lengthy process of rule promulgation before its PELs are approved and become law. As a result PELs are often out-of-date, while NIOSH values may be more current. Exposure values set by NIOSH are as follows:

Recommended Exposure Limit (REL)—This value is defined as a TWA for up to a 10-hour day during a 40-hour workweek, to which nearly all workers may be repeatedly exposed without adverse effects. RELs are generally equal to or lower than PELs. For example, both OSHA and NIOSH use 1.0 ppm as PEL and REL for hydrogen peroxide. But OSHA uses a PEL 0.75 ppm for formaldehyde while NIOSH recommends an REL of 0.016 ppm for the same substance.

Short Term Exposure Limit (STEL)—This is defined as a 15-minute TWA that should not be exceeded at any time during the workday. The REL for sulfur dioxide is 2 ppm while the STEL is 5 ppm. The STEL is higher than the REL and is sometimes comparable to the PEL-P set by OSHA.

REL-Ceiling (REL-C)—This is a value that should not be exceeded at any time. It is comparable to the ceiling values set by OSHA and, like those values, is uncommon. Calcium arsenate has a PEL-C of 0.010 mg/m3, but NIOSH sets a REL-C of only 0.002 mg/m3 because of its carcinogen status. Chlorine has a PEL-C of 1 ppm with an REL-C of 0.5 ppm.

Immediately Dangerous to Life or Health (IDLH)—As defined by OSHA, the IDLH is an atmospheric concentration of any toxic, corrosive or potentially asphyxiating substance that poses an immediate threat to life, or that would interfere with an individual's ability to escape from a dangerous atmosphere. As defined by NIOSH, it is a condition that poses a threat of exposure to airborne contaminants when that exposure is likely to cause death or immediate or delayed permanent adverse health effects or prevent escape from such an environment. IDLH environments are life threatening and must be approached with extreme care and a high level of personal protection. While IDLH values are set by NIOSH and are not enforceable, most agencies treat these values as if they were enforceable. Trichloroethane has an IDLH of 100 ppm, while the IDLH of styrene is 700 ppm.

Carcinogens (Ca)—NIOSH recommends that occupational exposure to carcinogens be limited to the lowest feasible concentration. This is very general and covers only those substances that are classified "Ca." More materials may be placed in this category in the future.

4.3.3 ACGIH EXPOSURE LIMITS

The ACGIH is a professional organization comprised primarily of industrial hygienists working for federal, state, or local governments, or in academia. This group recommends exposure limits for over 700 chemicals with updates published every year. For 2001, ACGIH made a notice of intended changes for 47 substances. ACGIH sets Threshold Limit Values (TLVs) defined as airborne concentrations of substances to which it is believed nearly all workers may be repeatedly exposed day after day without adverse health effects. As with NIOSH, ACGIH values generally are not legally enforceable. However, if no PEL values exist, OSHA has the option of enforcing a REL or TLV at a hazardous site under the General Duty Clause. Three categories of TLVs are specified.

Threshold Limit Value (TLV)—This is the TWA concentration for a normal 8-hour workday and a 40-hour work-week, to which nearly all workers may be repeatedly exposed, day after day, without adverse effects. In most cases, calculation of TLV is for an 8-hour workday, but in some cases it is permissible to use a 40-hour work-week. These values are based on the same concept as RELs and PELs. For naphthalene the PEL, REL, and TLV are all 10 ppm, while for benzene the PEL is 1.0 ppm, the REL is 0.1 ppm, and the TLV is 0.5 ppm.

TLV-Ceiling (TLV-C)—This is a concentration that should not be exceeded during any part of the work period. The ceiling value is set because there is a group of substances which is generally fast-acting and must be carefully monitored. The 1998 list of substances includes over 50 products with this characteristic. Most are in the form of gases or vapors. All three organizations list iodine as having a ceiling of 0.1 ppm. The ceiling values listed by OSHA and NIOSH are similar in concept.

TLV-STEL—The TLV-STEL is the maximum concentration of potentially hazardous material to which workers can be exposed for a period of up to 15 minutes continuously without suffering from irritation, chronic or irreversible tissue damage, or narcosis of sufficient degree to increase the likelihood of accidental injury or impairment of self-rescue. It is a TWA exposure that should not be exceeded at any time during a workday, even if the 8-hour TWA is within the TLV. Exposures of the STEL should not be longer than 15 minutes and should not be repeated more than four times daily. There should be at least 60 minutes between successive exposures at the STEL. The STEL is not a separate, independent exposure limit, but rather supplements the TLV-TWA limit when there are recognized acute effects from a substance that has primarily chronic effects. The STELs are recommended only when toxic effects in humans or animals have been reported from high short-term exposures, such as with heptane, which has a STEL of 500 ppm.

Carcinogen (Ca)—ACGIH recognizes two categories of human carcinogens. Confirmed human carcinogens are substances recognized as having the potential to produce cancers. This is based on evidence from epidemiological studies of disease in a population or clinical evidence in exposed humans. Examples of those 19 currently listed in this category are asbestos, beryllium, coal tar pitch volatiles, uranium and vinyl chloride.

Suspected human carcinogens are able to cause cancers in experimental animals and are suspected of causing cancers in humans. In most cases, the existing data is insufficient to suspect a cancer risk to humans. The suspected carcinogens list includes vinyl bromide, acrylonitrile, and lead chromate. ACGIH is the only professional group to set TLVs for some carcinogens in both categories and recommends that exposure to all confirmed or suspected carcinogens be kept to a minimum.

4.3.4 AIHA EXPOSURE LIMITS

The American Industrial Hygiene Association is the world's largest association of occupational, environmental health, and safety professionals. Included within AIHA is the Academy of Certified Industrial Hygienists. Unlike ACGIH and NIOSH, AIHA is directly involved in submitting recommendations and comments in matters affecting occupational health and safety regulations. The exposure limits which have been set as guidelines by AIHA are the Emergency Response Planning Guidelines (ERPGs) and Workplace Environmental Exposure Level Guides (WEELs).

ERPGs–Air exposure guidelines for more than 200 chemicals have been established by AIHA. ERPGs are one-hour planning guidelines developed by the AIHA to protect the general public (and workers) from the consequences of accidental chemical releases. The AIHA evaluates chemicals for their human and animal toxicity and places these chemicals in hazard categories depending on the degree of exposure hazard, as follows:

ERPG 3 is the maximum airborne concentration below which, it is believed, nearly all individuals could be exposed for up to one hour without experiencing or developing life-threatening health effects.

ERPG 2 is the maximum airborne concentration below which, it is believed, nearly all individuals could be exposed for up to one hour without experiencing or developing irreversible or other serious health effects or symptoms that could impair their abilities to take protective action.

ERPG 1 is the maximum airborne concentration below which, it is believed, nearly all individuals could be exposed up to one hour without experiencing other than mild transient adverse health effects or perceiving a clearly defined objectionable odor.

WEELS—These guidelines represent the workplace exposure levels to which, it is believed, nearly all individuals could be exposed repeatedly without experiencing adverse health effects. The AHIA WEEL Committee typically concentrates on chemicals for which there are no existing guidelines. Nearly 100 chemicals are included.

4.4 Limitations of Exposure Values

When evaluating exposure limits for specific situations, a variety of limitations must be kept in mind. Some are discussed below.

The three organizations (OSHA, NIOSH, and ACGIH) that establish numerical exposure limits often do not agree on what is a safe limit. AIHA WEELS includes chemicals that are typically not evaluated by OSHA, NIOSH, or ACGIH. Therefore, the limits set by all four organizations should be evaluated when establishing exposure limits for a specific situation. Whenever possible, the most conservative (or lowest) limit should be used. In addition, no single expression of exposure is adequate in protecting worker health, so a variety of values including TWAs, STELs, and ceiling values must be used (see Table 4-1).

When exposed to chemicals, there is no sharp line

Table 4-1 **Examples of the Comparative Values of Several Chemicals As Specified by OSHA, NIOSH, ACGIH, and AIHA**

	OSHA				NIOSH				ACGIH			AIHA	
Chemical	**PEL**	**ST**	**C**	**Peak**	**REL**	**STEL**	**C**	**IDLH**	**TLV**	**STEL**	**C**	**ERPG-1**	**WEEL (8-hr)**
Benzene	1 ppm	5 ppm	—	—	Ca 0.1 ppm	1 ppm	—	500 ppm	Ca 0.5 ppm	2.5 ppm	—	50 ppm	—
Formalin as Formaldehyde	0.75 ppm	2 ppm	—	—	Ca 0.016 ppm	—	0.1 ppm	20 ppm	Suspect Ca	—	0.3 ppm	—	—
Glutaraldehyde	—	—	—	—	—	—	0.2 ppm	—	—	—	0.05 ppm	—	—
Hydrogen Chloride	—	—	5 ppm	—	—	—	5 ppm	50 ppm	—	—	5 ppm	3 ppm	—
Isophorene	25 ppm	—	—	—	4 ppm	—	—	200 ppm	—	—	5 ppm	—	—
n-Pentane	1000 ppm	—	—	—	120 ppm	—	610 ppm	1500 ppm	600 ppm	—	—	—	—
Styrene	100 ppm	—	200 ppm	600 ppm; 5 min. max. peak in any 3 hours	50 ppm	100 ppm	—	700 ppm	20 ppm	40 ppm	—	50 ppm	—
Toluene	200 ppm	—	300 ppm	500 ppm; 10 min. max. peak	100 ppm	150 ppm	—	500 ppm	50 ppm	—	—	50 ppm	—
Aldicarb	—	—	—	—	—	—	—	—	—	—	—	—	0.07 mg/m^3
Urea	—	—	—	—	—	—	—	—	—	—	—	—	10 mg/m^3

between safe and unsafe. In the same way, there is no safety in numbers alone. Chemical exposure values are similar to speed limits. When the speed limit is exceeded, the risk of injury or death is increased. However, many people successfully exceed the limit. Conversely, traveling below the speed is no guarantee that injury or death will not occur.

Likewise, meeting a chemical exposure limit does not guarantee a person's well being, nor does exceeding the limit guarantee that a person will be injured. However, by not exceeding the exposure limit, the risk of injury is reduced. Exposure values are not a fine line between safe and unsafe conditions. Therefore, it is to the benefit of workers that a meter reading mentality to exposure limits be avoided.

The lack of safety in numbers also relates to how toxic effects are measured. Most exposure values are based on acute effects that result from exposure to high concentrations of chemicals over a short period of time. Acute is defined as a short-term toxic effect that occurs within 24 hours after exposure. Such effects may be relatively easy to measure, and cause-and-effect relationships can be established. Effects may range from eye irritation to nausea or to severe skin burns from acid, to coma, and to death.

Conversely, workplace exposure generally involves low exposure to chemical concentrations over an extended period of time. Health effects from such exposures may not be evident for many years. These delayed responses are termed chronic effects. The difficulty of determining what specific substance caused a specific problem ensures that cause-and-effect relationships for chronic illnesses are still hotly debated in science and industry. For this reason, only about 20 percent of the exposure values are based on chronic effects. Time elapsed between exposure and the onset of health problems, and the multitude of chemicals most people are exposed to, make documentation of cause-and-effect difficult. Asbestos, silica and cotton dust are examples of substances that produce chronic effects, yet it took decades of research to firmly establish the link between exposure and disease. Varying degrees of human sensitivity to chemicals also contribute to vagueness in exposure values.

To help attempt to distinguish safe from unsafe, the use of risk assessment is becoming more common. Risk assessment includes a qualitative judgment about the toxicity of a substance based on the existing evidence. This evidence includes dose-response assessment of animal experiments, human toxicology studies, an estimate of the likely degree of human exposure to a specific chemical, and a description of the nature and magnitude of human risk. Arguments have been made for and against these procedures, most dealing with the value of the human life saved from exposure to hazardous substances and the cost to industry to keep people protected from such substances.

Finally, chemical exposure studies are based largely on animal research. The difficulty is in relating animal response to human response. For example, guinea pigs are very sensitive to dioxins, while hamsters can tolerate a dosage 900 times greater. When setting toxic limits for humans, the decision of whether we are more like guinea pigs or hamsters can make a 900-fold difference in the allowable exposure level, and a tremendous difference in compliance costs.

4.5 Sensitivity

The PEL, REL, TLV, and WEEL are examples of exposure limits for workers. These are not acceptable exposure limits for the general population. Also, some people, workers and others, have adverse responses to chemicals at exposures well below the PEL, REL, or TLV, or WEEL while others appear immune to high levels of chemicals. Human sensitivity to chemical exposure is highly variable and depends on factors such as prior exposure, genetic differences, gender and hormonal differences, nutritional factors and medications, age and maturity, and smoking.

4.5.1 PRIOR EXPOSURE

When first exposed to a chemical, some people may show no adverse effects. However, upon subsequent exposures, even to extremely low doses, they may show a severe reaction. Formaldehyde, toluene, 2,4 diisocyanate, and chromium are chemicals that elicit such responses.

In other cases, prior exposure to a chemical may increase tolerance. Ammonia, alcohol and some pain and antibiotic medications are examples. Either type of reaction is difficult to predict.

4.5.2 GENETIC DIFFERENCES

Genetic variations in humans allow for wide differences in susceptibility to toxic chemicals. The genetic makeup of an individual determines the prevalence (or absence) of key enzymes used to process toxic chemicals in the liver and other cells. Additionally, several genetic variations in metabolism affect the response of the body to a toxin.

4.5.3 GENDER AND HORMONAL DIFFERENCES

Gender and hormonal status appear to be factors in susceptibility to certain chemicals. There is not an across-the-board situation where one sex is more or less sensitive to chemicals; it depends on the nature of the toxin. For example, female mice show little response to chloroform exposures that are lethal to male mice. Yet female rats and rabbits are more susceptible to the toxic effects of parathion and benzene than are the males. Pregnancy has been shown to markedly increase the susceptibility of mice to some types of pesticides, and similar effects have been reported for lactating animals exposed to heavy metals.

4.5.4 NUTRITIONAL FACTORS AND MEDICATIONS

Generally, people who lack essential minerals in their diet will absorb and retain these minerals along with an increased absorption of any toxic metals such as cadmium and barium. Malnourished children are particularly susceptible to harm following exposure to lead. Deficiencies in dietary minerals, especially calcium and phosphorous, speed the absorption of lead by the GI system. Recent research has shown that milk sugar (lactose) increases lead absorption. Another common chemical, benzene, seems to have its toxicity enhanced by too much fat in the diet. Research suggests that a vitamin C deficiency increases the toxic effects of DDT. A person on a low-calorie, low-protein diet is usually more sensitive to a number of toxic chemicals.

Certain medications also increase a person's susceptibility to many toxic chemicals. Over-the-counter products containing alcohol may increase the effects of exposure. Knowledge of prescription or over-the-counter products taken by workers should, therefore, be part of a comprehensive baseline medical evaluation in any business operating under OSHA regulations.

4.5.5 AGE AND MATURITY

Some chemicals are more toxic to infants and children than to adults. Lead ingestion has more severe effects on the nervous systems of infants and children than on adults. Also, the very young are less able to metabolize and detoxify chemicals since they have not fully developed the body mechanisms that can effectively detoxify certain chemicals. In addition, certain types of toxic effects can only occur during growth so adults would not be susceptible. For example, the full development of the brain takes from six to seven years after birth. Children younger than this are susceptible to toxic chemicals that can affect brain development, such as learning ability or behavior.

The very old may have lost this same ability to some degree, causing greater absorption of toxic materials in bones, tissues and target organs. Detoxifying capacity of the liver and the excretory capacity of the kidneys decrease with age.

4.5.6 SMOKING

It is well documented that smoking causes or contributes to respiratory and cardiovascular diseases. In addition, smoking acts with asbestos and other chemicals to cause severe damage to the respiratory system. Cigarette smoking also causes increased exposure to carbon monoxide. If workers are frequently exposed to carbon monoxide as part of their job, the additional levels from smoking can cause cardiovascular changes, dangerous especially to people with coronary heart disease. Other chemicals are found in tobacco that workers might be exposed to on the job. These include acetone, aldehydes, arsenic, cadmium, hydrogen cyanide, hydrogen sulfide, ketone, lead, phenol and polycyclic aromatic compounds. Additionally, heat generated by burning tobacco can change chemicals found in the workplace into more harmful substances. Smoking greatly affects the ability of lung tissue to detoxify or remove inhaled contaminants. The toxicity of chemical carcinogens in tobacco may be increased if pre-existing viruses, such as herpes simplex, or particulates, such as iron, are present.

4.6 Chemical Interaction

Many substances act on the body simultaneously, causing the effect of each chemical to be different, either enhanced or diminished, as compared to the effects of each chemical alone. Chemical mixtures have the potential to produce effects that are different from those caused by the single compounds.

4.6.1 ADDITIVE

The effects of two or more chemicals are additive when the combination of those chemicals produces an enhanced effect that is no greater than the sum of their individually measured effects (2+3=5). The ingestion of aspirin and antibiotics usually has an additive effect.

4.6.2 SYNERGISM

Two or more chemicals may work together to produce a synergistic or enhanced effect (2+3=15). Records show that asbestos workers who smoke are 50 times more likely to develop lung cancer than asbestos workers who do not smoke. Enhanced effect has also been found between alcohol and carbon tetrachloride. In one study, chronic alcoholics exposed to carbon tetrachloride experienced greater liver damage than did exposed non-alcoholics. Certain medications may also react with workplace toxins to produce an unwanted synergistic response.

4.6.3 POTENTIATION

Potentiation is the effect of two chemicals, one toxic and the other nontoxic, that combine to cause drastically increased toxic effects (0+3=15). Potentiators do not have a particular effect themselves but exaggerate the effects of other chemicals. The combination of isopropanol, which is not a liver toxin, and carbon tetrachloride, which is a liver toxin, combine to cause significantly more liver damage than carbon tetrachloride alone. Imbibed alcohol, an inhaled narcotic (trichloroethene), and food enhancers such as MSG also work this way.

4.6.4 ANTAGONISM

Antagonism may occur when one chemical actually interferes with another chemical and reduces the damage usually caused by that chemical (4+6=2), such as a base neutralizing an acid. This relationship forms the basis for antidotes. There may be a total counterbalance in which the resultant product is "0" or there may be the production of a less toxic product overall. For example, when taken together, barbiturates and norepinephrine may counteract the other effects on blood pressure since one decreases it while the other increases it; histamines will cancel the effect of antihistamines; people poisoned with methanol are sometimes given ethanol to reverse the reaction.

4.7 Acute and Chronic Toxicity

It is important to realize that chemicals may cause adverse health effects with a single, large exposure or through multiple, low exposures over a period of time.

4.7.1 ACUTE TOXICITY

Acute toxicity is the result of exposure to a chemical over a short period of time, usually within 24 hours. If the victim survives an acute exposure, the effects are often reversible. However, in many cases, the shorter the time period over which the body receives a dose, the greater the damage because the body's mechanisms for repairing or buffering the toxic damage will be overwhelmed. Acute exposures to chemicals are more common in transportation accidents, fires, or releases at chemical manufacturing and/or storage facilities. Many acute exposures result from being splashed by chemicals or from being near ruptured containers. First responders or workers present at the time of an accident or spill are likely to be affected. At abandoned waste sites, personnel engaged in sampling may have an increased risk of acute exposures from splashes, mists, gases, or particulates, in uncontrolled amounts and unknown concentrations.

4.7.2 CHRONIC TOXICITY

Chronic toxicity often results from small exposures over a long period of time (greater than 24 hours by regulatory definition). Chronic effects are not usually reversible and may result in permanent injury or death. Examples of chronic effects are asbestos-linked lung cancer and liver cancer from long-term exposure to vinyl chloride.

Chronic toxic effects are difficult to establish. Humans are exposed to thousands of chemicals, and trying to pinpoint the long-term effect of one while living in a smorgasbord of chemicals is difficult and tedious.

4.8 Other Terms Used in Toxicology

Following are terms, not previously defined, that are commonly used in toxicology. When reviewing these and other terms, keep in mind that toxicology often requires additional expert input from occupational physicians, chemists, and industrial hygienists. It may be difficult for the typical worker to understand chemical information. Because of this, much information can be gleaned from the MSDS. These sheets should be presented in training, read, understood and questions answered as applicable to each product. They should be readily accessible to workers, included as part of the site Health and Safety Plan, and available at key locations in and around the facility.

Asphyxiants are chemicals that interfere with oxygen transport to the body. Simple asphyxiants such as nitrogen, hydrogen, and methane displace the ambient air, leading to suffocation. Other chemical asphyxiants, such as carbon monoxide, inhibit oxygen transfer from the lungs to the cells. Carbon monoxide can cause chemical asphyxiation even when a normal quantity of oxygen is present in the air, since it prevents the body from using the oxygen.

Anesthetics are chemicals that induce a loss of sensitivity in a specific part of the body. Anesthetics can cause impaired judgment, dizziness, drowsiness, headache, unconsciousness, and even death. Anesthetized individuals experience a partial or total loss of the sense of pain. Examples are ether and gasoline.

Neurotoxins are substances that have the potential for slowing or reducing the nerves' ability to control various parts of the body such as peripheral muscles, hearing, and sight. Many of the organic phosphates used in insecticides, such as parathion, are neurotoxins. Some neurotoxins produce a depression of the central nervous system marked by stupor, a decrease in motor functions, and unconsciousness. Examples of neurotoxins are mercury and carbon disulfide.

One of the most striking studies in the effects of neurotoxins was Minamata Disease, named for the concentration of mercury poisoning cases in coastal areas of the Minamata Bay and Shiranui Sea, Japan. A company manufacturing acetaedehyde and vinyl chloride used large amounts of inorganic mercury. In-house chemical processes converted the mercury from an inorganic into a methylated form that was then dumped into the local bay. Here it made its way to the food chain through fish and into humans. Many adults and children eating contaminated fish and shellfish showed signs of bioaccumulation of methyl mercury and of mercury poisoning. This manifested itself in the degeneration of the central nervous system. Unborn children were exposed across the placental membrane, with severe retardation resulting. By the end of the formal study in 1972, nearly 800 cases of mercury intoxication had been verified, with some 2,800 others applying for verification. Cases had been identified in other villages and on-going studies are still being conducted.

Irritants are substances that can cause discomfort or pain to the eyes, skin, or respiratory system. They may injure the tissues of the respiratory system and lungs, thereby causing inflammation of these passages. The mechanism of irritation is generally corrosion of tissue at the site of contact. Tearing of the eyes caused by onions, hydrochloric acid mists or sodium hydroxide dust, or the burning sensation in the nose caused by breathing ammonia are examples of irritant effects.

Carcinogens are substances that cause malignant tumors. Some carcinogens may have a "safe" threshold level below which no cancer should develop, while others may have a "zero tolerance" threshold. Cancer resulting from exposures to carcinogens is considered a chronic effect since considerable time is involved between the exposure and onset of the disease. An example of several generally accepted human carcinogens and related diseases are listed on Table 4-2. A comparison of the definition of carcinogen by OSHA, NIOSH and ACGIH is also noted. The National Toxicology Program (NTP) and the International Agency for Research on Cancer (IARC) are other organizations involved in research and identification of carcinogens.

Reproductive Toxins include substances that lower the fertility of a species, the percent of conceptions leading to live birth, and/or the survival of the fetus. This type of toxicity may lead

Table 4-2 **Examples of Carcinogens and Their Effects**

Carcinogen	Target Organ(s) of Cancer	Other Toxic Effects
*+°Arsenic	Lungs, skin, liver	Spontaneous abortion; low birth weight; nasal ulceration; heart arrhythmia and disease; reduced kidney function; blood vessel spasms; liver cirrhosis; nausea; vomiting; abdominal pain; diarrhea; skin lesions; seizures; arm and leg weakness.
*+°Asbestos	Lungs	Irritation of larynx, eyes, gastrointestinal tract; pulmonary lung disease (asbestosis); finger clubbing.
*+°Benzene	Blood, Bone Marrow (Leukemia)	Anemia; changes in bone marrow; irritant of eyes, skin, nose, respiratory and central nervous system; staggering; headaches; fatigue: anorexia; general weakness; exhaustion.
*+°Beta-naphthylamine	Bladder	Dermatitis and skin hemorrhaging; breathing difficulty; inflammation of the bladder; inability to control motor functions; inability of the blood to carry oxygen; blood in the urine and painful urination.
*+°Vinyl chloride	Liver	Gastrointestinal bleeding; abdominal bleeding; male impotence; liver disease; blood disease; central nervous system degeneration; lymphatic alterations.

*Carcinogen as defined by OSHA +Carcinogen as defined by NIOSH °Carcinogen as defined by ACGIH

to birth defects and may create reduced birth weight or size. Common toxins are tobacco and alcohol. At least 50 chemicals, including lead, cadmium, glycol ethers, ionizing radiation, organohalide pesticides, and organic solvents have produced reproductive impairment in animals.

Mutagens. Substances that cause changes in the genetic makeup (deoxyribonucleic acid (DNA) changes in the chromosomes) of cells are termed mutagens. Mutagenetic changes may also affect later generations if the changed genetic material is passed on to offspring. Human mutagens include mercury and lead compounds, benzopyrene, mustard gas, UV light, and x-rays. Suspected human mutagens are DDT, sodium arsenate, dioxin 2,4D, cadmium sulfate, the pesticide 2,4,5-T, and nitrites, ozone, and benzene.

Teratogens. These are substances that cause abnormalities in the fetus during its development in the uterus. Such materials may be absorbed into a pregnant woman's body, travel to the unborn child via the bloodstream, and cause a variety of congenital malformations. Teratogenic agents are often cell specific so that a particular chemical is related to a specific type of birth defect. Arsenic may increase eye damage; heavy metals such as cadmium, lead, and zinc produce abortions and still births; and mercury can produce brain damage.

Hepatotoxins. Certain substances are toxic to the liver and produce liver damage. Symptoms include jaundice and liver enlargement such as cirrhosis. Examples are bromobenzene, carbon tetrachloride, kepone, methylene chloride, trichloroethene, and vinyl chloride.

Nephrotoxins. Certain substances that produce kidney damage are classified as nephrotoxins. Examples include lead, halogenated hydrocarbons, uranium, cadmium, and mercury.

Systemic. This pertains to the whole body rather than to a localized region or specific organ or to a system of tissues with a common function; for example, alcohol poisoning is systemic, as opposed to corneal damage to the eye which is localized. Hypothermia affects the entire body, whereas frostbite may be confined to toes, fingers, and the nose.

Lethal Dose. This term refers to the amount of a substance required to kill an exposed organism (such as laboratory rats, mice, or rabbits). When extrapolated from laboratory animals to humans, the determined value may not be exact. As was mentioned previously, many factors influence an individual's or species' tolerance or susceptibility to a toxin. Lethal doses of substances are generally expressed in milligrams (mg) of substance per kilogram (Kg) of body weight.

Lethal Dose-50% (LD50). LD50 is the dose administered by absorption or ingestion of a substance that causes death in 50 percent of the exposed population. This is the simplest and most commonly used toxicity test. It is a single-exposure study, with death as the sole criterion of toxicity. A single dose is administered to a limited number of animals and the number of deaths after 14 days is noted. Using rats, approximate acute LD50 exposures for some chemicals are shown below:

Material	*LD50 (mg/Kg)*
Sugar	29,700
Alcohol	14,000
Malathion	1,200
Aspirin	1,000
Lindane	1,000
Ammonia	350
DDT	100
Heptachlor	90
Arsenic	48
Dieldrin	40
Strychnine	2
Nicotine	1
Dioxin	0.001

Lethal Concentration–50% (LC50). This refers to a dose taken in from an airborne concentration (or waterborne for aquatic species) of a substance that causes death in 50 percent of the exposed population. It is similar to the LD50 described above. As with the LD50, the lower the number, the more toxic the substance.

4.9 Sources of Information

American Conference of Governmental Industrial Hygienists: *2001 TLVs and BEIs, Threshold Limit Values for Chemical Substances and Physical Agents.* Cincinnati, OH: American Conference of Governmental Industrial Hygienists, 2001.

American Industrial Hygiene Association: *Emergency Response Planning Guidelines and Workplace Environmental Exposure Level Guides.* Fairfax, VA: American Industrial Hygiene Association Press, 1999.

Anderson, Kenneth N., et al (eds.): *Mosby's Medical Nursing and Allied Health Dictionary.* 5th ed. St. Louis, MO: Mosby, 1997.

Boraiko Allen, A.: Storing Up Trouble: Hazardous Waste. *National Geographic 167, Number 3* (March 1985).

Cheremisinoff, Nicholas P., J.A. King, and R. Boyfo: *Dangerous Properties of Industrial and Consumer Chemicals.* New York: Marcel Dekker, Inc., 1994.

Environmental Protection Agency: *National Priorities List Sites: Colorado.* Washington, DC: U.S. Environmental Protection Agency, September 1990.

Environmental Protection Agency: *National Priorities List Sites: Pennsylvania.* Washington, DC: U.S. Environmental Protection Agency, September 1991.

Freudenthal, Ralph I. and S.L. Freudenthal: *What You Need To Know to Live with Chemicals.* Green Farms, CN: Hill and Garnett Publishing, Inc., 1989.

Harte, John, et al: *Toxins A to Z, A Guide to Everyday Pollution Hazards.* Berkeley, CA: University of California Press, 1991.

Kamrin, Michael A.: *Toxicology—A Primer on Toxicology Principles and Applications.* Chelsea, MI: Lewis Publishers, Inc., 1990.

Koren, Herman: *Handbook of Environmental Health and Safety, Principles and Practices.* 2nd ed., Vols. I and II. Michigan: Lewis Publishers, 1991.

Labar, Gregg: Safety in Numbers? *Occupational Hazards,* pp. 52–54 (August 1990).

LaDou, Joseph (ed.): *Occupational and Environmental Medicine.* 2nd ed. Stamford, CN: Appleton and Lange, 1997.

Lappé, Marc: *Chemical Deception—The Toxic Threat to Health and the Environment.* San Francisco: Sierra Club Books, 1991.

Matthews, Bonnye L.: *Chemical Sensitivity, A Guide to Coping with Hypersensitivity Syndrome, Sick Building Syndrome, and Other Environmental Illnesses.* Jefferson, NC: McFarland & Company, Inc., 1992.

Sackheim, George I. and D.D. Lehman: *Chemistry for the Health Services.* 7th ed. New York: Macmillan Publishing Company, 1994.

U.S. Department of Health and Human Services: *NIOSH Pocket Guide to Chemical Hazards.* In www.niosh.gov Washington, DC: August, 2001.

FIVE

Physical and Environmental Hazards

5 | CONTENTS

The distinguishing trait of hazardous waste sites is the presence of uncontrolled, and often unknown, chemicals. Protecting workers from these chemicals is the primary focus of this book. However, physical hazards are just as important as chemical hazards, and physical injury is more common among workers at hazardous waste sites than is chemical exposure. Ninety five percent of all fatalities and injuries incurred at a hazardous waste site are not due to chemicals, but rather to physical, mechanical or kinetic factors. This unit discusses common "non-chemical" hazards that are as important and dangerous as chemical hazards. Physical and environmental hazards include explosion and fire, oxygen deficiency or excess, biological organisms, dangerous surfaces, electrocution, heat stress, cold stress, and related conditions.

5.1 Explosion and Fire

Some of the many potential causes of explosions and fires at hazardous waste sites include:

- chemical reactions that produce explosion, fire, or heat (organic chemicals that readily vaporize, white phosphorous);
- chemical degradation that produces unstable products (picric acid), dioxanes;
- ignition of explosive or flammable chemicals (acetone, benzene, etc.);
- ignition of materials due to oxygen enrichment (oxidizers);
- agitation of shock- or friction-sensitive compounds (nitroglycerin);
- sudden release of materials under pressure (pressurized drums, containerized gases); and
- ignition of dry grass or shrubs from catalytic converters on vehicles.

Explosions and fires may occur spontaneously. However, they more commonly result from site activities, such as moving drums, accidentally mixing incompatible chemicals, or introducing an ignition source (such as a spark from equipment) into an explosive or flammable environment. At hazardous waste sites, explosions and fires not only pose the obvious hazards of intense heat, open flame, smoke inhalation, and flying objects, but may also cause the release of toxic chemicals into the environment. Such releases can threaten both on-site personnel and members of the general public living or working nearby.

To protect against these hazards, qualified personnel should monitor fuel sources for fire, explosive atmospheres, flammable vapors, and oxygen levels. All potential ignition sources should be kept away from explosive or flammable environments by using nonsparking, explosion-proof equipment such as brass tools and spark arresters on mechanical equipment. Finally, safe practice must be followed when performing any task that might result in the agitation or release of chemicals.

Due to the variety of chemicals, mixtures, and interactions which could be involved, it is not possible to include them all in this section. (Careful reference to CHEMTREC, NIOSH, and related informational sources should be made before entering sites or beginning the removal of substances.)

5.2 Oxygen Deficiency

The oxygen content of standard air is approximately 21 percent by volume. While the percentage of oxygen in the air remains the same as elevation increases, the density of the air decreases. Therefore, air "thins out" as elevation increases and less oxygen is available for respiratory functions and flammability hazards. Physiological effects of oxygen deficiency in humans are readily apparent when the oxygen concentration in the air decreases to 16 percent by volume. These effects include impaired attention span, judgment and coordination, and increased breathing and heart rate. Oxygen concentrations lower than 16 percent can result in nausea, vomiting, brain damage, heart damage, unconsciousness, and death. Oxygen levels less than five percent by volume are fatal.

Individual physiological responses, air density variations from sea level conditions, and variations in measurement by instruments which may be inaccurately calibrated, must be taken into account when determining the quality of air on a work site. Concentrations of 19.5 percent oxygen or lower are considered to be indicative of oxygen deficiency. OSHA mandates that such environments cannot be entered without a supply of air that has normal oxygen levels.

Oxygen deficiency may result from the displacement of oxygen by another gas, or from the consumption of oxygen by a chemical or biological reaction. Confined spaces and low-lying areas such as trenches and pits are particularly vulnerable to oxygen deficiency and should always be monitored prior to entry. Personnel must use atmosphere-supplying respiratory equipment when oxygen concentrations drop below 19.5 percent by volume and the cause of oxygen deficiency must be determined since this is a potential IDLH situation.

Excessive levels of oxygen may also have physiological effects. Lung damage may result from several days of breathing a highly enriched oxygen atmosphere. The primary danger of an oxygen enriched atmosphere is not its health effect, but its effect on combustion. Enriched oxygen environments may be caused by oxidizers or compressed oxygen gas and are fire hazards. In 1967 during a pre-flight training exercise, three astronauts were killed when a minor fire broke out in their Apollo craft. The oxygen enriched atmosphere changed the minor fire into a deadly conflagration.

OSHA has determined 25 percent concentration by volume to be the upper limit of safety for working in most environments. Due to additional hazards associated with confined spaces, OSHA has set an upper level of oxygen in confined spaces at 23.5 percent.

5.3 Biologic Hazards

Wastes from hospitals and medical research facilities can contain disease-causing organisms that could infect site personnel. Like chemical hazards, pathogens may be dispersed in the environment by water and wind. Other biologic hazards that may be present at a hazardous waste site include poisonous plants, insects, spiders, ticks, animals, and indigenous pathogens such as Hepatitis A and B, bubonic plague, tetanus, Rocky Mountain spotted fever, and hantavirus. Protective clothing and respiratory equipment can help reduce the chances of exposure. Thorough washing of any exposed body parts and equipment will also help protect against infection. Sometimes, even if a worker is adequately protected by clothing, the element of surprise when encountering a biologic hazard, such as a snake, large lizard, or rat under a drum, may be enough to cause a person to slip, fall, tear their protective clothing, or even hyperventilate. Any of these responses, especially in confining PPE, may become hazardous to the individual's safety.

Animals can bite and scratch, especially if cornered or threatened. Such actions may produce puncture wounds allowing toxins to come into contact with the skin. They could also lead to infection with an animal-borne disease such as rabies. In addition, animals carry fleas and ticks that may transfer diseases to humans, such as lyme disease, plague, Rocky Mountain spotted fever, and hantavirus. For example, the often fatal hantavirus pulmonary syndrome is caused by a virus carried by the deer mouse. The virus is present in the saliva, urine, and feces of infected mice. People can become infected by breathing in the virus during direct contact with rodents or from disturbing dust and feces from mice nests or surfaces contaminated with mouse droppings or urine.

Plants, such as poison oak, poison sumac, or poison ivy, may cause skin irritation. However, some individuals are so sensitive that inhaling smoke from burning these plants may lead to the severe irritation and swelling of breathing passages. These individuals may suffocate if the situation is not corrected in time. Stinging nettle is another plant that causes severe irritation on contact.

Perhaps the greatest threat from biologic hazards are those caused by pathogens, or disease-causing organisms, including bacteria, viruses, and fungi. HBV can be contracted by workers dealing with medical waste, especially waste that contains needles and other sharp instruments containing blood, saliva, semen, and vaginal fluids from hepatitis-infected individuals. Therefore, all medical waste must be considered as infected with biological agents and handled appropriately. Legionnaire's disease, anthrax, and hantavirus can pose greater threats to life than the chemical contents of a drum. Some of their danger lies in their inability to be seen.

5.4 Topographic and Physical Hazards

Topographic and physical hazards may pose a greater threat than chemicals or biological hazards. Such hazards include:

- holes or ditches which may be fully or partially hidden by vegetation, or must be crossed while wearing bulky protective equipment;
- precariously positioned objects such as drums, pallets, cylinders, and boards, which may shift suddenly and fall;
- sharp objects such as nails, metal shards, broken glass, and general landfill garbage including bedsprings, snow fence, and machinery parts;
- slippery surfaces such as damp grass, plastic, and clay soils can turn slick with moisture. (Rainfall or the use of water at a remediation site can change the footing from day to day. A stable walking surface on Monday may be a slip-and-fall hazard on Tuesday);
- negotiating steep grades can be hazardous, especially for workers in protective equipment that restricts their vision or mobility;
- steep slopes may also harbor unsafe footing with sand grains or pebbles acting as loose ball bearings;
- uneven terrain, including animal diggings, marshy soil, snow, ice, and partially buried rocks may be difficult to detect wearing protective equipment; and
- unstable surfaces, such as walls that may collapse or flooring that may give way, pose dangers. (Some workers, such as those with fire fighting experience, are more cognizant of these dangers since burning buildings exhibit this instability. However, in abandoned or non-burning structures, care is sometimes not taken to protect against this hidden hazard.)

Some safety hazards are a function of the work itself. For example, heavy equipment creates an additional hazard for workers in the vicinity of the operating equipment. Protective equipment can impair a worker's agility, hearing, and vision, thus increasing the risk of accident.

Accidents involving physical hazards can directly injure workers and can create other hazards such as increased chemical exposure due to damaged protective equipment or danger of explosion caused by the mixing or spillage of chemicals. Site personnel should constantly look for potential safety hazards and immediately inform their supervisors of any new hazards so that mitigating action can be taken. Site entry plans (Unit 15) should include any known physical hazards and workers should be briefed accordingly.

5.5 Electrical Hazards

Overhead power lines, downed electrical wires, and buried cables pose a danger of shock or electrocution if workers contact or sever them during site operations. Electrical equipment used on-site (such as generators), may also pose a hazard to workers. Common sense,

throughout site characterization to identify electrical hazards, and following OSHA's electrical hazards regulations (29 CFR 1910.137) are the best techniques to manage this hazard.

Insulation and grounding are two recognized means of preventing injury during electrical equipment operations. With the wide use of portable tools on construction and hazardous waste sites, the use of flexible cords often becomes necessary. Hazards are created when cords, cord connectors, receptacles, and cord- and plug-connected equipment are improperly used and maintained. Regulations require the use of GFCIs (Ground Fault Circuit Interrupter) on receptacle outlets that are in use and not part of the permanent wiring of the building and structure. Alternately, a scheduled and recorded equipment "grounding conductor program" may be in place on construction sites, covering all cord sites and receptacles that are not part of the permanent wiring of the building or structure, and equipment connected by cord and plug that are used by workers in the field.

GFCIs are used to reduce electrical hazards on hazardous waste sites and on construction sites. Tripping of GFCIs—interruption of current flow—is sometimes caused by wet connectors and tools. It is good practice to limit exposure of connectors and tools to excessive moisture by using watertight or sealable connectors. Providing more GFCIs or shorter circuits can prevent tripping caused by the cumulative leakage from several tools, or by leakage from extremely long circuits. The GFCIs must be checked regularly prior to use to ensure that they are working properly. If the GFCI is not operational, then it should be immediately removed from service.

Lightning is a hazard during outdoor operations, particularly for workers handling elevated drilling equipment. To eliminate this hazard, weather conditions should be monitored and work should be suspended during electrical storms. Although no rule exists as to what constitutes a safe distance from lightning and storms, a general rule of thumb might be to discontinue work and seek shelter if thunder is heard less than five to 10 seconds following a lightning flash. Shelter should not be sought under tall, vertical objects (including trees), under drilling rigs with elevated masts, or near active electrical equipment.

5.6 Heat Stress

Heat stress is possibly the most common hazard on a hazardous waste site, especially for workers wearing protective clothing. The same protective materials that shield the body from chemical exposure also limit the dissipation of heat and moisture from the body. Personal protective clothing can, therefore, create a hazardous condition. In addition, workers often sweat and become dehydrated while performing work. Depending on the ambient conditions and the work being performed, heat stress can occur very rapidly—within as little as 15 minutes. It can pose as great a danger to worker health as chemical exposure.

Related safety problems are fairly common in hot environments. Workers suffering from heat stress may be more accident-prone due to fatigue, dizziness, or fogging of protective facial equipment. Heat lowers an individual's mental alertness and physical performance and may promote irritability, anger, and extreme emotional reactions to situations, which, in turn, may lead to accidents.

In its early stages, heat stress can cause rashes, cramps, general discomfort, and drowsiness, resulting in impaired functional ability that threatens the safety of both the individual and co-workers. Continued heat stress can lead to heat stroke and death. Avoiding overprotection, careful training, frequent monitoring of personnel who wear protective clothing, judicious scheduling of work and rest periods, and frequent replacement of fluids can protect against this hazard. Certain protective equipment, such as ice vests, and heat exchange cooling suits are available for extreme conditions.

5.6.1 THE BODY'S HEAT EXCHANGE MECHANISM

Most heat problems in the workplace are caused by an imbalance of heat production and heat loss in the body. Humans are able to regulate their internal body temperature within narrow limits. This is accomplished by heat being carried by blood from muscles and deep tissues to the surface of the body for dispersal. The initial heating of the body is environmentally affected by the air temperature, humidity, air flow, and radiant temperature, all of which may vary widely during a daily work shift. The human body uses sweat as a means of evaporative cooling. The effective temperature an individual actually "feels" is a combination of the surrounding environment, work activity level, and the clothing that is worn. The efficiency of the body's cooling mechanisms can be affected by the work environment, including the tasks being performed, psychological stress levels, clothing, the surrounding environment, and the health, age, and physical condition of the worker.

Workplace heat exposures are different between hot-dry or warm-moist heat. In a hot-dry situation, the body releases heat readily through sweating and evaporation, but fluid-loss rates may be extreme enough to produce dehydration of body tissues. Warm-moist heat may retard the ability of sweat to evaporate and cool the body. In either situation, the body may absorb more heat than it has the ability to release through sweating. Depending on the individual, any type of heat may be a hazard resulting in one or more of the following symptoms.

5.6.2 SIGNS AND SYMPTOMS OF HEAT STRESS

Human reaction to heat is divided into several levels of severity. In general, these reactions are classified as follows:

Heat rash is a skin irritation that may result from continuous exposure to heat or humid air and is caused by rubbing, chafing, or even a reaction to a person's own sweat with its included salts and minerals. In most cases the sweat glands become blocked and infected.

Heat cramping is a phase that signals the start of physical danger and indicates that the worker should rest or slow down. Heat cramps are caused by heavy sweating with inadequate electrolyte replacement or redistribution. Signs and symptoms include:

—muscle spasms; and/or
—pain in the hands, feet, and abdomen.

However, not all workers suffer heat cramps before more serious forms of heat stress occur.

Heat exhaustion produces a definite warning that the body is being stressed by excessive heat. It occurs from dehydration and increased heat stress on various body organs. Signs and symptoms may include:

—heavy sweating (this is a universal sign);
—pale, cool, moist skin;
—dizziness, fainting, light-headedness;
—nausea;
—headaches;
—irritability;
—slightly elevated body temperature;
—weak, rapid, and/or irregular pulse;
—decreasing blood pressure; and/or
—general fatigue and poor feeling.

Heat stroke is a potentially life-threatening condition that may result in death. Temperature regulation fails, and the body temperature rises to critical levels. The defining symptom of heat stroke is a lack of or reduced perspiration, resulting in hot, dry skin. Other symptoms may include:

—nausea;
—level of consciousness change (confusion or disorientation);
—strong, rapid pulse;
—sharp drop in blood pressure;
—strong heart rate from adrenaline rush;
—fatigue; and/or
—coma.

When wearing protective clothing that inhibits sweat evaporation, the skin is usually moist in heat stroke victims, thus masking this important symptom. Thus, behavior monitoring is often used. Reduced mental alertness, excessive fatigue, illogical actions, and inappropriate behavior often indicate heat stroke.

Exertional heat stress (hyperpyrexia) may occur when workers wearing chemical protective clothing (CPC) demonstrate symptoms of both heat exhaustion and heat stroke, making diagnosis of the actual clinical situation difficult. Enclosed protective equipment curtails the body's ability to reduce heat through the processes of evaporation, radiation and convection. Sweat collects inside the suit, yet the actual time when it was produced is masked. The skin of the face may become discolored, but whether as a result of heat stroke or from the heat trapped in the suit, is difficult to tell. Wet, red skin does not fit into either set of classic symptoms. Additionally, blood pressure may rise or fall, the pulse may vary from strong to weak and thready, body temperature will elevate and a level of consciousness change will occur. Under any circumstances, rather than try to guess at the clinically defined condition, find and administer medical treatment.

No scientific studies have been made of the long-term results of heat exposure and ensuing chronic effects. Questions have been raised about the effect on the lungs of breathing hot, dry air, but no conclusions have been reached. Undue stress on the body's heat exchange system may affect the kidneys and heart. In any event, exposure to heat and other environmental factors may use up part of the total energy reserves of the body, making the individual more susceptible to attack from organic or chemical exposures.

5.6.3 TREATMENT OF HEAT STRESS

5.6.3.1 Heat Cramps

As soon as a worker experiences a heat cramp, he or she should leave the work site immediately, seek a cooler environment, and drink liquids, especially those with electrolytes. To relieve pain, gently stretch the affected area. Massaging may or may not help. Medical attention is rarely needed.

5.6.3.2 Heat Exhaustion

An individual suffering from heat exhaustion should leave the work site immediately and rest in a cool area. Remove protective clothing and loosen other clothing. Drink cool (not iced), noncaffeinated and nonalcoholic liquids, unless nausea is present. Cool, wet compresses may be applied to the skin. It is usually recommended that those suffering from heat exhaustion lie down with their feet eight to 12 inches off the ground. Medical assistance may be required for those suffering from heat exhaustion.

5.6.3.3 Heat Stroke

Heat stroke is a life-threatening condition requiring immediate action. Contact emergency medical services immediately. Remove the victim from the work area, and remove protective clothing and as much of the patient's clothing as possible. Very slowly cool the victim by fanning, wet towels, or compresses. If the patient starts to shiver, slow the cooling process.

5.6.4 TOLERANCE TO HEAT STRESS

The physiological factors that may affect workers' ability to tolerate heat stress include their physical condition and health practices, their level of acclimatization, their age, their gender and their weight.

The ACGIH, NIOSH, and other organizations have produced recommendations and guidelines for monitoring pulse rate, body temperature and weight loss for individual workers to protect against heat stress. Calculations and equations are given based on factors such as the environment (temperature/humidity), the workload, a work-rest regimen, water and electrolyte supplementation, protective clothing, acclimatization, fitness, and adverse health effects of heat on individuals such as pregnant women.

5.6.4.1 Physical Condition/Health Practices

Physical fitness is a major factor influencing a person's ability to perform work under hot conditions. At a given level of work, a fit person, compared to an unfit person, will have:

- less physiological strain;
- a slower heart rate;
- a lower body temperature, which indicates less retained body heat (a rise in internal temperature precipitates heat injury);
- a more efficient sweating mechanism;
- slightly lower oxygen consumption; and
- slightly lower carbon dioxide production.

In addition to physical fitness, personal habits may influence an individual's reaction to heat. It has been determined that individuals who drink an excessive amount of alcohol within a day or two prior to heat exposure have less tolerance to heat-related disorders since alcohol suppresses a hormone that prevents dehydration. The physician conducting annual medical checks should evaluate tolerance to heat stress and PPE usage as part of the examination protocol. In addition, medication for chronic problems such as asthma, allergies, and headaches should be checked with a company physician prior to entering a site where heat exposure could produce additional stress on the body. Smoking may raise internal body temperatures and lower extremity temperatures, as well as make breathing in protective gear more difficult. In addition, unusual or sporadic physical problems such as diarrhea will have negative effects on tolerating a hot environment.

5.6.4.2 Level of Acclimatization

The degree to which a worker's body has physiologically adjusted or acclimatized to working under hot conditions affects his or her ability to do work. Acclimatized individuals generally have lower heart rates and body temperatures than nonacclimatized individuals, and sweat sooner and more efficiently. This enables them to maintain lower skin and body temperatures at a given level of environmental heat and work load than nonacclimatized workers. Sweat composition also becomes more dilute with acclimatization, which reduces salt loss.

Acclimatization can occur after just a few days of exposure to a hot environment. NIOSH recommends a progressive six-day acclimatization period for workers before allowing them to do full work on a hot job. Following this regimen, the first day of work on-site begins with only 50 percent of the anticipated workload and allows for 50 percent exposure time. Ten percent is added each day through day six. With fit individuals, the acclimatization period may be shortened two or three days. However, workers can lose acclimatization in a matter of days, and work regimens should be adjusted to account for this.

5.6.4.3 Age

Generally, maximum work capacity declines with increasing age, but is not always the case. Active, well-conditioned older individuals often have performance capabilities equal to or greater than young, sedentary individuals. However, there is some evidence, indicated by lower sweat rates and higher body core temperatures, that older individuals are less effective in compensating for a given level of environmental heat and work loads. At moderate thermal loads, however, the physiological responses of "young" and "old" are similar, and performance is not affected.

5.6.4.4 Gender

The literature indicates that females tolerate heat stress at least as well as their male counterparts. Generally, a female's work capacity averages 10 to 30 percent less than that of a male. The primary reasons for this are the greater oxygen-carrying capacity and the stronger heart in the male. However, a similar situation exists as with aging: not all males have greater work capacities than all females.

ACGIH has found that during the first trimester of pregnancy, if a female worker's core temperature exceeds 102.2°F for extended periods, there is an increased risk of malformation of the fetus. Additionally, core temperatures above 100.4°F may be associated with temporary infertility in both males and females.

5.6.4.5 Weight

The ability of the body to dissipate heat depends on the ratio of its surface area to its mass (surface area/weight). Heat loss (dissipation) is a function of surface area, and heat production is dependent on mass. Therefore, heat balance is described by the ratio of the two.

Since overweight individuals (those with a low ratio) produce more heat per unit of surface area than thin individuals (those with a high ratio), overweight individuals should be given special consideration in heat stress situations. However, when wearing impermeable clothing, the weight of an individual is not always a critical factor in determining the ability to dissipate excess heat.

5.6.5 MONITORING

The incidence of heat stress depends on a variety of factors; therefore, all workers, even those not wearing protective equipment, should be monitored. General monitoring requirements fall into two categories as follows:

- workers wearing permeable clothing (e.g., standard cotton or synthetic work clothes) should follow recommendations for monitoring requirements and suggested work/rest schedules in the current ACGIH TLVs for heat stress. If the actual clothing worn differs from the ACGIH standard ensemble in insulation value and/or wind and vapor permeability, change the monitoring requirements and work/rest schedules accordingly; and

- workers wearing semipermeable or impermeable encapsulating ensembles should follow corrective standards that have been established for additional personal protective clothing and equipment. Such workers should also be monitored when the temperature in the work area is above 70°F.

Monitoring the worker for heat stress usually employs one or more of the following strategies:

- **Heart rate**—Count the inside wrist pulse during a 30-second period as early as possible during the rest period.
  —if the heart rate exceeds 110 beats per minute at the beginning of the rest period, shorten the next work cycle by one-third and keep the rest period the same length of time.
  —if the heart rate still exceeds 110 beats per minute during the next rest period, shorten the following work cycle by one-third.
- **Oral temperature**—Use a clinical thermometer (three minutes under the tongue) or a similar device to measure the oral temperature at the end of the work period (before drinking). Ear canal or other measurement techniques may be incorporated as long as they accurately relate to oral temperature.
  —if oral temperature exceeds 99.6°F, shorten the next work cycle by one-third without changing the rest period.
  —if oral temperature still exceeds 99.6°F at the beginning of the next rest period, shorten the following work cycle by one-third.
  —do not permit a worker to wear a semipermeable or impermeable garment when his/her oral temperature exceeds 100.6°F.
- **Body water loss (if possible)**—Measure the worker's weight on a scale accurate to ±0.25 lb. at the beginning and end of each work day to see if enough fluids are being taken to prevent dehydration. Weights should be taken while the employee wears similar clothing or, ideally, is nude. The body water loss should not exceed 1.5 percent total body weight loss in a work day (or approximately 2.25 pounds for a 150 pound person).
- **Monitor worker behavior**—Heat stress often affects the central nervous system. Workers should be monitored for illogical actions, reduced alertness, and inappropriate behavior.

Initially, the frequency of physiological monitoring depends on the air temperature adjusted for solar radiation and the level of physical work. The length of the work cycle will be governed by the frequency of the required physiological monitoring.

Although the need for monitoring during elevated heat periods is recognized, physically accomplishing the task can be cumbersome, especially when PPE is involved. For any of the above criteria to be evaluated, the individual may have to withdraw from the work site to a location that is safe enough for him or her to remove or open protective gear to allow measurements.

5.6.6 PREVENTION

The best way to deal with heat-related disorders is prevention. Proper training and preventive measures can help avert serious illness and loss of work productivity. Preventing heat stress is particularly important because once a person suffers from heat stroke or heat exhaustion, that person may be predisposed to additional heat injuries. To avoid heat stress, management should take the following steps:

- Adjust work schedules;
  —modify work/rest schedules according to monitoring requirements;
  —mandate work slowdowns as needed;
  —rotate personnel and alternate job functions to minimize overstress or overexertion during one task.;
  —add additional personnel to work teams; and
  —perform work during cooler hours of the day, if possible, or at night if adequate lighting can be provided.
- Provide shelter (air-conditioned, if possible) or shaded areas to protect personnel during rest periods;
- Maintain workers' body fluids at normal levels. Daily fluid intake must approximately equal

the amount of water lost in sweat. The normal thirst mechanism is not sensitive enough to ensure that enough water will be ingested to replace lost sweat. When heavy sweating occurs, workers should be encouraged to drink more. The following strategies may be useful:

—maintain water temperature at 50° to 60°F (cool enough to be refreshing but not too cold to chill the body);

—provide small disposable cups that hold about four ounces. This makes measuring consumption easier, especially if cups are labeled with the worker's initials;

—have workers drink 16 ounces of fluid (preferably water or dilute drinks) before beginning work;

—urge workers to drink at least eight ounces of water at each monitoring break. A total of one to 1.6 gallons of fluid per day are recommended, but more may be necessary to maintain body weight;

—do not drink caffinated or alcoholic beverages; such beverages are ineffective at replacing body fluids; and

—weigh workers before and after work to determine if fluid replacement is adequate.

- Encourage workers to maintain an optimal level of physical fitness;
  - —where indicated, acclimatize workers to site working conditions: temperature, protective clothing, and workload; and
  - —urge workers to maintain normal weight levels.
- Provide cooling devices to aid natural body heat exchange during prolonged work or severe heat exposure. Cooling devices include:
  - —field showers or hose-down areas to reduce body temperature and/or to cool off protective clothing; and
  - —cooling jackets, vests, or suits.
- Train workers to recognize and treat heat stress. As part of training, have workers identify the signs and symptoms of heat stress.

5.7 Cold Stress

Cold stress (frozen extremities and hypothermia) and impaired ability to work are dangers at low temperatures, when the wind chill factor is low, or when working with cryogenics. PPE is usually not designed to prevent cold stress since PPE often enhances cold stress by trapping moisture that can rapidly cool the body under the right conditions.

The body has two mechanisms for maintaining its internal temperature when external temperatures drop. First, blood vessels to the body surface constrict so less blood approaches the surface and less internal heat is released. In this instance, the hands and feet may become numb and cold. A person with cold, stiff, numb, or painful hands may have problems performing manual tasks; accidents can result from losing one's grip on tools, dropping items, or tripping. Second, the body produces rapid muscle contractions (shivering) which generate heat and help maintain the inner temperature of the body. Shivering makes it difficult to do work requiring manual dexterity.

The human body does not acclimatize to cold as it does to heat, so humans do not become more efficient at withstanding cold temperatures. Generally, people who must work in cold workplaces get mentally used to the cold, may feel less uncomfortable and, therefore, work somewhat more effectively. While cold stress can be manifested in several ways, frostbite and hypothermia are the most common problems.

5.7.1 FROSTBITE

The basic sign of frostbite is the white or grayish-yellow color of the skin. Unfortunately, when wearing PPE, the skin is usually covered and not readily accessible to check for discoloration. While pain is often associated with frostbite, it is usually only felt early in the process, and is not a reliable indicator. In addition, once frostbite sets in, the affected area is usually numb. Some medical authorities recognize three stages of frostbite:

1. **Incipient frostbite (or frost nip)** is characterized by discolored skin, but the skin is still soft.
2. **Superficial frostbite** is characterized by hard, discolored, and cold skin that is usually numb. However, there may be a sharp aching pain that results from ice crystals forming in the tissues. Superficial frostbite is reversible because only the surface of the skin is affected.

For some people the affected area may have chronic symptoms such as excessive sweating, pain, numbness and/or abnormal color. As the skin thaws, the flesh becomes red and painful.

3. **Deep frostbite** is characterized by extremities being discolored (from white to grayish-blue) and hard throughout. In such cases, the flow of blood to the affected area stops and blood vessels may be damaged. Medical attention is essential and amputation may be required.

5.7.2 HYPOTHERMIA

Hypothermia is the cooling of the body's core and is caused by prolonged cold exposure and heat loss. This condition is initially characterized by uncontrolled shivering, which may be followed by drowsiness, general numbness, and illogical actions, slurred speech, abnormally slow breathing, loss of coordination, and often death.

If a person becomes overly tired during physical activity, he or she will be more likely to have heat loss. As exhaustion approaches, the blood vessel constriction mechanism overloads and a massive dilation of the vessels takes place, with a rapid loss of internal heat and subsequent cooling of core organs. Alcohol and sedative consumption increase the danger of hypothermia.

This condition is likely to develop slowly with a gradual loss of mental acuity and physical ability. Many deaths occur when temperatures are above freezing. For example, immersion in cold rivers or springs during summer hiking may cause hypothermia.

5.7.3 PREVENTION

When working in cold temperatures, warm-up periods should be scheduled at least once every two hours. During the break, workers should drink noncaffeinated beverages since water retention will help retain body heat. Do not become overheated when working, or remove or open PPE during the break to dry clothing. Appropriate clothing should be worn under the PPE. Monitor your buddy for signs of cold stress, such as clumsiness, shivering, and illogical behavior. Work should be designed to minimize cold exposure, with workers being provided warm areas or protection from direct weather influences such as wind.

ACGIH has promulgated standards to protect workers from cold stress. These standards include proper clothing, medical monitoring, environmental monitoring, establishment of a work-warming regimen, shift flexibility, the buddy system, and special recommendations for site specific workplaces such as refrigerator rooms.

As with excessive heat, few studies have been done to show the chronic health effects of cold stress. Besides obvious tissue loss or amputation of extremities, some research has shown that chronic lung disease and sinus irritation may be lasting effects.

5.7.4 TREATING COLD STRESS

5.7.4.1 Frostbite

For those suffering from incipient or superficial frostbite, immediately bring the victim into a warm environment, remove protective clothing, place affected part in warm water (100–105°F) until completely thawed (or normal color returns), and give the individual warm drinks that are noncaffinated and nonalcoholic. Do not rub the affected area. If the feet are affected, do not have the individual walk on them, but allow them to dangle while being treated.

Do not initiate the thawing process if there is any danger of refreezing; keeping the tissue frozen is less dangerous than submitting it to refreezing. Do not use other heat sources for warming since the desensitized tissues could be burned. Do not allow the individual to use tobacco since nicotine constricts the blood vessels and may further curtail circulation. While medical attention may not be necessary for incipient frostbite, it is required for superficial frostbite due to tissue damage.

5.7.4.2 Hypothermia

This life-threatening condition requires immediate attention and professional medical help. While waiting for assistance, bring the victim into a warm area and remove protective clothing and any damp or wet clothing. Wrap the victim in blankets or any other warm insulating material being certain that the head is covered and that the person is insulated from the cold ground. Do not give the individual liquids; instead, slowly warm the victim from the outside.

5.8 Noise

Noise is described as unwanted sound, independent of loudness, that may produce an undesired physiologic or psychological effect. It is considered to be a health hazard since its obvious effects include:

- workers being startled, annoyed, irritated or distracted;
- physical damage to the ear that results in pain and temporary and/or permanent hearing loss; and
- communication interference that may increase potential hazards due to inability to warn of danger and the necessary proper safety precautions.

OSHA has established regulations for workplace noise and 29 CFR 1910.95 deals with specific industry programs and safeguards. Equipment coming on to a site should be monitored for noise output and be installed with appropriate noise controls. Noise warning signs may also be posted.

Noise is measured in decibels on a logarithmic scale similar to that used for earthquake intensity. The numbers are dimensionless, in that they do not represent a tangible volume or number of items. For example, "1" on the scale represents the threshold of acute hearing which may vary within a population, but generally represents the most sensitive. Values for certain noises when rated on this scale follow:

1—threshold of hearing-acute

15—threshold of hearing-average

39—quiet office

46—household refrigerator

60—conversation

90—heavy city street traffic

120—auto horn

130—pneumatic rock drill

The major outcome of intense or long-term noise exposure is NIHL (Noise Induced Hearing Loss) that damages the structure of the hearing organ. Hearing loss may take two forms. A Temporary Threshold Shift (TTS) is caused by a brief exposure to a high level sound such as a short tour through a noisy manufacturing plant. The greatest effect comes immediately after exposure. The second loss is Permanent Threshold Shift (PTS) where recovery from exposure to noise is less than total. This is likely to occur when an individual is exposed to noise levels over 85 decibels for extended periods of time, such as several years of operating noisy equipment.

The control of NIHL is accomplished primarily through prevention. Controls can take many forms, including modifying equipment to decrease noise at the source, enclosing the source of the receiver, and increasing the distance between the source and the receiver or providing shielding. Managers may protect workers by rotating jobs or limiting the number of machines operating at one time. Ear plugs and muffs that are rated for noise reduction are the most common form of protection. Equipment ratings are also available for most situations.

5.9 Other Potential Hazards

Other conditions and situations are currently under evaluation by OSHA, ACGIH, AIHA, and NIOSH in an attempt to determine what, if any, acute and chronic effects may be felt by workers. Some of these include:

- motion;
- light and nonvisible radiation;
- electromagnetic and other radiation frequencies;
- pressure variations;
- impulse noise (blast over-pressure); and
- combined effects with chemical substances (noise and chemical exposure interaction).

5.9.1 MOTION

In Unit 2 the field of ergonomics was discussed, along with the relationship of various syndromes to repetitive motion. While program standards are being handled by OSHA under 29 CFR 1910 Subpart Y, the ACGIH has set up standards and TLVs for Hand-Arm (Segmental) Vibration Syndrome (HAVS). This refers to excessive vibrations received through the hand and dissipated primarily into the arm but not generally beyond there. Occupational tasks involving mining (drills), forestry (chain saws), metal working (grinders, etc.), and some jack-hammers are examples of this exposure. Overexposure to vibration while using hand tools may cause bone and muscle deterioration in the fingers, hands, and arms, with the most extreme condition being Raynaud's syndrome or dead fingers. The circulation may be impaired on a temporary or permanent basis creating a condition similar to frostbite.

OSHA has initiated an ergonomics program standard to address the significant risk of work-related musculoskeletal disorders (MSDs) encountered by employees in a variety of jobs in general industry workplaces. General industry employers covered by the standard are required to establish ergonomics programs containing some, or all of the following elements:

- management leadership and employee participation;
- job hazard analysis and control;
- hazard information and reporting;
- training;
- MSD management; and
- program evaluation.

These elements may differ from site to site as well as within individual companies depending on the types of jobs in that workplace and whether a musculoskeletal disorder covered by the standard has occurred.

The standard requires all general industry employers whose employees perform manufacturing or manual handling jobs, to implement a basic ergonomics program in those jobs. If an employee in a manufacturing or manual handling job experiences and OSHA-recordable MSD that is determined by the employer to be covered by the standard, the employer must implement a full ergonomic program for that job and for all other jobs in the establishment involving the same physical work activities. This full program includes, in addition to the elements in the basic program:

- a hazard analysis of the job;
- the implementation of engineering, work practice, or administrative controls to eliminate or materially reduce the hazards identified with that job;
- training the employees in the job and their supervisors; and
- the provision of MSD management, including, where appropriate, temporary work restrictions and access to a health care provider or other professional if a MSD occurs.

General industry employers whose employees work in jobs other than manual handling or manufacturing and experience a MSD that is determined by the employer to be covered by the standard are also required to implement an ergonomics program for those jobs. However, following presidential and congressional changes, the OSHA Ergonomics Standard of 2000 was repealed in March 2001 by the Congressional Review Act. Therefore, at the time this manual was published, there is no OSHA ergonomics standard, and all references to this standard should be considered good practice guidelines and not regulatory requirements.

5.9.2 LIGHT AND LIGHTING

Light may have negative effects on workers since poor lighting is a major cause of fatigue and accidents. OSHA sets specific minimum illumination values for generic work situations (29 CFR 1910.120(m)) but does not elaborate on the effects of poor lighting. Factors such as glare, shadows, and the quality and quantity of light are being evaluated since morale can suffer in extreme lighting conditions. The ACGIH has established TLVs for visible wavelength radiation and the Illuminating Engineering Society has also established standards for industrial lighting.

5.9.3 ELECTROMAGNETIC AND NONIONIZING RADIATION FREQUENCIES

Besides broad-spectrum visible light, infrared and UV exposures that range from visible to nonvisible rays may pose problems for workers. Infrared radiation may cause thermal burns and extreme exposure to eyes results in cataracts. UV rays may also cause severe sunburn, skin cancer, and damage to the lens of the eye. Certain types of lamps, such as black lights, ozone-producing lights, mercury vapor lamps, germicidal lamps, and flash tubes, should carry warnings, and workers should be made aware of potential problems. The ACGIH has set TLVs for both infrared and UV light. The AIHA has published informational technical documents on the evaluation and control of nonionizing radiation.

Another light source of concern to worker safety and health is the laser. This device amplifies and concentrates a beam of light in one direction without spreading it. This point source of great brightness is a hazard to individuals. The laser may injure the eye, skin, or possibly, internal organs. In addition, the energy-intense beam of light may vaporize or fragment harmful substances or oxidize chemicals in the ambient air, producing airborne health hazards.

Energy produced by the motion of electromagnetic energy is widespread and varied. It includes cosmic rays, gamma rays, x-rays, UV rays, and the visible spectrum through infrared, microwaves, and radio waves. Nonionizing radiation is under study by several worker safety organizations. Radio waves, including microwaves, are used in major industrial processes such as heating, melting, and curing plastic, rubber, or glue. Other uses include radar installations, and satellite communications systems. Possible problems incurred by humans when exposed to these energy sources may include damage to the eye, central nervous system, heart, blood, and immune system. Effects on reproduction and the development of offspring of women exposed during pregnancy have also been suspected. Finally, workers exposed to excessive amounts of radio frequency/microwave radiation may develop adverse thermal effects from the heating of deep body tissues, especially those organs having a high water content.

Protection from such energy sources can be accomplished by minimizing the time of exposure, keeping a maximum distance from the source, and using proper shielding. The ACGIH has set TLVs for radio frequencies, microwave radiation, and static magnetic fields. The organization is also studying electromagnetic pulses and radio frequency radiation. The AIHA has published technical documents on the evaluation and control of ionizing radiation.

5.10 Sources of Information

American Conference of Governmental Industrial Hygienists: *2001 Threshold Limit Values for Chemical Substances and Physical Agents.* Cincinnati, OH: American Conference of Governmental Industrial Hygienists, 2001.

American Industrial Hygiene Association: *General Concepts for Nonionizing Radiation Protection* (Publication No. 352-EA–98). Fairfax, VA: American Industrial Hygiene Association Press, 1998.

American Industrial Hygiene Association: *Radio Frequency in Microwave Radiation* (Publication No. 182-EA–94). Fairfax, VA: American Industrial Hygiene Association Press, 1994.

American Industrial Hygiene Association: *Ultraviolet Radiation* (Publication No. 136-EA–91). Fairfax, VA: American Industrial Hygiene Association Press, 1991.

Bennet, Claude J., M.D., and F. Plum, M.D. (eds.): *Cecil Textbook of Medicine.* Vol. 1. Philadelphia, PA: W.B. Saunders Company (Division of Harcourt Brace Company), 1996.

Bureau of Labor Statistics in cooperation with State and Federal agencies: *Census of Fatal Occupational Injuries, 1992–1999—Fatal Occupational Injuries by Event or Exposure, 1992–1999.* In www.osha.gov [Occupational Safety and Health Administration (OSHA) World Wide Web homepage]. Washington, DC: Bureau of Labor Statistics, 2000.

"Ergonomics Program," *Code of Federal Regulations* Title 29, Part 1910. In www.osha.gov [Occupational Safety and Health Administration (OSHA) World Wide Web homepage]. August 2001.

"Ground Fault Circuit Interrupters," *Code of Federal Regulations* Title 29, Part 1926.404. 2001.

Koren, Herman: *Handbook of Environmental Health and Safety, Principles and Practices.* 2nd ed., Vols. I and II. Chelsea, MI: Lewis Publishers, 1991.

Larson, David E., M.D. (ed.): *Mayo Clinic Family Health Book.* 2nd ed. New York: William Morrow and Company, Inc., 1996.

Martin, William F., J.M. Lippitt, and T.G. Prothero: *Hazardous Waste Handbooks for Health and Safety.* 2nd ed. Stoneham, MA: Butterworths-Heinemann, 1992.

Mooney, Blake: *1993 Altitude-Rated Places: A Medical Atlas (United States).* Vol. 1. New Orleans, LA: McNaughton & Gunn, 1993.

National Institute for Occupational Safety and Health: *Working In Hot Environments.* Cincinnati, OH: National Institute for Occupational Safety and Health, 1992.

Stellman, Jeanne M. and S.M. Daum: *Work Is Dangerous to Your Health.* New York: Pantheon Books, 1973.

Tiernery, Lawrence M. Jr., M.D., S.J. McPhee, M.D., and M.A. Papadakis, M.D.: *Current Medical Diagnosis and Treatment, 1999.* 38th ed. Stamford, CN: Appleton and Lange, 1999.

SIX

Radiation Hazards

6 | CONTENTS

Radioactivity, the spontaneous release of energy by a nucleus as it changes form, has been a part of the universe since the beginning of time. It was not until science had matured in the late 1800s and early 1900s with the development of detection devices that the presence of radioactivity was recognized and defined. Today, radioactive materials are used regularly in a number of applications such as medicine, nuclear power, detection of underground pipeline leaks, agriculture, biology, astrophysics, art, and archeology. In spite of beneficial features, certain radioactive materials may also be used in ways that could adversely affect the environment. Nuclear fuel such as plutonium is highly radiotoxic and ranks with botulism (food poisoning bacterium) as one of the world's most poisonous materials. The inadvertent release of radioactive waste products, such as those generated from the operation of nuclear reactors and defense industry projects, still poses major environmental problems and the transportation and disposal of these materials are the subject of extensive discussion.

Numerous regulatory agencies are involved in attempts to minimize hazards associated with radioactive materials. The Nuclear Regulatory Commission (NRC) licenses and oversees any facility using radioactive materials for commercial use, as well as regulating the construction and operation of commercial nuclear reactors and nuclear disposal sites. OSHA has regulations for protecting workers from unnecessary exposure to radiation or radioactive materials. When radioactive materials are shipped, their transportation is regulated by the DOT. The DOE oversees the operations of nuclear defense facilities.

Radioactivity results from the decay or disintegration of the nuclei of certain elements. Most elements with more than 82 protons in their nuclei are unstable and therefore, radioactive. These unstable nuclei emit energy and/or particles to become more stable. The particles released are dominated by protons, neutrons, and electrons, while the energy released is usually in the form of gamma rays. Both particle and energy release are examples of ionizing radiation. This means that as each of these particles or energy waves pass through matter (such as skin, organs, or bone), the matter involved is ionized or excited. Ionization is a process involving the removal of an electron from an atom or molecule, leaving them with a positive charge. Excitation is the addition of energy to an atom or molecule, changing them from stable to excited states. Either change may cause cell damage by upsetting the cell chemistry and by disrupting the ability of the cell to repair itself.

6.1 Types of Radiation

6.1.1 ALPHA RADIATION

Alpha radiation consists of positively charged particles comprised of two neutrons and two protons. Alpha particles have high energy but limited penetrating ability (only about 0.1 mm, the thinness of a sheet of paper) because of large mass. These airborne particles are usually not hazardous to humans unless they lodge in the lungs, enter through abraded skin, or are swallowed on particulates. The alpha emitter can enter the body and then release more alpha particles. When an alpha particle enters the lining of the lungs, cells in the vicinity will receive an enormous blast of energy as the alpha particle loses its energy. This energy can cause tissue damage.

Radon, a gaseous uranium decay product, is the most common alpha emitter. Radon naturally occurs in a wide variety of materials. The means of entry is inhalation and lung cancer may be a result of prolonged or high exposure to radon gas.

6.1.2 BETA RADIATION

Beta radiation originates from negatively charged electrons that are emitted when a neutron decays. These particles have less energy than alpha radiation but their smaller mass gives them more penetrating ability. Such particles can penetrate 5–10 mm into human skin. Beta particles can cause skin burns and are hazardous when they enter the body through inhalation or ingestion. Such particles have a variety of sources, such as carbon 14 used in medical research, cosmic radiation and certain minerals.

6.1.3 GAMMA RADIATION

Gamma radiation consists of short-wave-length electromagnetic energy comprised of photons, or packets of energy that are given off from an excited nucleus. Gamma rays travel in straight paths at speeds close to that of light. Gamma radiation has a high penetrating ability but may not be as hazardous as other forms of radiation because it can pass directly through certain portions of the body without depositing any energy. Harm may result if an intervening mass such as bone or a dense organ absorbs the energy. Gamma radiation is common and originates from cosmic and terrestrial sources, such as uranium ore.

6.1.4 NEUTRON RADIATION

A neutron is one of the elementary particles found within the nucleus of the atom and consists of one proton and one electron. It is without a charge and is emitted when atoms decay or break apart. Neutrons have a high penetrating ability and a high potential to cause biological damage. They travel long distances in air and can induce radioactivity in soils, buildings, metals and foods. These particles are found at nuclear facilities and as by-products of nuclear arms devices, and can produce long-lasting effects to the environment when they are released.

6.2 Toxicology of Radiation Exposure

Everyone is exposed daily to a certain amount of low-level ionizing radiation from radioisotopes occurring naturally in our environment. Many people are also intentionally exposed to radiation with the use of medical or dental x-rays or in the treatment of cancerous conditions. Neither of these types of exposures generally results in the development of long-lasting health problems, since the dose of the radiation is relatively low.

Body tissue can be exposed to a certain amount of radiation without adverse effects. Only unusually intense quantities of radiation are accompanied by any physical sensations. Very high doses of radiation will, ultimately, alter body tissues and result in death or ulceration of the tissue. The amount of radiation producing these effects depends on the type of radiation absorbed, its energy, the specific area of the body that has been exposed, the duration of exposure and the age of the individual. Both acute and chronic effects can result from radiation exposure.

When radioisotopes are ingested or placed in the body, they can be difficult to eliminate. These materials tend to collect in body regions where they may cause damage to the molecular components of the cells. Radioisotopes of elements that are naturally found in the body tend to accumulate in places where the natural element is located. For example, when radioactive calcium 45 is ingested, it is deposited and concentrated in bone matter. In this way, such radioactivity can be said to bioaccumulate.

6.2.1 CELLS AND TISSUE

When radiation passes through tissue, it causes ionization within tissue cells. Radiation passing through the cell is most likely to interact with water molecules producing either direct or indirect effects:

Direct Effects—breaking apart of the hydrogen and hydroxyl ions in the cell, creating free radicals.

Indirect Effects—formation of chemically active species of molecules such as hydrogen gas (H_2), hydrogen dioxide (HO_2), and hydrogen peroxide (H_2O_2).

Many of the ionizing events cause changes in the cell exactly like those which occur naturally within the cell and have no noticeable effects. In the presence of oxygen, however, some of the ionizing events result in compounds not normally found in the body, such as hydrogen peroxide and hydrogen dioxide.

The body is amazing in that it possesses the ability to repair limited damage, either by repairing the damaged cell or by replacing the damaged cell. As long as the body is able to repair the damage faster than the damage is produced, the immediate effects of radiation exposure are minimal. If there is severe damage to a group of cells with an important biological function, more significant biological effects may occur. For example, if cells that create white blood cells are damaged, the body will lose the ability to fight diseases resulting in serious consequences.

Radiation can also cause genetic changes. Genetic codes are transmitted from one generation to the next by protein molecules called DNA. If DNA is altered by radiation, this change can be passed on to daughter cells. Cells whose genetic material has been altered due to radiation can continue to divide and grow. The degree of change in the characteristic of the cell produced by the damage can vary. Such changes in cell characteristics are called mutations and can be carried from one cell generation to the next. These mutations can range from minor to serious physical defects, or to mutations which are recessive or not apparent in the new individual. Mutations which effect future generations are termed genetic mutations, while mutations which effect only the individual exposed to the radiation are called somatic mutations. A major somatic mutation may result in a fetal malformation.

6.2.2 CELL RADIOSENSITIVITY

Different cells in the body differ in their sensitivity to radiation exposure. Thus, an amount of radiation which is lethal or produces severe damage to one type of cell may not damage another cell type. Several factors influence a cell's radiosensitivity:

- reproductive capacity of the cell—rapidly dividing cells are very radiosensitive;
- stage of cell division—cells in the process of cell division have maximum radiosensitivity;
- cell activity—the higher the metabolic rate of a cell, the greater the radiosensitivity; and
- blood and food supply—undernourished cells are generally less radiosensitive than normal cells.

Highly radiosensitive cells include white blood cells, bone marrow, and eye cells. Cells which exhibit low radiosensitivity include red blood cells, bone cells, muscle cells, and cells of the nervous system.

6.2.3 CRITICAL ORGAN AND BODY BURDEN

An organ may concentrate a radionuclide. The particular organ receiving the greatest dose from a particular radioactive material is designated the critical organ. The critical organ depends both upon the chemical nature of the radioactive material and the radiosensitivity of the organ. For example, the thyroid gland accumulates radioactive iodine.

The body burden of a particular radioactive material is simply the total amount of radioactive material within the body at any one time. The body burden is affected by the effective half life of the radioactive material and whether the material is soluble or insoluble in the body. Effective half life is the time required for half of the radioactive material to be eliminated.

6.2.4 EFFECTS ON AND RISKS TO HUMANS

Biological damage caused by radiation can be grouped into classifications of effects.

Chronic Effects—exposure to radiation received over years to a life-time. A body can better tolerate radiation exposure over an extended period of time than it can tolerate the same exposure delivered in a large single dose. Some of the effects of chronic exposure to radiation include life-shortening cancer and leukemia.

Threshold Effects—no permanent biological damage unless the total dose to the body or an organ of the body exceeds some value.

Acute Effects—effects due to high levels of radiation exposure to the body over a short period of time (< 24 hours). Acute effects are shown in the table below.

Dose	*Effects*
< 25 rem	No readily noticeable, visible effects.
25–100 rem	Minor blood changes.
100–200 rem	Some sickness is observed.
200–450 rem	Radiation sickness increases.
450 rem	LD 50/30d (50 percent of population exposed will die within 30 days)
450–1000 rem	Severe sickness and death; loss of hair; internal bleeding; nervous system impairment; survival possible with medical treatment.
> 1000 rem	Survival improbable; nervous system shutdown and coma.

Most activities of life carry risk of injury or death. There are hazards at work, home, and in the public environment. An acute exposure to 10 mrem whole body radiation exposure carries a lifetime risk of death equivalent to riding a bicycle for 10 miles, driving a car 300 miles, smoking 2 cigarettes, or eating 100 charcoal-broiled steaks.

6.3 Radiation Dosage and Related Terms

What amount of radiation exposure is safe? There is no unqualified answer to this. Guidelines have been proposed and regulations specified for various types of exposures. The terms by which these quantities of radiation are defined are varied and at times difficult to grasp. The following terms and definitions are based on physics, chemistry, and environmental interpretations of the types and energy levels associated with radiation.

The ability of radiation to produce carcinogenic and/or genetic damage in humans is expressed as the energy delivered per unit weight of tissue. Logically, we divide the total energy delivered to a sample of tissue material (both the energy and tissue being carefully measured in the lab) by the mass in grams or kilograms of the tissue. Energy units are generally given in ergs, the unit of work equal to the work done by a force of one dyne (the force that would give a mass of one gram an acceleration of one centimeter per second) acting through a distance of one centimeter. The problem with this system is that the area of actual exposure is generally many times smaller than that being measured. Perhaps only a few cells would actually be involved. Therefore, the actual event may be a large extrapolation of the answer that is derived. For example, if 100 ergs of energy are deposited in 1 gram of tissue, the tissue is said to have received 1 rad of radiation or 10^2 ergs per gram. This definition does not take into consideration an important fact of radiation exposure: the time it takes to deliver this amount of energy to the tissue in question. If the rate of exposure is required, then the doses would be expressed in rads per minute, hour or day..

The term rad also says nothing about the type of radiation being considered. Section 6.1 described the differences in energy concentrations among alpha, beta, gamma, and neutrons. Other forms of ionizing radiation such as x-rays may be used.

For specific effects where the energy and its relative transfer into tissues must be known, the rad unit must be modified to a unit that demonstrates the relative biological damage caused by one type of radiation over another. The RBE, or relative biological effectiveness, is used for this expression. Consider a situation in which alpha particle radiation causes twenty times as much health effect to a lung cell per rad as does beta particle radiation. The RBE for that specific health effect would be 20. A unit has been designated for this combined effect, the rem.

$$\text{rem} = (\text{rads}) \times (\text{RBE})$$

The RBE for differing effects of various types of ionizing radiation may change significantly from one biological effect to another. It may be 10 for lung tissue, yet only 1 or 2 for skin. The RBE for one radiation compared to another is not a fixed quantity, so it is not possible to use a single RBE value to describe alpha particle radiation, for example, in comparison with x-rays or beta particle radiation.

6.4 Radiation Controls

The amount of radiation to which a worker may be exposed in the workplace is regulated usually by NRC or DOE. Regulations will apply if no other agency has jurisdiction over specified amounts in specified situations. Besides setting levels of exposure for safety, regulators also stress shielding, distance, and exposure times as three major ways to control the harmful effects of radiation. A concerted effort by all personnel at a facility can ensure that personal safety is maintained, public safety is not threatened, and the environment is not adversely affected.

6.4.1 OCCUPATIONAL EXPOSURE LIMITS

Regulatory agencies establish limits for exposure to ionizing radiation, based on two criteria:

- any amount of exposure to ionizing radiation may be hazardous; and
- no adverse biological effects are observed at or below the limits established.

A brief summary of the occupational limits imposed by various regulatory agencies is as follows:

Occupational Dose Limits

Organ/ Affected Area	*DOE rem/yr*	*NRC/OSHA rem/qtr*
Whole Body, Head, and Trunk	5*	1.25**
Lenses of the Eyes	15	1.25
Extremities	50	18.75
Skin of the Whole Body	50	7.5

*DOE has established an Administrative Control Level of 2 rem/yr to challenge management of personnel to limit radiation exposure to well below regulatory dose limits.

**Personnel may exceed the limit for whole body exposure so long as the exposure does not exceed 3 rem per calendar quarter and other requirements of the applicable regulations are followed.

6.4.2 POSTING REQUIREMENTS

Regulatory agencies require various warning signs when radiation hazards exist. The NRC posting requirements are found in 10 CFR, Chapter 20. The DOE requirements are found in DOE order 5480.11. Many state requirements can be found in the health department regulations. The NRC requirements are summarized as follows:

Radiation Area is an area where radiation levels are greater than 5 mrem/hr, or where a person could receive 100 mrem in any 5 consecutive days.

High Radiation Area is an area where radiation levels are greater than 100 mrem/hr. High radiation areas are also controlled through increased security and monitoring requirements.

Very High Radiation Area is an area where radiation levels are greater than 500 rem/hr.

Airborne Radioactivity Areas are areas where airborne radioactive material exists in concentrations in excess of 25 percent of the amounts specified in 10CFR20 for a particular radionuclide.

Radioactive Materials Area is any demarcated area in which radioactive materials are used or stored in an amount exceeding the quantities specified in a regulation.

6.4.3 ALARA

ALARA is an acronym for "As Low As Reasonably Achievable." This principle applies to all aspects of radiation protection. Though limits are established by regulatory agencies, it is advisable that employers minimize exposure of personnel to levels as low as possible. Exposure of personnel is minimized by using three basic protection practices: time, distance, and shielding.

6.4.4 TIME, DISTANCE, AND SHIELDING

Time, distance, and shielding are practices used to protect radiation workers from unnecessary exposure to ionizing radiation. Understanding and properly implementing these radiation practices are essential for maintaining radiation exposure ALARA. To minimize exposure, personnel can minimize exposure time, can

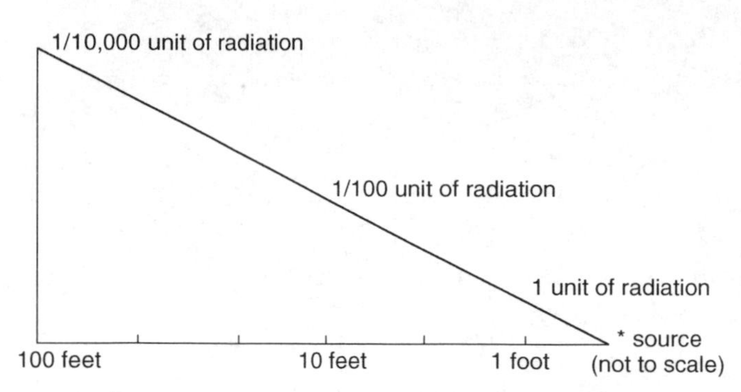

Figure 6-1 Radiation Exposure Calculations

increase their distance from the source of radiation, and can place shielding between themselves and the source.

6.4.4.1 Time

A radiation dose is directly proportional to the exposure time. For example, if the time spent in a particular radiation field is doubled, the worker's dose is doubled. Site supervisors can limit the exposure by reducing the time spent in radioactive environments by rotating workers out of the radioactive work area.

6.4.4.2 Distance

Exposure to radiation can be minimized by maximizing the distance between personnel and the radioactive source. Exposure rates can be calculated at various distances from the source. A radiation dose is inversely proportional to the square of the distance of separation from the source (Figure 6-1). For example, a person standing 10 feet from a source will receive only 1/100 of the radiation received by a person standing 1 foot from the source.

6.4.4.3 Shielding

The type of shielding needed to stop radiation depends on the type and form of radiation. If alpha and beta radiation are present as air-borne particles, the use of a particulate filtering cartridge would help capture these particles before they are inhaled. In other situations, alpha shielding can be accomplished by paper and beta particles can be shielded using aluminum foil, plastic, or similar materials. Gamma shielding utilizes metals (steel and lead) or water. Plastic, water, and specialized synthetic materials can provide shielding from neutrons.

6.5 Personal Monitoring

Since humans cannot detect radiation with their senses, various types of instruments must be used. Some methods of detection include gas ionization, photographic emulsions, scintillation, semiconductors, chemical decomposition, and radiophotoluminescent. These methods work on the principle of radioactive materials interacting with detection media. Although the types of detection instruments sound exotic, they differ only in the substance in which detection takes place and in the method with which detection is registered.

Survey instruments are designed to measure the rate of ionization in milliroentgen per hour or counts per minute. Personal monitoring devices (generally worn by the worker) are designed to measure the total cumulative radiation exposure in units which are related to the absorbed dose. Monitoring is generally done for two specific zones, the working environment and the individual. Some common detection devices used for monitoring sites and working environments for occupational exposure are listed below.

6.5.1 THERMOLUMINESCENT DOSIMETER

The thermoluminescent dosimeter (TLD) is a personal monitoring badge worn by on-site personnel who may potentially be exposed. It detects cumulative exposure to ionizing radiation and is read through a readout device which indicates exposure to radiation in millirems. It is worn externally on the body to best indicate whole body radiation exposure. A control badge is read at the same time as the personnel badges to check for accuracy.

6.5.2 GEIGER-MUELLER INSTRUMENTS

Geiger-Mueller Instruments are detection devices which consist of a gas filled tube. When radiation ionizes the gas, an electronic pulse is generated. This pulse is then electronically amplified and registered on a meter. It works well in the surveying of alpha, beta, and gamma.

6.5.3 SCINTILLATION SURVEY INSTRUMENTS

Scintillation counters operate by transferring radiation energy into light. The light is then translated into an electrical pulse by a photosensitive vacuum tube. These instruments are very sensitive and can be used to detect alpha, beta, and gamma radiation.

6.5.4 OTHER RADIATION SURVEY INSTRUMENTS

Specific instruments are available to measure particles or waves. Such dedicated devices are generally not used in waste-site survey activity unless preliminary site investigations or record evaluations indicate that radioactive material may be present and pose a threat to workers. The instruments can measure special characteristics such as the type of radionuclides or the strength of nonionizing radiation such as microwaves.

6.6 Radiation in the Environment

Common occupational sources of radiation include medical and dental sources (currently the largest artificial source of radiation exposure), nuclear power plants, radioactive static eliminators, isotope-tagged products, radioactive luminous markers, and continuous inspection by x-rays fluoroscoping. In addition, large doses of radiation are used in industry to sterilize food and drugs. Radiation also kills insects in seeds, toughens polyethylene materials, and activates chemical reactions in petroleum processing.

Workers in special occupations may have relatively high risks of radiation exposure. For example, workers in underground mines may be "bombarded" from all directions with natural elemental decay products of uranium, even though they are not specifically mining what would be termed radioactive materials. Firefighters may be exposed to radiation from improperly stored or spontaneously released materials. Airline pilots and service personnel may be exposed to high levels of cosmic radiation.

6.7 Transportation Controls

When radioactive materials are moved from one place to another, the potential hazard comes not so much from the chemical properties of the materials as it does from the radioactive nature of the materials. It is necessary to have all materials labeled in such a manner as to make their radioactive nature readily identifiable. Some regulations concerning transportation are presented in a later unit on the transportation of hazardous materials.

For general precautions, it is necessary to remember certain basics when dealing with radioactive shipments. First, if drums containing radioactive materials exhibit radiation levels above the normal background, a health physicist should be contacted immediately. A worker should not handle any drums that are determined to be radioactive until experts in this area have been consulted. Next, the degree of hazard will vary depending on the type and quantity of material involved. The DOT has set a specific class, Class 7, to describe shipping requirements (container type, labels, etc.) and emergency response guidelines for different categories of radioactive materials. Other regulations or guidelines for the transportation, storage, disposal, and use of radioactive materials are addressed by the EPA, the NRC, the Interstate Commerce Commission, DOE, the Coast Guard, and the Civil Aeronautics Administration.

6.8 Sources of Information

Gofman, John W.: *Radiation and Human Health*. San Francisco: Sierra Club Books, 1981.

Gollnick, Daniel A.: *Basic Radiation Protection Technology*. 3rd ed. Pacific Radiation Corporation, 1994.

Koren, Herman: *Handbook of Environmental Health and Safety Principles and Practices*. Vols. I and II. Chelsea, MI: National Environmental Health Association/Lewis Publishers, 1991.

Meyers, Eugene: *Chemistry of Hazardous Materials*. 3rd ed. Englewood Cliffs, NJ: Prentice-Hall, 1997.

U.S. Department of Transportation, Research and Special Programs Administration, Transport Canada, Safety and Security, Dangerous Goods Secretariat of Transport and Communications, Transport Dangerous Goods Directorate, Mexico, Secretaria de Communicaciones y Transportes: *2000 North American Emergency Response Guidebook*. Washington, DC: U.S. Government Printing Office, 2000.

SEVEN

Medical Monitoring

7 | CONTENTS

A medical program is essential to assess and monitor worker health and fitness both prior to and during the course of employment, to provide emergency and other treatment as needed, and to keep accurate records for future reference. Information from a site medical program may be used to conduct future epidemiological studies, to adjudicate claims, to provide evidence in cases of litigation, and to report worker medical conditions to any agencies requiring them by law.

Medical surveillance is designated in specified operations and for specified employees, each covered under a variety of legal sections. OSHA regulation 29 CFR 1910.120(f), Hazardous Waste Operations and Emergency Response: medical surveillance, designates which employees are covered for medical attention. Employees may be monitored and/or placed under surveillance for having contact with, or for working in the presence of, hazardous substances. In general, those workers covered by the medical surveillance program are listed under OSHA law 29 CFR 1910.120 (a)(1)(i) through (a)(1)(iv), those not covered by (a)(2)(iii) exemptions, and employers of employees specified in paragraph (q)(9).

Section	*Designation*
(a)(1)(i)	Clean-up operations required by a governmental body such as federal, state, local or other, involving hazardous substances that are conducted at uncontrolled hazardous waste sites (including, but not limited to, the EPA's NPL, state priority site lists, sites recommended for the EPA NPL, and initial investigations of government identified sites which are conducted before the presence or absence of hazardous substances has been ascertained;
(a)(1)(ii)	Corrective actions involving clean-up operations at sites covered by the RCRA of 1976 as amended;
(a)(1)(iii)	Voluntary clean-up operations at sites recognized by federal, state, local or other governmental bodies as uncontrolled hazardous waste sites;
(a)(1)(iv)	Operations involving hazardous waste that are conducted at TSD facilities regulated by 40 CFR Parts 264-265 pursuant to RCRA;

or by agencies under agreement with the U.S.E.P.A. to implement RCRA regulations;

(a)(2)(iii) Operations within the scope of (a)(1)(iv) must comply only with the requirements of paragraph (p) of this section. This includes regulations pertaining to "Certain Operations Conducted Under the Resource Conservation and Recovery Act of 1976." It presents the specific criteria for RCRA site health and safety, medical monitoring, and all aspects of operations; and

(q)(9) Emergency response to hazardous substances releases: medical surveillance and consultation; this covers employers whose employees are engaged in emergency response no matter where it occurs, except for employees engaged in operations specified in (a)(1)(i) through (a)(1)(iv).

Employees covered by a medical surveillance program are those engaged in specified operations (section (a)) and/or are specified under 29 CFR 1910.120 (f)(2)(i) through (f)(2)(iv). Workers falling within these classifications are:

(f)(2)(i) All employees who are or who may be exposed to hazardous substances or health hazards at or above the established permissible exposure limit, above the published exposure levels for these substances, without regard to the use of respirators, for 30 days or more a year;

(f)(2)(ii) All employees who wear a respirator for 30 days or more a year or as required by 1910.134;

(f)(2)(iii) All employees who are injured, become ill or develop signs or symptoms due to possible overexposure involving hazardous substance or health hazards from an emergency response or hazardous waste operation; and

(f)(2)(iv) Members of HAZMAT teams.

Frequency and type(s) of medical examinations and consultations are:

Section	*Medical Surveillance*
(f)(2)(i) (f)(2)(ii) (f)(2)(iv)	Prior to assignment; At least once every 12 months unless the attending physician believes a longer interval (not longer than biennially) is appropriate; At termination of employment or reassignment to an area where the employee would not be covered if the employee has not had an examination within the last six months; As soon as possible upon notification by an employee that the employee has developed signs or symptoms indicating possible overexposure to hazardous substances of health hazards, or that the employee has been injured or exposed above the permissible exposure limits or published exposure levels in an emergency situation; and At more frequent times, if the examining physician determines that an increased frequency of examination is medically necessary.
(f)(2)(iii) (a)(1)(iv)	For employees covered under these paragraphs, who may have been injured, received a health

impairment, developed signs or symptoms which may have resulted from exposure to hazardous substances resulting from an emergency incident to hazardous substances at concentrations above the permissible exposure limits or the published exposure levels without the necessary PPE being used, medical surveillance will be:

As soon as possible following the emergency incident or development of signs or symptoms; and

At additional times, if the examining physician determines that follow-up examinations or consultations are medically necessary.

Medical examinations required by paragraph (f)(3) shall include a medical and work history (or updated history if one is in the employee's file) with special emphasis on symptoms related to the handling of hazardous substances and health hazards, and to fitness for duty including the ability to wear any required PPE under conditions (such as temperature extremes) that may be expected at the work site.

1910.134(e) Minimum requirements for medical evaluation that employers must implement to determine the employee's ability to use a respirator prior to the employee being fit tested or required to use the respirator in the workplace. The employer may discontinue an employee's medical evaluations when the employee is no longer required to use a respirator. Specifics for respirator testing are found in 1910.134 (e)(2) through (e)(7). Appendix C of 1910.134 includes a mandatory Respirator Medical Evaluation Questionnaire.

This unit presents general guidelines for designing a medical program for personnel at hazardous waste sites, including information and sample protocols for pre-employment screening, periodic medical examinations, emergency and nonemergency treatment, recommendations for program recordkeeping, and reviews of procedures.

Recommendations in this unit assume that workers will have adequate protection from exposures through administrative and engineering controls as well as access to appropriate PPE and decontamination procedures described elsewhere in this manual. Medical surveillance should be used to complement these other controls.

7.1 Developing a Program

A medical program must be developed for each work site based on the specific needs, work tasks, and potential exposures of employees. It should be designed by an experienced occupational health physician or other qualified occupational health consultants in conjunction with the site Health and Safety Officer. The director of an occupational medical program should be a physician, board-certified in occupational medicine, or a medical doctor who has had extensive experience managing occupational health services. Since many waste sites are located in rural or remote areas where a physician with such qualifications is hard to find, the site medical program may be managed by a local physician with assistance from an occupational medicine consultant. Some functions may also be performed by a qualified Registered Nurse (preferably an occupational health nurse) under the direction of a suitably qualified physician who has responsibility for the program.

Blood test analyses, such as those for exposure to airborne lead and zinc as well as for bloodborne pathogens and other toxins, must be performed by an accredited laboratory that meets OSHA accuracy requirements in proficiency testing. OSHA, the College of American Pathologists, and the Wisconsin State Laboratory of Hygiene analyze and report blood testing data. A list of Laboratories Approved for Blood Lead

Analysis and licensed and accredited under the Clinical Laboratory Improvement Act (CLIA) and Medicare is updated and distributed by OSHA.

A site medical program should provide the following elements:

Surveillance
- pre-placement screening
- periodic medical examinations
- exposure specific examinations
- termination examination

Treatment
- emergency
- nonemergency (as warranted on a case-by-case basis)

Recordkeeping

Program review

Table 7-1 outlines a recommended medical program. Screening and examination routines are described in following sections. These routines are based on potential health risks for hazardous waste workers, a review of available data relating to actual exposures, and an assessment of several established medical programs. Conditions and hazards vary considerably from place to place at each site; therefore, only general guidelines are given.

The effectiveness of a medical program depends on appropriate content and surveillance, active worker involvement, and management commitment. Management commitment is shown through medical surveillance, treatment, and management directives. All occupational medical examinations, procedures, and monitoring required by the job description or government agencies must be provided to employees at no cost to the employee. To this end, management should urge employees to provide the examining physician with a complete, honest, and detailed occupational and medical history. They should assure their workers of the confidentiality of their medical records and require workers to report any suspected exposures, regardless of the amount. Finally, management should require workers to alert the site physician to any unusual physical or stressful conditions which may be new to them. Employee training should emphasize that even vague or seemingly minor complaints (such as skin irritation or headaches) may be important.

When developing an individual program, site conditions must be considered and the monitoring needs of each worker should be determined based on the worker's past exposure history, age, health status and the potential exposures on-site. The routine daily tasks of each worker should also be considered. For instance, a heavy equipment operator exposed to high noise levels would require a different monitoring protocol from a field sample collector with minimal noise exposure.

Potential exposures that may occur at a site must also be evaluated during an exposure assessment and discussed in the Site Health and Safety Plan. Certain chemicals such as lead, PCBs, mercury, or pesticides may be part of medical surveillance as determined by the exposure assessment. The presence of these materials should be measured in the worker at periodic medical screenings to determine whether it is necessary to continue testing for such substances.

7.2 Pre-Placement Screening

The pre-placement screening has several goals. One is to determine an individual's fitness for duty, including the ability to work while wearing protective equipment. Another is to establish a baseline against which any future changes in health or physical well-being can be evaluated. Pre-placement screening should identify any underlying illness or medical conditions that may be aggravated by exposures or job activities. It should recognize any abnormalities or prescription drug use that may affect the ability to work safely so that corrective measures may be taken.

7.2.1 DETERMINATION OF FITNESS FOR DUTY

Workers at hazardous waste sites are often required to perform strenuous tasks and wear PPE, such as respirators and protective clothing, that may cause heat stress and other problems. To ensure that prospective employees are able to meet work requirements, the pre-placement screenings should focus on any condition that increases the risks of aggravating these situations.

Table 7-1 **Recommended Medical Program**

Type of Medical Service	Typical Components
Pre-placement Screening	• procure and record medical history (detailed and comprehensive) • record occupational and exposure history • perform physical examination • determine fitness to work wearing respirator or special protective clothing and equipment • set up and perform baseline monitoring for specific exposures
Periodic Medical Evaluations	• produce update of medical and occupational exposure history • execute physical examination based on 1) examination and interview results, 2) exposure, 3) job class and task, 4) health status, 5) age
Exposure Specific Examination	• perform when a known hazardous substance exposure has occurred or when symptoms seem to indicate such an exposure • perform when an exposure has occurred without benefit of PPE • perform as soon as possible • determine need for possible follow-up evaluations
Emergency Treatment	• provide emergency first aid at site • develop liaison with local hospital and medical specialists • arrange for decontamination of victims
Nonemergency Treatment	• develop mechanism for nonemergency health care; includes colds, flu, headaches, etc., including referral to personal physician
Termination Exam	• determine medical status of worker at termination of employment
Recordkeeping and Review	• maintain and provide access to medical records in accordance with OSHA and state regulations • report and record occupational injuries and illnesses • review Site Health and Safety Plan regularly to determine if additional testing is needed • review program periodically; focus on current site hazards, exposures, and industrial hygiene standards

7.2.1.1 Occupational and Medical History

This is a detailed health, occupational, and exposure history of the worker including a worker-completed questionnaire which is carefully reviewed by the physician prior to the physical examination. The document should also be discussed with the worker, with special attention paid to occupational exposure and to other potential chemical and physical hazards. It should be conveyed to the worker that truthfully listing all potential problems is important since it gives the physician the basis for specific testing as well as for future monitoring procedures.

The detailed history should include a review of past illnesses and chronic diseases such as diabetes, high blood pressure, allergies, asthma, and lung and cardiovascular diseases. Other medical issues, such as shortness of breath, labored breathing on exertion, chest pain, high blood pressure, and heat intolerance must be evaluated. Individuals who are exposure sensitive such as someone with a history of severe asthmatic reaction to a specific food, drug or environmental agent, should be identified.

Finally, it is necessary to record relevant lifestyle habits including exercise, cigarette smoking, alcohol use, pharmaceutical use, and hobbies such as crafts, home repairs, and automobile maintenance. Such pastimes pose the possibility of exposing the individual to materials as glues, solvents, cleaners, welding fumes, heavy lifting, and/or noise which may harm the individual without his or her immediate knowledge.

7.2.1.2 Physical Examination

The physical examination should include a comprehensive examination of all major body organs focusing on the pulmonary, cardiovascular, skin and musculoskeletal systems which are most likely to be effected by exposures and generally show the most pronounced evidence of past exposures (Table 7-2). Obesity and lack of exercise usually indicate to physicians that an individual may be more susceptible to heat stroke, while missing or arthritic fingers, facial scars, dentures, poor eyesight, or perforated eardrums may signal a potential worker inability to effectively use respirators and/or other necessary types of protective equipment.

7.2.1.3 Ability to Work While Wearing Protective Equipment

This medical evaluation includes a review of the worker's written occupational and personal histories as well as medical tests based on physical and mental criteria. These evaluations help to disqualify individuals who are clearly unable to perform based on physical problems such as severe lung disease, heart disease, back/orthopedic problems, or who are claustrophobic. They also identify workers with limiting conditions (can only wear contact lenses or have breathing difficulties such as asthma) who may not be able to use full-face respirators or on-demand respirators. The evaluation may include additional testing of lungs and heart. Sometimes the physician may conduct a stress test (for example, a treadmill test) in order to ensure a proper final diagnosis.

The final determination of worker ability to use and to tolerate proper protective equipment is made on the basis of the individual worker's profile, medical history, and the physical examination, as well as age, previous exposures, and testing. A written assessment is necessary if wearing respiratory equipment and working on a hazardous waste site is a job requirement. OSHA respirator standard (29 CFR 1910.134 (b)(10)) states that no employee should be assigned to a task that requires the use of a respirator unless it has been determined that the person is physically able to perform under such conditions.

7.2.2 BASELINE DATA FOR FUTURE EXPOSURES

Data collected from all available sources forms the baseline history. This is used to later measure the effectiveness of protective measures and to determine if exposures have adversely affected the worker. Reliable biological monitoring tests, such as blood tests for lead levels, are useful in determining pre-exposure levels of specific substances to which the worker may be exposed.

Testing of specific toxins is typically conducted by the physician for workers who may receive significant exposure from these agents. For example, serum PCB testing can monitor long-term exposure during cleanup of a polychlorinated biphenyl waste facility. Standard procedures are also available for determining levels of lead, cadmium, arsenic, and organophosphate pesticides. In some cases, pre-placement blood specimens and serum are frozen for later testing when PCBs and some pesticides are to be monitored. Finally, a battery of tests based on the worker's past occupational and medical history and an assessment of significant potential exposures are used to determine a baseline history.

7.2.3 PHYSICIAN'S WRITTEN OPINION

Following the pre-placement examination, the employer must obtain a copy of a written opinion from the examining physician. The employee may, upon request, be furnished with a copy. This opinion should contain:

- an explanation concerning the detection of any medical conditions likely to place the employee at an increased risk in the occupational environment;
- the risk of material impairment of the employee's health from work in hazardous waste operations or from respirator use;
- recommendations for limitations on the employee's assigned work;
- the results of medical examinations and tests if requested by the employee; and

Table 7-2 **Tests Frequently Performed by Occupational Physicians**

Body Systems and Organs	Test	Example
Liver:		
General	Blood Tests	Serum aminotransferase, albumin, globulin, total bilirubin (direct bilirubin if total is elevated)
Obstruction	Enzyme test	Alkaline Phosphate
Cell Injury	Enzyme tests	Gamma glutamyl transpeptidase (GGTP), lactic dehydrogenase (LDH), serum glutamic-oxaloacetic transaminase (SGOT), serum glutamic-pyruvic transaminase (SGPT)
Kidney: Urinary	Blood tests Urinalysis Dip sticks	Blood urea nitrogen (BUN), creatinine, uric acid; Including color, appearance; specific gravity; pH, quantitative glucose, total protein, bile, and acetone; microscopic examination of centrifuged sediment
Digestive: Multiple Systems and Organs	Hands-on Examinations	Occult blood in stool evaluation; palpation of gastro-intestinal organs
Blood-Forming Function	Blood tests	Complete blood count (CBC) with differential and platelet evaluation, including white cell count (WBC), red blood count (RBC), hemoglobin (HGB), hematocrit or packed cell volume (HCT), and desired erythrocyte indices. Reticulocyte count may be appropriate if there is a likelihood of exposure to hemolytic chemicals
Lungs	Pulmonary Function Spirometry Stethoscope X-ray	Rate and volume of air discharged from the lungs is measured and related to normal values based on age, height, race and gender; X-rays can reveal scars, tumors, or other lung abnormalities
Cardiovascular (heart)	Blood pressure Stethoscope Blood sugar Cholesterol	Blood pressure monitoring; stress test on treadmill or stairs; blood tests for diabetes risk assessment; blood tests to determine cholesterol risk factor
Hearing (ears)	Audiogram	Noise-proof booth with earphones
Eyes	Acuity Color Blindness Peripheral vision	Sneller chart Color blindness number diagrams Eye tracking; side vision
Musculo-skeletal and Skin (dermal)	Examination of joints, limbs, skin color, features	Visual and hands-on examination of joint movement and flexibility; strength tests; reflex action of joints; visual examination of skin
Nervous System	Oral history; hands-on exam of reflexes; Potential exposure to neurotoxins may demand additional, sophisticated tests such as: electrical stimulation for nerve conduction; Magnetic Resonance Imagery (MRI); Computed Tomography (CT); Position Emission Tomography (PET)	Reflex action of joints; quantitative sensory testing (touch, thermal, vibratory, electrical stimuli); scans for tumors using opaque or tumor-absorbing dyes, nerve pinching, incorrect bone and spine alignment

- a statement verifying that the employee has been informed by the physician of the results and any medical conditions which may require further examination or treatment.

Specific findings or diagnoses unrelated to occupational exposures are not part of this written decision.

7.3 Periodic Medical Examinations

Periodic medical examinations should be used in conjunction with pre-placement exams and medical history for comparison with baseline data to determine biological trends that may indicate early signs of the adverse health effects of exposure. From the findings, appropriate protective or remedial measures may be taken to protect the employee. Periodic medical exams are usually given at intervals of 12 months, unless the occupational physician recommends a longer interval that does not exceed 24 months. The frequency and content of periodic exams will vary depending on the nature of the work being performed, the type and number of exposures and other factors discussed in previous paragraphs. Once again, the worker's individual medical profile will help to determine examination frequency. For example, a worker participating in the cleanup of a PCB contaminated building may be examined monthly for PCB serum level while a worker drilling ground water monitoring wells at a former municipal landfill may have a yearly exam. In the case of the PCB worker, if the monthly tests find no change in PCB levels, indicating no appreciable exposure, the frequency of the testing may be reduced.

It is important that the workers know whether and how their general health has changed since the last evaluation. They should advise the physician of symptoms such as a greater incidence of colds, sickness, coughs, eye irritation, skin rash abdominal pain, etc. Although these may indicate non-life-threatening situations, all symptoms should be considered significant where exposures to possible hazardous substances are concerned.

Periodic screening exams can include any of the following.

- interval medical history focusing on changes in health status or acute exposures at the work site. The physician should have information about any worker exposure as well as exposure monitoring information from the job site;
- a full physical examination is generally conducted annually;
- additional medical testing may be required depending on available exposure information, medical history, and examination results. In this case, testing should be more specific for possible medical effects of the worker's exposure. Multiple testing for a wide range of potential exposures is not always useful;
- pulmonary function tests should be administered if the individual uses a respirator, has been or may be exposed to pulmonary irritants or toxins, or if the individual has breathing difficulties, especially when wearing a required respirator;
- annual audiometric tests are required for persons subjected to high noise exposures;
- annual vision tests are recommended to check for vision degradation; and
- blood and urine tests are indicated when there is definite exposure to potential toxins such as PCBs or materials as determined by on. exposure assessment.

7.4 Exposure Specific Exams

Special medical examinations may be required if employees are injured, develop signs or symptoms which may have resulted from exposure to hazardous substances during an emergency incident, or have been exposed to hazardous substances at concentrations above the PELs without benefit of PPE. These tests should be performed as soon as possible following the incident or the development of symptoms and may require follow-up examinations or consultations.

7.5 Termination Examination

At the completion of a job at a hazardous waste site, all personnel should undergo an exit medical examination as described in the previous sections.

The exam may be limited to obtaining an interval medical history of the period since the last full examination, if three conditions are met.

- the last full medical evaluation was within the last six months;

- no exposure occurred since the last examination; and
- no symptoms associated with exposure have occurred since the last examination.

If any of these criteria are not met, a medical evaluation is legally required at the termination of employment.

7.6 Emergency Treatment

Provisions for treatment of emergency situations (including biological waste exposures, chemical warfare agents, explosions, punctures, and heat stroke) and for acute nonemergency treatment (such as falls, headache, frostnip, snake bites, bee stings, and plant poisoning) should be made at each site. Not only site workers, but also contractors, visitors, and other personnel (particularly emergency responders) may require emergency treatment.

Emergency medical treatment should be integrated with the overall site emergency response program. Several guidelines have been recommended for establishing an emergency treatment program.

- Train a team of site personnel in emergency first aid including a certified course in cardiopulmonary resuscitation (CPR) and first aid, emphasizing treatment for explosion and burn injuries, heat stress, and acute chemical toxicity. The team should include an emergency medical technician (EMT), if possible;
- Train personnel in emergency decontamination procedures;
- Be certain personnel know what roles and responsibilities each would assume during an emergency;
- Establish an emergency first aid station onsite with the capability of stabilizing patients requiring off-site treatment as well as treating those who need general first aid for minor conditions. The station should be located in the support zone adjacent to the decontamination area, with plans in place for emergency decontamination. A standard first aid kit and supplies should be available plus additional items such as emergency showers, stretchers, potable (drinkable) water, ice, emergency eyewash, decontamination solutions, and fire-extinguishing blankets. Supplies should be checked regularly and restocked immediately after each use;
- Arrange for a physician (and possibly a backup physician) who can be paged on a 24-hour basis;
- Arrange for an on-call team of occupational health specialists for emergency consultations (toxicologist, dermatologist, hematologist, industrial hygienist, allergist, ophthalmologist, cardiologist, and/or neurologist);
- Establish a protocol for monitoring heat or cold stress, and train all workers to recognize symptoms in themselves and in others; and
- Make advance arrangements for transportation of injured workers to a nearby medical facility. This includes alerting the facility to possible treatments required as well as contamination control procedures which may be needed. Assisting the facility in developing procedures for site-related emergencies will help protect hospital personnel and patients, and will minimize delays due to concerns about hospital safety or contamination. For specific illnesses or injuries, provide details of the incident including information on the toxic agent and the worker's past medical history to the appropriate staff. Such information is critical when specific medical treatment is required as in the case of exposure to carbon monoxide, cyanide or organo-phosphate pesticides. Knowing a medical history will help the physicians select medicines that will not complicate existing conditions or allergies.

It is necessary to conspicuously post lists containing the names, phone numbers, addresses, and procedures for contacting all medical personnel, facilities, and emergency services that might be required. These lists should be situated at multiple locations and include maps and directions to the site as well as to the medical facility since individuals responsible for communications may not be available to talk directly with the transporting service. At large sites, radio communication systems should be available. Designated leaders should review emergency procedures daily with all site personnel at safety meetings prior to starting each work shift. Finally, a copy of the worker's medical records should be readily available. Sometimes, for long-term work, medical records are placed at the site and, when appropriate, at a nearby hospital.

7.7 Nonemergency Treatment

Arrangements should be made for the nonemergency medical care of site workers experiencing adverse health effects resulting from exposure to hazardous substances or physical hazards, such as eye irritation, skin rash, bug bites, slipping, etc. These conditions and exposures should be noted in the worker's medical records as part of the on-going medical surveillance program. If nonrelated personal physicians are treating a worker for non-job-related illnesses, such treatment should be reported for inclusion in the site medical records. Failure to do this may put the worker at risk because of specific work requirements. For example, a bad cold or flu may interfere with respirator use, or a change of diet or water intake may upset the GI system, making it difficult to work any distance from a toilet.

7.8 Medical Records

Proper recordkeeping is essential at hazardous waste sites because of the nature of the work and the risks involved. Employees may work for several employers at a number of geographically separate sites over the course of their careers, and adverse effects of long-term exposure may not become apparent for many years. Records enable subsequent medical care providers to be informed about workers' previous and recent exposures.

OSHA regulations mandate that unless a different time period is specifically called for under OSHA standards, the employer must maintain and preserve medical records on exposed and all workers for 30 years after they leave employment. The results of medical testing, and full medical records and analyses must be made available to workers, their authorized representatives, and authorized OSHA representatives. Finally, the employer must maintain records of occupational injuries and illnesses for at least five years, and post a yearly summary report of these instances.

7.9 Program Review

The medical program must be constantly evaluated to ensure its effectiveness. Maintenance and review of medical records and test results helps medical personnel, site officers, and parent company managers to assess the effectiveness of the health and safety program. Evaluations also help determine whether the correct concerns are being addressed through examinations, training, and protective equipment use.

Once a year, the site safety officer, medical consultant, and/or management representative should evaluate whether each accident or illness was promptly and properly investigated to determine the cause. Following such investigations, they should ascertain if necessary changes were made in the health and safety procedures to prevent a recurrence. They should evaluate the effectiveness of specific medical testing as it relates to specific site exposures, and add or delete medical tests as necessary. They must keep current on field activities and exposure assessments from industrial hygiene and environmental data that could drive the addition or deletion of medical tests. Employers must review potential exposures and Site Health and Safety plans at all sites to determine if additional testing is required. Finally, they must review emergency treatment procedures and update all lists of emergency contacts, being careful to replace all dated material with current information.

7.10 Sources of Information

Bronstein, Alvin C. and P.L. Currance: *Emergency Care for Hazardous Materials Exposure*. 2nd ed. St. Louis, MO: Mosby-Year Book, Inc., 1994.

National Institute for Occupational Safety and Health/Occupational Safety and Health Administration/U.S. Coast Guard/Environmental Protection Agency: *Occupational Safety and Health Guidance Manual for Hazardous Waste Site Activities*. Washington, DC: U.S. Government Printing Office, 1985. pp. 5–1 to 5–10.

National Institute for Occupational Safety and Health/Occupational Safety and Health Administration: *Occupational Diseases: A Guide to Their Recognition* (Publication No. 77–181). Washington, DC: National Institute for Occupational Safety and Health/U.S. Department of Health, Education and Welfare/Centers for Disease Control, June 1977.

Occupational Safety and Health Administration: *Blood Lead Laboratories, Program Description and Background*. In www.osha.gov [Occupational Safety and Health Administration (OSHA) World Wide Web homepage]. Washington, DC: OSHA, September 1999.

Occupational Safety and Health Administration: *Occupational Safety and Health Standards for General Industry*. Chicago, IL: CCH Incorporated, January, 2001.

EIGHT

Air Monitoring

8 | CONTENTS

Air monitoring is required by OSHA 29 CFR 1910.120(h) whenever there is the potential for employees to be exposed to:

- airborne contaminants in levels that may present a health hazard;
- an IDLH environment;
- oxygen deficiency; or
- a flammable or explosive atmosphere.

It is also a necessary component of the site safety plan. Air monitoring demonstrates the presence of specific chemicals or classes of chemicals, and allows the Health and Safety Officer at a site to make appropriate choices regarding respiratory protection and protective clothing. It is used to determine the need for medical monitoring, to assess potential health effects of exposures, to verify chemical exposures, and to help determine appropriate response and evacuation actions. This unit discusses some of the common portable equipment used in the field to detect, identify, and monitor airborne contaminants in the working environment.

Most of the air monitoring equipment discussed in this unit provides information in "real-time" or direct read-out, with results given at the time monitoring is performed. The advantage of this is time saved not having to send samples to an off-site analytical laboratory for results. A disadvantage is that the data usually does not have the accuracy or specificity of laboratory results. Most air monitoring programs require some laboratory analysis to identify specific vapors or to verify field measurements. Some equipment also has the capability of storing data for later downloading into computer data bases. Such information may be used for medical surveillance, exposure record documentation, or as part of an over-all site analysis.

Direct-reading instruments are often complex electronic devices. Prior to use, workers must be trained in the proper use, calibration, and interpretation of the data generated, as well as in understanding the instrument's limitations. Additionally, instruments are designed for specific functions, and provide limited information. Instruments cannot make decisions; only an informed and trained worker can do this.

8.1 Selection Criteria

It is essential that the correct monitoring instrument be used for each specific scenario. Following are guidelines for proper equipment selection:

Easy to Use. The simplest instrument capable of collecting the necessary data should be selected. Complicated field instruments are prone to operator error or mechanical failure and should be avoided whenever possible.

Accurate. Field instruments must be accurate enough to meet objectives that can range from worker protection to contaminant identification.

Reliable. Instruments that quickly lose their battery charge, require frequent calibration, or are affected by temperature, humidity, barometric pressure, or other weather conditions may not be reliable, and their use in the field may be limited.

Fast Response. Field instruments should be able to generate data quickly. This is obviously important for worker protection, but is also important for efficiency. The conditions represented by samples collected for field analysis can change quickly; thus, a rapid measurement is essential.

Easy to Interpret. The data generated by a field instrument should take only a second to read and understand. Instruments that generate graphs and charts requiring interpretation have limited use in most field situations.

Sensitive and Selective. Sensitivity refers to the lowest concentration that an instrument can accurately analyze. Selectivity describes the instrument's capability of measuring the target contaminant(s) of interest, even in the presence of other contaminants. Instruments best suited for field use should have high sensitivity, a wide range of readings, and high selectivity.

Reproducible. Results registered by instruments should be reproducible. When a reading in the same area is taken several times within a brief interval, the results should not vary greatly, assuming that the contaminant concentration is relatively constant. Some instruments require time to readjust their sensors following measurements. It is possible that damage may occur during contaminant surges, and that the instrument will not be able to produce accurate or precise readings until recalibration or other adjustments.

Portable. The use of heavy or awkward instruments is generally limited to areas accessible by vehicles. Portable units should be easy to move, rugged, and able to withstand rigorous use and movement. They should also be quick to assemble, and have a short check-out and calibration time.

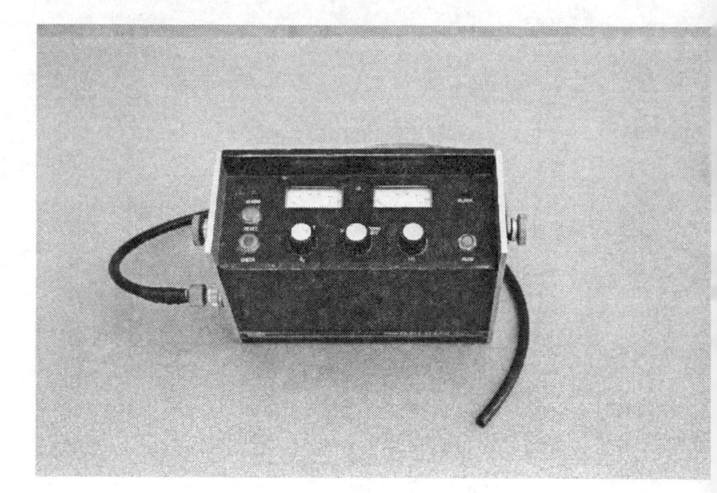

Figure 8-1 Combustible Gas Indicator

Withstand Weather Conditions. Field instruments should be able to withstand heat, cold, wind, dust, and humidity.

Approved for Hazmat Use. Portable field instrumentation used at hazardous sites must be demonstrated to be safe to use in chemical environments. NIOSH has an approval procedure for field instruments used to monitor conditions for worker safety. Only instruments having this approval for air monitoring use should be taken into the field to detect respiratory hazards. Instrumentation used in areas with combustible or flammable atmospheres must be certified by the Mine Safety and Health Administration (MSHA) for use under those conditions.

8.2 Combustible Gas Indicators

It is important to be able to anticipate the potential for flammable or explosive environments. The combustible gas indicator (CGI) can determine if the air contains sufficient levels of vapors to cause an explosion or to support combustion. This instrument is also known as an explosion meter or explosimeter, and is one of the most widely used field instruments in the hazardous materials industry. Most of these devices have a readout unit that gives the percent of the LEL, ranging from 0 percent to 100 percent of the LEL.

8.2.1 OPERATION

For most CGIs, gases are detected by allowing air to diffuse into a sensor or by pumping air into a sensor unit. In either case, the air moves through a metal filter and comes into contact with two hot filaments inside the sensor. Both filaments are initially at the same temperature and have the same electrical resistance. One filament is coated with catalyst, and combustible gases are burned on this filament. The burning increases the temperature of the catalyst-coated filament, thus changing its electrical resistance. The resistance between the two filaments is now different, and this imbalance is translated into percent LEL. The greater the amount of combustible gases in the sampled air, the greater the burn on the catalyst-coated filament, the greater the temperature differential, and the greater the LEL reading.

CGI readings are relative to the calibration gas. When measuring a different gas, the instrument works on the same principle, but the LEL reading may not be accurate. When burned, other gases generate more or less heat than the calibration gas, giving inaccurate LEL readings. For example, when a pentane calibrated CGI is used in a methane environment, a reading of 50 percent LEL is actually only approximately 30 percent LEL because methane burns substantially hotter than pentane, and a lower resistance and higher LEL are recorded. However, the magnitude of differences is specific to the instrument; instruction manuals must be consulted for specifics. In contrast, when a pentane calibrated CGI is used in a xylene environment, a reading of 50 percent LEL will be substantially higher because xylene burns at a cooler temperature than pentane, giving a higher resistance and lower LEL reading. This latter condition can lead to a potentially deadly instrument interpretation on the part of the unsuspecting technician. However, calibration discrepancies only occur at concentrations below 100 percent LEL. At 100 percent LEL, both filaments are burning vapors and safety devices are activated that give the correct reading.

The CGI operates by burning gases, so an appropriate level of oxygen is required for accurate results. Most models require at least 16 percent oxygen for accurate readings. Low oxygen concentrations will produce low values that are false. Oxygen-enriched environments (oxygen concentrations greater than 25 percent) will enhance combustion and will result in high readings that are false. Due to this property, the CGI is generally used in tandem with an oxygen meter.

Substances commonly associated with hazardous waste sites can damage equipment and distort readings. The catalytic filament of the CGI can be damaged by contaminants such as sulfur compounds, heavy metals (especially lead), and silicon compounds. These materials often form fumes that coat the filament and interfere with the burning process. Enclosed spaces, such as storm sewers or trenches, may alter oxygen levels, concentrate combustibles, or confine airborne vapors. Oxygen levels may stratify, producing a variety of readings at different levels without the worker being aware of a change in hazard in the work location. High humidity can reduce readings due to water vapor cooling the filaments.

8.2.2 INTERPRETATION OF DATA

If a CGI is calibrated to methane and used to test for a natural gas (methane) leak, a reading of 50 percent LEL would mean that half of the methane needed to reach an explosive environment is present in the air. If the CGI reads 100 percent LEL, there is sufficient methane in the air to explode once an ignition source is introduced.

The CGI can be used to estimate contaminant concentration if a single compound is present. Using the above example, a CGI reading of 25 percent LEL means that 25 percent of the concentration of methane needed to reach an explosive environment is present in the air. The LEL (100 percent) given for this gas is 5.3 percent methane in air, so one quarter of this value would be 1.3 percent or 13,000 ppm of methane in the air (13,000 = 1.3 percent of 1,000,000).

When an atmosphere at or above 100 percent LEL is sampled, the catalyzed filament will respond first, followed by the other filament that will also burn the vapors. The result is that resistance will balance. With some CGIs this results in the instrument first reporting 100 percent LEL then returning to 0 percent LEL as the circuits balance. Therefore, CGI operators must watch the readings and not simply listen for the manufacturer's alarm. Some manufacturers have incorporated microprocessors in their instruments to measure the resistance of both filaments. A latching device will lock

Figure 8-2 Oxygen Meter

the display or produce a warning if a change in the uncoated filament is monitored.

OSHA recommends that whenever the CGI indicates 25 percent of the LEL, personnel should evacuate the area and not return until the LEL is less than 25 percent. However, an LEL reading of less than 25 percent does not necessarily mean the environment is safe for breathing; it simply means the environment is safe from explosion. Again, the oxygen may be so low that a false LEL reading is indicated, and the area is actually filled with a potentially explosive or toxic gas. Due to the additional hazards posed by confined spaces, OSHA requires evacuation at 10 percent LEL. Some industrial guidelines also use 10 percent LEL as an action guideline for confined and non-confined spaces. Site-specific health and safety plans may use any stable reading above 0 percent LEL as an action limit requiring investigation and control of the source of the combustible vapor.

8.3 Oxygen Meters

Oxygen meters determine the percentage of oxygen in the air and are usually calibrated to detect concentrations between 0 and 25 percent. The normal concentration of oxygen in the air is 20.9 percent, but a variety of chemicals and physical conditions can increase or decrease the oxygen to dangerous levels. Oxygen meters are commonly used in conjunction with a CGI to verify that there is sufficient oxygen for accurate operation of the CGI.

8.3.1 OPERATION

Air is drawn into a detector cell by a pump or by diffusion. The oxygen molecules diffuse through a membrane of the oxygen detection cell which contains an electrolyte (potassium hydroxide paste). A chemical reaction between the oxygen and the electrolyte within the cell produces a small current that is proportional to the air's oxygen content. The current passes through an electronic circuit, and the resulting signal is displayed on a readout.

Certain vapors can damage the detection cell. Carbon dioxide concentrations greater than 0.5 percent can permanently affect the detector cell due to its interference with the chemical reaction. When oxidizers are monitored in the atmosphere, they react with the sensor and to the oxygen, resulting in a false high response. Airborne particles can clog the detection unit, thus slowing air transport into the sensor and producing an unusable "real-time" reading.

Elevation above sea level affects both the efficiency and the accuracy of the meter, due to atmospheric pressure changes on the membrane. Lower air pressure reduces both the rate and amount of diffusion, thus giving both slow and false readings. It is important to calibrate these instruments for local conditions, including very high and very low pressure weather systems.

The electrolyte in the oxygen meter begins to wear out from the time the unit is first used. In most cases, the sensor should be replaced every six months.

8.3.2 INTERPRETATION OF DATA

OSHA regulations state that oxygen concentrations between 19.5 percent and 25 percent are safe for normal working conditions. However, in certain situations, such as confined spaces, the upper limit for safe working decreases to 23.5 percent. When oxygen levels are lower than 19.5 percent, air containing adequate oxygen must be supplied to the worker using an air line or self-contained breathing apparatus (SCBA). An oxygen enriched environment must be vented before workers can enter.

Low oxygen environments can have several causes. In some cases, the oxygen is displaced by another gas that can be toxic or flammable. However, toxic conditions may exist before oxygen becomes deficient. Low oxygen readings can also be caused by bacterial

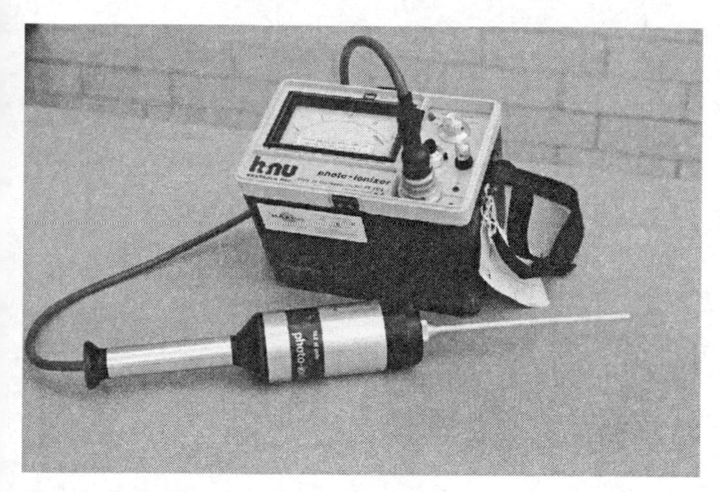

Figure 8-3 Photoionization Detector

consumption in confined spaces, or by chemical reactions, such as combustion, that consume atmospheric oxygen.

High oxygen conditions are generally caused by the presence of oxidizers or by leaks of compressed oxygen gas. For a more detailed discussion of oxygen, see Unit 5.

8.4 Photoionization Detectors

When sufficient energy is added to molecules, the molecules can break apart (ionize) into charged particles called ions. The energy needed for this reaction is referred to as the ionization potential (IP) and is measured in electron volts (eV). A photoionization detector (PID) provides this energy for ionization by using ultraviolet (UV) light.

A PID is used to determine the concentration of ionizable vapors in the air. In most cases, the ionizable vapors are organic. These instruments are commonly used for general monitoring and detecting situations where protective equipment is needed.

8.4.1 OPERATION

Air is pumped into an ionization chamber that is flooded with UV light. The light provides the energy to split uncharged molecules into charged ions. The ions are attracted to a metal grid within the ionization chamber. The grid conducts a small amount of current, and the ions attracted to the grid produce a change in current. This change is converted to a visual display, expressed in parts per million. The readout is proportional to the change in current that, in turn, is dependent on the number of ions attracted to the grid following ionization by the UV light.

A PID is usually calibrated to isobutylene, a nontoxic and organic vapor. The instrument's ability to accurately detect other gases is dependent on how closely those other gases resemble isobutylene in terms of ionization. Measurement may also be made by applying a correction factor to the PID reading to account for ionization differences between isobutylene and the contaminant gas or vapor. Most PID units have a "span" setting that can electronically change the reading to correspond to the calibration gas concentration. The span control adjusts the amplification of the current change and acts somewhat like the volume control knob on a radio.

Anything that interferes with light transmission can affect PID readings. When water vapor is present in the ionization chamber, it acts like fog and scatters and reflects the UV light back toward the source. For example, consider a worker who is monitoring vapors as a drum is opened for sampling. Suddenly the PID plunges to a reading of 0.00 ppm (or less). One possibility is that water vapor released from the drum scattered that the UV light in the PID. Another possibility is that the drum released a vapor that the PID could not ionize. The PID acts the same in both situations, but one situation is benign and the other is potentially dangerous.

8.4.2 INTERPRETATION OF DATA

The PID measures all ionizable vapors within the range of the energy supplied by the UV light source. The instrument can ionize only those vapors that have an IP (in eV) equal to or less than the eV emitted by the UV lamp. For example, a PID having a UV lamp of 10.2 eV will not detect any vapors that have an IP greater than 10.2 eV. The eV emitted by the UV lamps in PIDs generally range from 9.6 to 11.7. Most organic vapors have a low IP, so the PID is an excellent instrument for detection of most organic vapors. There are important exceptions such as methane which has an IP of 12.98 eV.

While the PID detects all vapors within its ionization range, it cannot separate specific vapors or tell the operator which vapor is present. Therefore, the PID is of greatest value when the operator has a good idea which vapors are likely to be present in a given area. In situations where vapors from numerous chemicals might be present, instrument calibration and personnel response are focused on the most hazardous vapor. For example, if a drum yard was suspected of containing vinyl chloride, toluene, and xylenes, the PID would be calibrated to the most dangerous component (in this case vinyl chloride). Although the PID will detect all three vapors, response is usually based on the premise that all detected vapors are vinyl chloride.

When entering an unknown environment containing ionizable vapors within the range of the PID, a general field practice says that readings of 1 to 5 ppm or greater above background indicate the need for respiratory protection and further site evaluation. This rule is also followed when more than one type of ionizable vapor is present. This is only a general rule and must be applied with caution because certain organic vapors, such hydrazine are regulated at levels at or below 1 ppm. Also, the PID may or may not be indicating a direct one-to-one response for the gaseous contaminants present. Finally, the concentrations of vapors usually change dramatically within seconds. Therefore, a field management response is generally governed by the highest reading.

8.5 Flame Ionization Detectors

The flame ionization detector (FID), also called organic vapor analyzer (OVA), is similar to the PID in that it detects ionizable vapors and gases. A more complicated instrument to operate than the PID, it has a greater detection range, being able to discern organic vapors with IPs beyond the capacity of a PID.

8.5.1 OPERATION

The operational theory of a FID is similar to a PID, with one important difference. The energy source for a FID is a hydrogen flame that generates the equivalent of approximately 15.3 eV, enough energy to ionize any compound containing carbon, with the exception of carbon dioxide and carbon monoxide.

In the FID instrument, gases are pumped into a detection chamber containing a hydrogen flame. When organic compounds are burned, they produce ions that are attracted to a grid within the detector. An electrical current is generated proportional to the ionic concentration. The change is then converted to a visual display, expressed in parts per million.

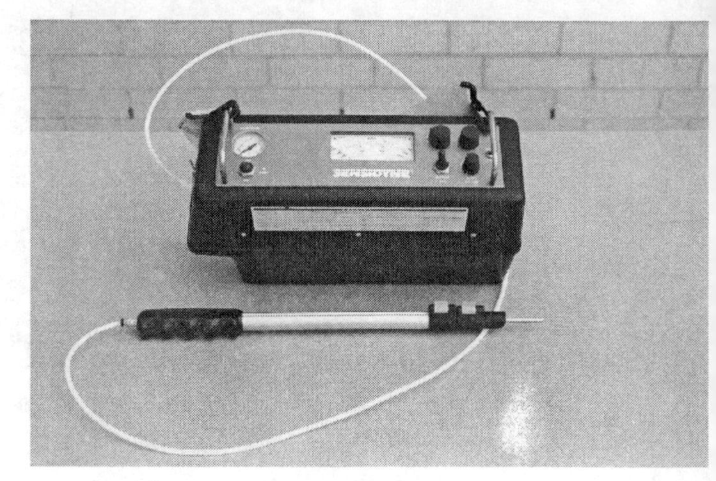

Figure 8-4 Flame Ionization Detector

The hydrogen fuel for the energy source must contain less than 1 ppm total hydrocarbon contamination or 0.0001 percent. If not, the hydrocarbons present in the fuel will be burned and measured by the FID, producing high background values. Additionally, there must be sufficient oxygen in the air to support combustion. Insufficient oxygen will smother the flame and the readout will drop to zero. Some halogenated compounds, such as carbon tetrachloride, may also extinguish the flame.

The FID is typically calibrated to methane. Like the PID, this instrument has a span setting that can electronically recalibrate the instrument to other specific gases.

In contrast to the PID, humidity does not effect the operation of the FID. However, it should not be used when the ambient air temperature is less than 40°F or in an oxygen-deficient atmosphere. In addition, if a FID is used in a potentially explosive environment, it must be an intrinsically safe model that will not act as an ignition source. In comparison to a PID, a FID is usually heavier and bulkier and is often used to detect semi-VOCs, such as oils and tars.

8.5.2 INTERPRETATION OF DATA

The FID measures the total ionizable vapors that the hydrogen flame can ionize. As previously mentioned, this includes nearly all carbon compounds. The instrument is unable to distinguish what type of vapor is detected. Therefore, the data is of limited use unless

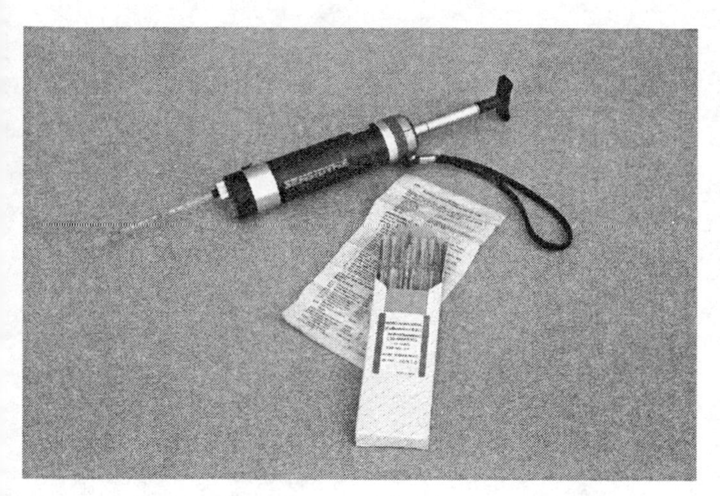

Figure 8-5 Sampling Pump and Colorimetric Tubes

the operator has an idea of the vapors likely to be present in the area under investigation. Some models have a gas chromatograph (GC) capable of specific gas identification.

8.6 Colorimetric Indicator Tubes

Chemically-reactive detector tubes are used to determine the presence and concentration of a known gas. The tubes are filled with an absorbent that chemically reacts with a specific gas to produce a color change. The length of the color change within the tube is related to concentration. Some kits have indicator tubes and flow charts that can identify unknown vapors into families of compounds such as alcohols, ketones, etc.

8.6.1 OPERATION

This air monitoring system consists of a hand-operated piston or bellows pump, a detector tube, and a stroke counter. Following selection of the appropriate tube, the glass tips at both ends are broken off and the tube is inserted into the pump in the correct direction.

A full pump stroke draws 100 cc of air through the tube. Each type of tube requires a certain number of pump strokes, with a minimum amount of time between each stroke. Manufacturer instructions must be followed carefully to ensure that the proper volume of air is drawn through the tube.

The gas reacts with a solid absorbent contained within the tube, resulting in a color change. The quantity of absorbent changing color is proportional to the concentration of the gas.

The pump must be periodically checked for leaking valves. This is often accomplished by inserting an unbroken detector tube into the pump orifice and attempting to aspirate air through the pump. If any air is drawn in, the pump leaks. Additionally, the pump should be volumetrically calibrated on a regular basis to be certain that 100 cc is still being drawn through per stroke.

Colorimetric tubes have a shelf life that is marked on the box containing the tubes. Expired tubes must not be used, and tubes may only be used once. A tube must be used only with the specific pump for that tube as designated by the manufacturer.

8.6.2 INTERPRETATION OF DATA

The length of the color change within the tube corresponds to the gas concentration. Tubes are marked in either percent or ppm. Often it is easier to read the tube by comparing the exposed tube to an unexposed tube of the same type. Unfortunately, the color change is commonly not a straight line but a jagged or faded edge, thus making it difficult to determine exactly where the color change ends. Additionally, a worker with color blindness or some sight disorder may be unable to read or see specific changes.

If detector tubes are used when more than one gas is present, cross-sensitivity may occur. This means that a gas other than the one of interest can cause a color change within the tube. For example, the tube used for detection of ethyl acetate is also sensitive to ketones. Another gas may increase or decrease the color change indicator, depending on colorimetric tube chemistry.

Humidity, temperature, and atmospheric pressure can also affect the measurements made by colorimetric tubes, with interferences increasing or decreasing tube response. Corrections for temperature, humidity, and pressure, if needed, are included in the instructions. Even with these corrections, the error factor can range from 25 percent to 50 percent.

The authors have experienced erroneous readings by colorimetric tubes received from a supplier. It is suspected that the tubes were either stored improperly by the supplier when received from the manufacturer, or had inadequate quality control from the manufacturer. Verification of colorimetric tube readings is therefore recommended, usually in a laboratory or other nonfield situation. For example, verification

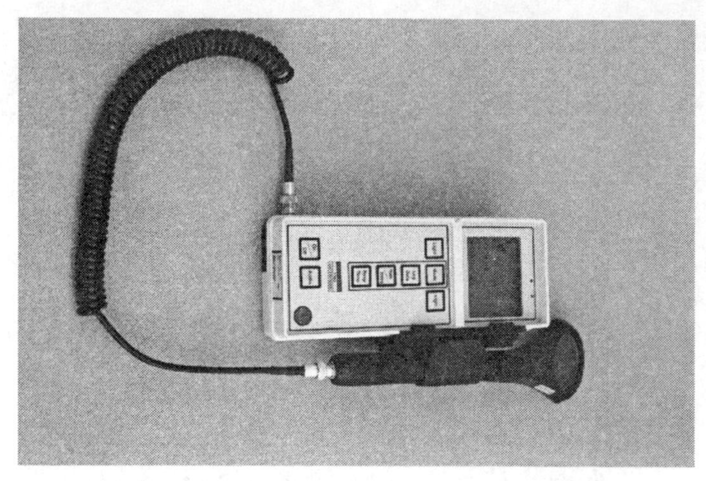

Figure 8-6 Geiger Counter

might be possible by using a calibration gas specific to the tube's contaminant of concern and applicable concentration range.

8.7 Radiation Meters

Whenever radioactive materials might be encountered, a radiation survey meter should be used to detect the type (generally alpha, beta, and gamma) and level of radiation. Using this information, protective equipment and safe work practices can be determined. In addition to survey instruments, personal dosimeters are commonly used in radioactive environments to determine an individual's dose or dose rate of radiation.

8.7.1 OPERATION

Radiation survey instruments work on the principle that radiation causes ionization that can be measured by a detecting media. The ions produced are counted electronically and a relationship established between the number of ions and the intensity of radiation present. Some units have interchangeable detectors used to determine the specific type of radiation.

The most common type of radiation meter is the Geiger-Mueller detection tube, or a Geiger Counter. The tube has a thin mica window through which radiation passes. The resulting ionization within the tube is sensed electronically and displayed.

Radiation survey instruments are factory calibrated and returned to the factory annually for recalibration since this process requires a radioactive source and a license to maintain the source. However they can be field checked with low-intensity sources before each use in the field.

8.7.2 INTERPRETATION OF DATA

Radiation is usually measured in milliroentgens per hour. This unit expresses an exposure rate, or the amount of radiation that an individual would be exposed to at the point of measurements. Many meters also have a scale that reads in counts per minute. Both counts per minute and milliroentgens are functionally related.

When interpreting the data, keep in mind that the Geiger Counter may also be sensitive to radio waves, microwaves, magnetic fields, and electrostatic fields. Additionally, some watches, smoke detectors, lantern mantles, pottery, and jewelry may emit radiation and interfere with accurate field measurements. Humidity may affect the readings by blocking the alpha radiation.

The radiation level is dependent on distance from the source. As stated in Unit 6, the decrease in radiation with distance is an inverse square relationship. Therefore, what is considered a safe level at one point in a work area may be unsafe at another point when the worker moves closer to the source. Thus, when radioactive materials are suspected, a systematic and thorough monitoring of the area must be conducted.

8.8 Miscellaneous Instruments

Charcoal tubes, silica gel tubes, filter cassettes and other specialized sampling media may be used in conjunction with battery-powered, portable sampling pumps. These portable air samplers are used by industrial hygienists to determine TWA air exposure levels. These methods typically require laboratory analysis which may take days or weeks to produce results. Specialized instruments can now detect specific gases such as methane, carbon monoxide, and hydrogen sulfide. In most cases, these instruments function in a manner similar to the oxygen meter. Chemically reactive sensors produce electrical signals that are converted to gas concentration.

Colorimetric test strips are available and can detect pH, specific chemicals, and chemical families such as oxidizers, hydrocarbons, halogenated compounds, chlorinated solvents, etc. Humidity, temperature, and age of the strips can affect their reactions to chemicals and result in false readings.

Other equipment sometimes used to monitor health and safety in field situations are infrared spectrophotometers specific for compounds; mixed oxide semiconductors capable of being calibrated to a variety of gases; and programmed thermal desorbers which extract contaminants from detector tubes for on-site analysis.

8.9 Limitations of Equipment

Limitations have been presented for each specific measuring device. In addition, other limitations must be pointed out.

Direct-reading instruments cannot quantify concentrations of individual compounds unless the instrument is specifically designed for quantitative analysis.

Some instruments show wide fluctuations in readings over a short time. Such readings should not be averaged; the highest value should be accepted as the actual response.

Operator interpretation is probably the most important limitation. Field personnel may mistakenly rely on instruments to determine levels of protective equipment while ignoring their senses. For instance, a PID will not show readings for hydrogen sulfide, yet the nose can detect its presence. Regardless of the instrument's response, knowledge of the site and associated chemical hazards, knowledge of instrument limitations, and common sense must be used.

8.10 Sources of Information

American Conference of Governmental Industrial Hygienists: *Air Monitoring Instrumentation* by C.J. Maslansky and S.P. Maslansky (Publication 93316). Cincinnati, OH: American Conference of Governmental Industrial Hygienists, 1993.

American Conference of Governmental Industrial Hygienists: *Air Sampling Instruments* (Publication 0030). 8th ed. Cincinnati, OH: American Conference of Governmental Industrial Hygienists, 1995.

American Conference of Governmental Industrial Hygienists: *Air Sampling Instrument Selection Guide: Indoor Air Quality*, edited by C. McCammon (Publication 9852). Cincinnati, OH: American Conference of Governmental Industrial Hygienists, 1998.

American Conference of Governmental Industrial Hygienists: *EPA Training Manual #4* (Publication 3080). Cincinnati, OH: American Conference of Governmental Industrial Hygienists.

American Conference of Governmental Industrial Hygienists: *Fundamentals of Air Sampling* by G. Wight (Publication 9424). Cincinnati, OH: American Conference of Governmental Industrial Hygienists, 1994.

American Industrial Hygiene Association: *Direct Reading Colorimetric Indicator Tubes Manual*, 2nd ed., edited by J.B. Perper and B.J. Dawson. Fairfax, VA: American Industrial Hygiene Association Press, 1993.

Code of Federal Regulations Title 29, Part 1910.120. 2001.

Garis, John N. and R.S. Prodans: Sampling Needs Dictate Choice of Instrumentation. *Occupational Health and Safety* (May 1992).

LaBar, Greg: Hazardous Air: Monitoring Common Gases. *Occupational Hazards* (May 1991).

Mustard, Timothy S. and M.J. Loshak: Direct-Reading Instruments Have Advantages, Limitations at Hazwaste Sites. *Hazmat World* (June 1992).

U.S. Department of Health and Human Services: NIOSH Recommendations for Occupational Safety and Health Standards, 1988. *Morbidity and Mortality Weekly Report 37, Number S–7* (26 August 1988).

NINE

Personal Protective Equipment

9 | CONTENTS

PPE is used to shield or isolate individuals from the chemical, physical, and biologic hazards that exist when working at hazardous waste sites. Careful selection and use of appropriate PPE should protect not only the respiratory system, but the entire body.

Most working situations at hazardous waste sites require skin protection for chemical or physical hazards. Hard hats and safety shoes are other forms of common protective equipment. In addition, respiratory protection is required in IDLH or oxygen deficient atmospheres, in concentrations of specific chemicals at or above the action limits, in confined space entry with unknown atmospheres, and in the presence of skin or eye absorption of irritation hazards.

Use of PPE is required by OSHA regulations throughout 29 CFR Part 1910 and 1926 and is reinforced by EPA regulations in 40 CFR Parts 156, 169, 211, 721, and 763, among others. This unit is based on these regulations and NIOSH publications.

No single combination of protective equipment and clothing is capable of protecting against all hazards. Thus, PPE should be used in conjunction with other protective methods. The use of PPE can itself create significant worker hazards such as heat stress, physical and psychological stress, impaired vision, impaired mobility, and impaired communication. In general, greater levels of PPE protection can cause the associated risks to increase. For any given situation, equipment and clothing should be selected that provides an appropriate level of protection. Overprotection as well as underprotection can be hazardous and must be avoided.

9.1 Types of Personal Protection

Three types of personal protection are available: body defenses, personal hygiene, and PPE.

9.1.1 BODY DEFENSES

The body has built-in defenses for protecting itself against some harmful substances. Skin provides a relatively thick, water resistant barrier that is highly effective in protecting the human body from a variety of contaminants. Eyes have tear ducts that can wash some physical contaminants from the surface of the eye. The respiratory system has several defenses against airborne particulates. These defenses include:

Nasal hair—A first-line defense that traps large, air-borne particles;

Cilia—Tiny hairs lining the respiratory tract. Beating 10 to 12 times per second, they move mucous and trapped particles towards the back of the throat and away from the lungs;

Mucous blanket—This sticky substance lines the respiratory tract and functions to trap particles which are carried to the back of the throat where they are swallowed or expelled by coughing; and

Cough reflex—This protective reflex helps to expel mucus and foreign particles.

While body defenses are normally sufficient against solids, liquids, and vapors usually found in the natural environment, concentrations of contaminants in a chemical environment necessitate the use of protective equipment.

9.1.2 PERSONAL HYGIENE

Along with the wearing of PPE, it is necessary to practice good personal hygiene to reduce the toxicological effects of any contamination. Typically, the longer the human body is exposed to harmful substances, the greater the chance for contaminant-related effects. Good personal hygiene practices reduce the time the body is exposed. Some of the good personal hygiene habits expected from hazardous waste workers include:

- taking full body showers before leaving site at the end of the day;
- washing hands and face after leaving work areas;
- never smoking in work areas;
- never eating, drinking, or chewing gum or tobacco or applying cosmetics in the work areas;
- changing into clean work uniforms daily;
- never wearing contaminated work clothing home, in personal vehicles, or off the work site;
- never laundering work uniforms at home;
- never directly entering the clean room area after working (always using the decontamination room first); and
- making good personal hygiene practices into good habits.

9.1.3 PERSONAL PROTECTIVE EQUIPMENT

Even with the healthiest body defenses and good personal hygiene, the use of specialized safety and protective equipment is still required for work at hazardous waste sites. OSHA requires employers to provide personal safety and protective equipment for their employees. Table 9-1 describes the OSHA standards for using PPE.

Table 9-1 **OSHA Standards for Use of PPE**

Regulation	Type of Protection
29 CFR 1910.132	General Requirements
29 CFR 1910.133	Eye and Face Protection
29 CFR 1910.134	Respiratory Protection
29 CFR 1910.135	Head Protection
29 CFR 1910.136	Foot Protection
29 CFR 1910.137	Electrical Protective Devices
29 CFR 1910.1000	Air Contaminant Protection
29 CFR 1910.1001–1045	Respiratory Protection
29 CFR 1910.120 Appendix B	Levels of Protection for Hazardous Waste Workers
29 CFR 1910.95	Hearing Protection

Other OSHA regulations address PPE for specific hazards such as asbestos, coal tar pitch, lead and other metals, coke oven emissions, bloodborne pathogens, and cotton dust.

Hearing damage is one of the most common occupational diseases in the country, affecting workers in manufacturing, construction, transportation, agriculture, and the military. Noise, or unwanted sound, is a by-product of many industrial processes. The National Institute for Occupational and Health's National Occupational Exposure Survey estimated in 1994, that 421,000 construction workers alone were exposed to noise above 85 dBA and that 15 percent of all workers exposed to noise levels of 85 dBA or higher would develop material hearing impairment. Occupational NIHL is a slowly developing hearing loss over a long period as the result of exposure to continuous or intermittent loud noise. Exposure to noise in excess of the current OSHA standards may also put workers at risk for developing hypertension and elevated blood pressure levels.

If employees are exposed to continuous noise levels in excess of 90 decibels averaged over an eight-hour work shift, then feasible engineering and/or administrative controls must be instituted. When an employee has exposure to noise levels in excess of 85 decibels averaged over an eight-hour work shift, then an effective hearing conservation program is necessary. The elements of the hearing conservation program include monitoring hearing hazards, engineering and administrative controls, audiometric testing, hearing protection provision, education and training of employees, record keeping, and program evaluation. When noise control measures are infeasible, or until such time as they are installed, hearing protection devices are the only way to prevent hazardous levels of noise from damaging the inner ear. Making certain that these devices are worn effectively requires continuous attention on the part of supervisors as well as noise-exposed employees. Comfort of fit has been determined as the main reason for employees to wear protective devices in a timely fashion. Careful fitting, training, attention to wear and disposal, and hygiene of devices help in assuring employees that management concern for their welfare is valid.

9.2 The Respiratory Protection Program

Under 29 CFR 1910.134, OSHA requires a formal written program for the selection and use of respirators. In organizing such a program, the two basic objectives should be to protect the worker from respiratory hazards and to prevent injury to the wearer from the incorrect use or malfunction of the respirator.

The program must also include protocols for the selection, training, fitting, use, storage, and maintenance of respirators. Managers should record training attendance and issue certificates for program completion. Periodic fit-tests and refresher classes should also be recorded and placed in the worker's file as well as in the general site file. Such a program should include the following:

- standard operating procedures for selection and use of respirators;
- proper selection of respirators on the basis of hazard;
- training of personnel in use and limitations;
- regular cleaning and maintenance;
- proper storage;
- routine monthly inspections of emergency use respirators and inspections before and after use;
- continual evaluation of respiratory program effectiveness after it is put into operation;
- determination of medical fitness of each potential user; and
- fit-testing of respirator users and use of only approved equipment.

9.2.1 FIELD EVALUATION

In addition to the above, the respiratory protection program shall be evaluated as needed to assess effective implementation and conduct the program. Surveillance of the workplace is to be conducted to assess work area conditions and degree of employee exposure or stress. Program effectiveness will be determined by ensuring that respirators are fitted and selected appropriately, are worn as necessary, and are properly maintained. Employees shall be consulted periodically to assess their acceptance of wearing respirators. Employee acceptance includes comfort, resistance to breathing, interference with vision, communication, job performance and confidence in the protection afforded by respirators. Additionally, random inspections of the work place are to be conducted to ensure that the provisions of the program are being implemented properly.

If a determination has been made for improvements in the program, the program must be revised. Revisions are also made based on employee input and also any new regulations or guidelines published.

9.3 Selection of Respiratory Protection

Before an effective respiratory protection program can be developed, the potential contaminants and exposure levels must be determined so that the proper respiratory selection, medical surveillance, and training can be performed.

Since air-purifying respirators (APRs) filter ambient air, the following conditions must be met if an APR is to be worn:

- atmospheric oxygen level is above 19.5 percent;
- chemical substance is known;
- chemical substance must be able to be filtered, absorbed, or neutralized by the APR;
- a change schedule established for replacement of APR cartridges;
- airborne concentration of chemical substance does not exceed the maximum use limit of the respirator and/or cartridge; and

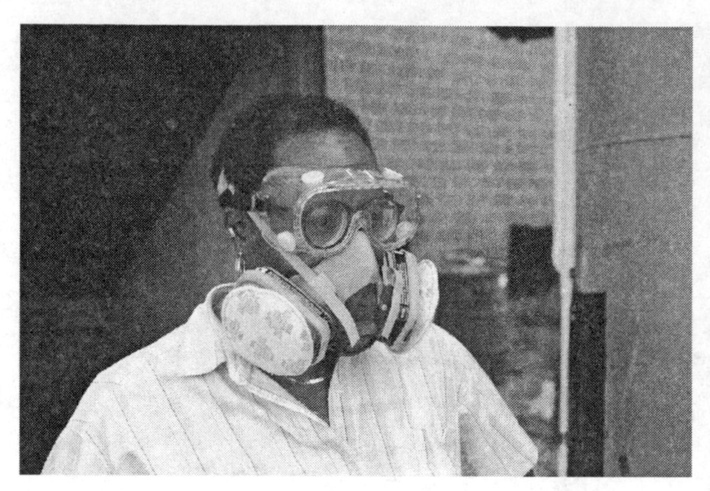

Figure 9-1 Example of an Air Purifying Respirator

Figure 9-2 Example of a Self-Contained Breathing Apparatus

- airborne concentration of chemical substance does not exceed the established IDLH.

If the environment does not meet these criteria, a SCBA (Figure 9-2) or supplied-air respirator (SAR) must be used. Pressure demand self-contained units must be used for IDLH, oxygen deficient, or uncharacterized contaminant atmospheres. A decision-making respirator selection flowchart is presented in Figure 9-3. A discussion of some of the conditions that affect respiratory selection follows.

9.3.1 OXYGEN-DEFICIENT OR ENHANCED ATMOSPHERE

OSHA defines an oxygen-deficient atmosphere as any atmosphere containing oxygen at a concentration below 19.5 percent. NIOSH certification of SARs or APRs is limited to those respirators used in atmospheres containing at least 19.5 percent oxygen.

The minimum requirement of 19.5 percent oxygen provides an adequate amount of oxygen for most work assignments and includes a safety factor. The safety factor is needed because oxygen-deficient atmospheres offer little warning of the danger, and the continuous measurement of an oxygen-deficient atmosphere is difficult.

Breathing oxygen concentrations below 16 percent can decrease mental effectiveness, visual acuity, and muscular coordination. At oxygen concentrations below 10 percent, loss of consciousness may occur, and below 6 percent oxygen, death will result. Often only mild subjective changes are noted by individuals exposed to low concentrations of oxygen, and collapse can occur without warning.

9.3.2 IDENTIFIED AIR CONTAMINANT

To determine the type of respirator to be used, the specific air contaminant and the airborne concentration must be determined. Then, an APR, SAR, or SCBA respirator can be selected based on the air contaminant's physical and chemical properties, the toxicological effects, and the respirator protection factor (PF).

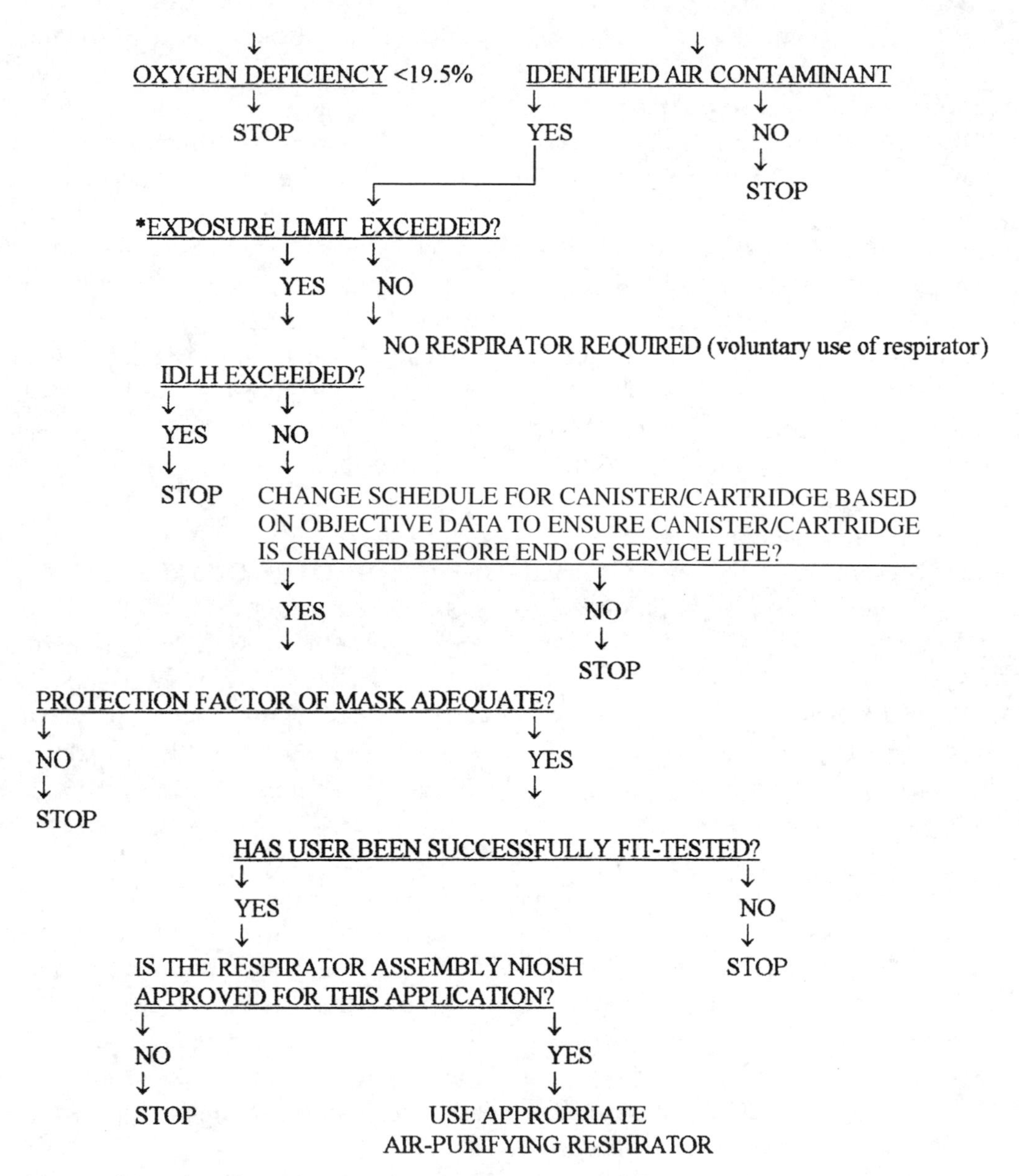

Figure 9-3 Decision-making chart for APR selection. The "STOP" decisions mean that an approprate atmosphere supplying respirator must be used, or that further training or administrative procedures are needed. *Exposure Limit (EL) may be a TLV, PEL, REL, WEEL or other reputable exposure limit. Adapted from: NIOSH et al., *Occupational Safety and Health Guidance Manual for Hazardous Waste Site Activities,* 1985.

9.3.3 KNOWN AIR CONTAMINANT CONCENTRATION

After the air contaminant is identified, the airborne concentration must be determined and compared to applicable exposure limits including PEL, TLV, REL, WEEL, IDLH, and ERPGs. The proper decision can then be made regarding exposure and respirator selection.

9.3.4 WARNING PROPERTIES

A warning property exists when a chemical vapor has a distinctive odor or taste, or when it causes respiratory tract or eye irritation. An "adequate" warning property for an air contaminant is found when these symptoms occur at concentrations below the PEL. For example, naphthalene has an odor threshold of 0.001 to 0.010

ppm while the PEL is 10 ppm, thus giving a good buffer between detection and danger. Recognition of an odor depends on a person's sensory ability to detect it. Since the range of odor recognition thresholds within a population is very wide, odor recognition may not be reliable as the only means for some individuals to determine if their APR system is providing proper protection. Similarly, individuals have varying sensitivity to taste and irritation quantities.

The Preamble to the revised Respiratory Protection Final Rule of 1998 no longer permits the use of warning properties as the sole basis for determining that an APR affords adequate protection against exposures to vapors and gases. For atmospheres which are not IDLH, APRs are considered acceptable by OSHA as long as appropriate precautions and changeout schedules are in place, even for use against substances with poor warning properties.

9.3.5 PROTECTION FACTORS

The level of protection that can be provided by a respirator is indicated by its PF as shown in Tables 9-2 and 9-3. This number, which is determined experimentally by measuring facepiece seal and valve leakage, indicates the relative difference in concentrations of airborne contaminants outside and inside the facepiece. For example, the PF for full-facepiece APRs is 50 in the asbestos and benzene examples shown by Tables 9-2 and 9-3, respectively. This means that workers wearing these respirators should be protected in atmospheres containing chemicals at concentrations that are up to 50 times higher than the appropriate exposure limits. However, this is a conservative and theoretical value that may not accurately reflect the protection offered by an APR. Another means of determining the PF is through a quantitative fit-test (see section 9.5.4.3.1). Fit-testing is used to determine the PF for a single-worker wearing a specific respirator unit. With quantitative fit-testing, a higher PF may be allowed for some work tasks.

At sites where the identity and concentration of chemicals in air are known, a respirator should be selected with a PF sufficiently high to ensure that the wearer will not be exposed to chemicals above the applicable limits. These limits include TLVs, PELs, WEELs, and the RELs. Such limits are designed to protect most workers who may be exposed to chemicals day after day throughout their lifetime in the working environment. The protection provided by a respirator can be compromised in several situations. For example, if a worker has a high breathing rate, he or she can "overbreathe" the respirator. This means that the worker's breathing rate exceeds the ability of the respirator to provide breathable air. If the ambient temperature is high, excessive sweat may cause a break in the face-to-facepiece seal. At very low temperatures, the exhalation valve and regulator may become ice-clogged due to moisture in the breath and air. If the worker has a poor facepiece-to-face seal, such as from excessive facial hair or from lifting the facepiece to speak, the seal is similarly compromised.

9.3.6 EMPLOYEE QUALIFICATIONS

Before an employee receives clearance to wear a respirator, she or he must undergo a medical examination with a licensed physician to help assure that the employee is capable of working with the added stress. Some latitude is offered by 29 CFR 1910.134, which recognizes a licensed health care provider.

Training should be site- and equipment-specific, and should include a thorough discussion of the risks and hazards inherent in the activities requiring PPE. Other elements that should be stressed are worker responsibilities to use the equipment correctly, cleaning, inspection, maintenance, storage, fit-testing, performance, and PPE regulations relevant to the site. At a minimum, the worker must have received the following to use the respiratory protection selected:

- current medical examination approved for respirator use;
- fit-test for the respirator selected; and
- training on the care and use of the respirator selected.

9.3.7 SERVICE LIFE AND CHANGE SCHEDULE

Not all canisters and cartridges are created equal. Each canister and cartridge is manufactured for a certain respirator facepiece system and for a range and number of types of contamination. This range must be compatible with the exposure occurring at the site. Canisters and cartridges are manufactured for certain contaminant categories (for example, organic vapors, acid gases, ammonia, and particulates) and are dated for shelf-life.

Table 9-2 **OSHA Assigned Protection Factor Classifications of Respirators for Protection Against Asbestos Fibers**[1]

Assigned Protection Factor[2]	Type of Respirator
10	Half mask air-purifying respirator, other than a disposable respirator, equipped with high-efficiency filters.[3]
50	Full facepiece air-purifying respirator, equipped with high-efficiency filters.[3]
100	Any powered air-purifying respirator, equipped with high-efficiency filters.
100	Any supplied air respirator operated in continuous flow mode.
1,000	Full facepiece, supplied air respirator operated in pressure-demand mode.
>1,000 or unknown	Full facepiece supplied-air respirator operated in a pressure-demand mode, equipped with an auxiliary self-contained breathing apparatus.

[1]Source: Table 1, Respirator Protection for Asbestos Fibers, 29 CFR 1910.1001 as amended, January 8, 1998, Federal Register, Volume 63. No 5, p. 1285 (also see 1910.134, Respiratory Protection Standards citations).

[2]The required protection factor is calculated by finding the PEL, REL, or TLV for the hazard of concern, and finding the concentration present in the air by monitoring the atmosphere. A simple formula utilizing the PEL is used to calculate the protection factor necessary, as shown below.

$$\text{Protection Factor (PF)} = \frac{\text{measured concentration of substance in air}}{\text{Exposure Limit (PEL)}}$$

If the asbestos fiber measurement is 5f/cc and the PEL = 0.1 f/cc, then 5 divided by 0.1 = 50. A respirator capable of protecting the wearer from the concentration present, must have a PF of 50 or higher from the above table.

Note that the exposure limit used in the denominator is usually the lowest published exposure limit, and may be the PEL, TLV, REL, or WEEL for the contaminant.

[3]A high efficiency filter means a filter that is at least 99.97 percent efficient against monodisperse particulates of 0.3 micrometers in diameter or larger. These are usually termed HEPA filters. Note, however, that the classification and naming of particulate respirator filters has changed, as described in part 9.3.8 of this unit.

Table 9-3 **Assigned Protection Factor Classifications of Respirators for Protection Against Benzene Gas/Vapor Exposures**[1]

Assigned Protection Factor[2]	Type of Respirator
10	Half mask air-purifying respirator equipped with organic vapor cartridges.
50	Full face air-purifying respirator equipped with organic vapor cartridges, or full facepiece gas mask with chin style appropriate organic vapor canisters.
100	Full facepiece powered air-purifying respirator with an appropriate organic vapor canisters.
1,000	Full facepiece supplied-air respirator operated in a positive-pressure mode.
>1,000 or unknown	Full facepiece self-contained breathing apparatus operated in positive-pressure mode.
>1,000 or unknown	Full facepiece supplied-air respirator operated in a positive-pressure mode with an auxiliary self-contained air supply.

[1]Source: Table 1, Benzene, 29 CFR 1910.1028 as amended, January 8, 1998, Federal Register, Vol. 63, No. 5, p. 1289 (also see 1910.134, Respiratory Protection Standard citations).

[2]$$\text{Protection Factor (PF)} = \frac{\text{measured concentration of material in atmosphere}}{\text{Defined Respiratory Limit (PEL)}}$$

For example, if the PEL for a chemical is 1.0 ppm, and air monitoring indicates a concentration of 10.0 ppm, the PF is 10.0 divided by 1.0 = 10. In this case, any of the above respirator systems may be used.

Users need to ensure the canister/cartridge they intend to use is appropriate for the hazard category and that the canister/cartridge is used prior to its expiration date.

A cartridge/canister change schedule must be established to ensure that the cartridge/canister is charged before the end of its service life. This schedule can be included in the site-specific health and safety plan where respiratory protection is needed. The change schedule can be determined by consulting one or more of the following:

- manufacturers' technical literature;
- manufacturers' respirator technical specialists;
- NIOSH or OSHA guidance documents;
- ANSI, AIHA, or ASSE technical documents; and/or
- Certified Industrial Hygienist or Certified Safety Professional.

The change schedule will consider environmental and work factors such as the temperature, humidity, work rate, duration of employee use of the respirator, and nature and concentration of the contaminant(s).

Particulate air filtering cartridges should be changed when breathing becomes difficult when fibers, particulates, or even high humidity clog the filter and restrict airflow. Changing the particulate cartridge prior to this point is often not recommended because the efficiency of this type of filter increases with use. Conversely, vapor cartridges should be changed at the first warning sign of breakthrough or at pre-determined intervals as set by the site safety officer.

Respiratory filtration works to capture and remove inhalable particles of a variety of sizes before they enter the worker's lungs. These respirators use negative pressure during inhalation to draw ambient air through the filters to remove particulates and are used in situations that are not immediately dangerous to life and health and that contain adequate oxygen to support life. The respirator standard (42 CFR 84) establishes performance criteria for nonpowered, air-purifying, particulate filter respirators. These are the most widely used types of respirators, supplying protection against hazardous particles sometimes in combination with an organic vapor cartridge in a variety of operations. These include welding, grinding, spray painting, sanding, dry pesticide spraying, and asbestos removal.

9.3.8 PARTICULATE FILTER CLASSIFICATION

Prior to the 1998 standard revision, respirator filters were classified according to the type of hazard being encountered (dust, mist, fumes, etc.). The revised rule establishes nine classes of particulate filters. This is accomplished by three series of particulate filters; N, R and P with each series having three levels of efficiency; 95 percent, 99 percent, and 99.97 percent, indicating the percentage of particles removed from the air. The N, R, and P series designations pertain to the presence or absence of oil particles in the work environment as follows:

- N series filters are *Not* resistant to oil;
- R series is Resistant to oil;
- P series is oil Proof;
- if no oil particles are present in the work environment, use a filter of any series;
- if oil particles such as lubricants, cutting fluids, glycerin, etc. are present, use a R or P series (N series cannot be used if oil particles are present); and
- if oil particles are present and the filter is to be used for more than one work shift, use only a P series filter.

The selection of filter efficiency, 95 percent, 99 percent and 99.97 percent, depends on how much filter leakage is acceptable, with higher efficiency meaning lower filter leakage. Upgrading from a half-face respirator to a full-face respirator increases the level of protection.

The new rule also eliminates the pesticide and spray paint respirator categories of the previous ruling. Filters used in conjunction with chemical cartridges, including those that might be used for spray paint and pesticides are now tested under the same provisions as particulate respirators.

9.4 Conditions Affecting Respirator Use

The following are special problems that may be encountered in the process of wearing and using respiratory protective equipment.

9.4.1 FACIAL HAIR/COSMETICS

Any material located between a worker's skin and the respirator sealing surfaces, such as beards, mustaches, "five o'clock shadow," cosmetics, suntan lotion, or face creams (including medications in a liquid base), should not be permitted, since a tight facepiece fit is necessary to provide expected protection. A negative-pressure respirator that seals poorly on the face, may allow leakage of outside contaminated air into the mask. A worker should not enter a contaminated work area when conditions prevent a good seal of the respirator facepiece to the face.

9.4.2 EYEGLASSES

Ordinary eyeglasses cannot be used with full-facepiece respirators since temple bars or straps that pass between the sealing surface of a full-facepiece and the worker's face will prevent a good seal. Special corrective lenses can be mounted inside a full-facepiece respirator and are available from all manufacturers of full-facepiece respirators. To ensure good vision, comfort, and proper sealing of the facepiece, these corrective lenses should be mounted by an individual designated by the manufacturer as qualified to install accessory items.

Eyeglasses or goggles may also interfere with half-masks. Often the solution is to use a full-facepiece with special corrective lenses.

9.4.3 CONTACT LENSES

The OSHA standard allows contact lenses to be worn by respirator users. However, several factors may restrict the use of contact lenses while wearing a respiratory device. This is especially true for air-supplying respirators. With full-facepieces, incoming air directed toward the eye can cause uncomfortable ocular dryness or discomfort from dirt, lint, or other debris lodging between the contact lens and the cornea.

If the facepiece seal is broken for any reason, irritating or toxic vapors or gases may enter the mask and be trapped under the contact lenses. Other hazardous materials may also react with the mucous membrane of the eye creating caustic by-products. The time necessary to exit the area for proper emergency decontamination may be long enough for acute or permanent eye injury to have occurred.

However, use of contact lenses in conjunction with respirator use is approved for some work situations, according to the American National Standards Institute publication on respiratory protection. A qualified health and safety professional, such as a Certified Industrial Hygienist or Certified Safety Professional should be consulted if use of contact lenses by a respirator user is being considered.

9.4.4 FACIAL DEFORMITIES

Facial deformities such as scars, deep skin creases, prominent cheekbones, severe acne, and the lack of teeth or dentures, can prevent a respirator from sealing properly.

9.4.5 LOW AND HIGH TEMPERATURES

Low temperatures may fog respirator lenses. Coating the inner surface of the lens with the anti-fogging compound normally available from the respirator manufacturer should help prevent fogging. Full facepieces with nose cups that direct the warm, moist exhaled air through the exhalation valve without its touching the lens, are available and also help control fogging. However, severe fogging can still occur because of work load (sweating), humidity and air flow factors. At very low temperatures, respirator valves may freeze due to moisture. In such situations, APRs should be replaced by SCBAs or SARs. Specifications for moisture content and other qualities of compressed breathing air are available from the Compressed Gas Association and OSHA's 29 CFR 1910.134, respiratory protection standard.

A person working in a high air temperature is under stress. Wearing a respirator causes additional stress, that could be minimized by using a lightweight respirator with low breathing resistance when technically feasible. As an alternative to an APR, an airline-type SAR can be equipped with a vortex tube which cools the supplied air to comfort the worker. Sweat on the face can also compromise the seal of the mask.

9.4.6 PHYSIOLOGICAL RESPONSE TO RESPIRATOR USE

Wearing any respirator, alone or with other types of protective equipment, will impose some physiological stress on the wearer. For example, the weight of the equipment increases the energy requirement for a given task. Therefore, selection of respiratory protective devices should be based on the weight of the respirator, the type and amount of protection needed, the individual's tolerance of the given device and the overall comfort and acceptance of the respirator of the user.

Use of respirators in conjunction with protective clothing can greatly affect human response and endurance, especially in hot environments or during heavy work, since the body relies a great deal on heat loss through the evaporation of sweat. With impermeable clothing, heat loss by sweat evaporation is not possible. Additionally, the weight of the respirator (up to 35 pounds for an SCBA) adds to the metabolic rate of workers, increasing the amount of heat the body produces. The net effect can substantially increase heat stress.

9.5 Air-Purifying Respirators

9.5.1 PARTICULATE FILTERING

Particulate-filtering cartridges are used for protection from aerosols such as dusts, fumes, and/or mists. A dust is a solid particle mechanically produced by milling, sanding, crushing, or grinding. Fumes occur in high heat operations such as welding, smelting, and industrial furnace work. They are generated when metal is heated and quickly cooled, thus generating fine airborne particles. Mists are tiny liquid droplets found near spraying, mixing, or cleaning operations. However, many mists can quickly vaporize, thus requiring chemical filtering rather than particulate filtering.

As an aerosol is drawn onto or into the filter media, it is trapped. The probability that a single aerosol particle will be trapped depends on such factors as its size, its velocity, and, to some extent, the composition, shape, and electrical charge of both particle and filter media. With current filter media, any filter designed to be 100 percent efficient in removing particles would be unacceptably difficult to breathe through.

Particulate filters are of two types: absolute and nonabsolute. Absolute filters use screening to remove particles from the air; that is, they exclude the particles which are larger than the pores. However, most respirator filters are nonabsolute filters, which means they contain pores which are larger than the particles to be removed. They use combinations of interception, sedimentation, inertial impaction, diffusion capture, and electrostatic capture to remove the particles. The exact combination of filtration mechanisms that comes into play depends greatly upon the flow rate through the filter and the size of particle.

9.5.2 VAPOR AND GAS REMOVING

The other major class of airborne contaminants are gases and vapors. Gases are substances that are airborne molecules at room temperature such as carbon monoxide which is an exhaust gas from internal combustion engines. Vapors are substances that evaporate from liquids or solids, and are commonly associated with solvent cleaning, painting, and refining.

Air-purifying cartridges or canisters are available for protection against specific gases and vapors, such as chlorine and mercury vapor, respectively, and classes of gases and vapors, such as acid gases and organic vapors. In contrast to filters, which are effective to some degree no matter what the particulate, the cartridges and canisters used for vapor and gas removal are designed for protection against specific contaminants.

Vapor and gas removing cartridges normally remove the contaminant by the interaction of the gas or vapor molecules with a granular, porous material, commonly called the sorbent. The general method by which the molecules are removed is called sorption. In addition to sorption, some cartridges use catalysts that react with the contaminant to produce a less toxic gas or vapor.

Three mechanisms are used in vapor and gas removing cartridges; adsorption, absorption, and catalysts.

9.5.2.1 Adsorption

Adsorption retains the contaminant molecule on the surface of the sorbent granule by physical attraction, like a magnet attracting iron. The intensity of the attraction varies with the type of sorbent and contaminant.

9.5.2.2 Absorption

In contrast to adsorption which is a physical process, absorption is a chemical process. Like adsorbents, absorbents are porous filters, but they do not have as large a surface area. Gas or vapor molecules that enter the absorptive material undergo a chemical reaction that changes the airborne contaminant to a nonhazardous chemical. For example, when an acid gas enters an absorptive cartridge, it reacts with potassium hydroxide (or a similar basic chemical) to produce a salt and water, both of which are harmless to the user.

9.5.2.3 Catalysts

A catalyst is a substance that influences the rate of chemical reaction between substances. For example, one catalyst used in respirator cartridges and canisters is hopcalite, a mixture of porous granules of manganese and copper oxides that speeds the reaction between toxic carbon monoxide and oxygen to form carbon dioxide which is expelled to the air. In most respirators, the foregoing processes are essentially 100 percent efficient until the sorbent's capacity to adsorb gas and vapor or catalyze their reaction is exhausted. Then the contaminant will pass completely through the sorbent, into the facepiece, and into the lungs.

Therefore, it is important to recognize when breakthrough occurs. The health and safety officer must carefully monitor for a suspected hazard, and set a time limit for work in a questionable area so that workers can withdraw from the area to change cartridges. This system is used more widely in the mining industry and seldom in hazardous materials work.

9.5.3 COMBINATION PARTICULATE, VAPOR, AND GAS-REMOVING

Cartridges and canisters are available to protect against both particulates and vapors and gases. These devices look much like the sorbent cartridge or sorbent canister alone. There are several methods of attaching a particulate filter to a typical chemical cartridge. One includes manufacture of the particulate filter inside the cartridge. Another method involved mounting the particulate filter to the outside of the cartridge using a snap-on cover. Other variations may be found, but the principle is the same. Where filters are used in combination with cartridges, the filter must always be located on the distal side (outside) of the cartridge relative to air flow direction.

It is not safe to attempt to assemble a combination filter by taping one filter to another. This is an unacceptable practice. Also, be certain that the cartridges which have been selected are compatible and certified for use with the air purifying respirator (APR) device since cartridges and APRs are not interchangeable among manufacturers.

9.5.4 APR OPERATIONS

9.5.4.1 Donning Procedure Guidelines

Half-face Respirator:

1. Respirators are available in several sizes. Be certain that the mask is not too large or too small to fit securely. The mask must comfortably fit the individual's face size and structure. Fit must be verified by fit-testing and by assessing the individual's acceptance of the respirator. Make certain that this is the same make and size of respirator used in the fit-tests.
2. Ensure that the respirator is assembled correctly and is equipped with the proper valves, gaskets, headgear and filter or cartridge for the work assignment to be performed. OSHA regulations require a respirator to be inspected by the wearer before and after each use to ensure that it is in good working condition.
3. Place the respirator under the chin and over the nose. The narrow part of the facepiece should be located over the nose.
4. Pull the crown strap over the head and adjust its placement until it is comfortable. The crown strap is adjustable in length for ease of fit.
5. Using both hands, hook the lower headband straps together behind the neck.
6. Adjust the upper headbands by pulling downward from the crown strap. Adjust the lower headbands by pulling sideways from the hook and loop. Do not adjust headbands and tension at the yoke. Continue adjusting headbands until a comfortable fit has been obtained.
7. Do not overtighten the respirator to the face. Ensure that any protective eyewear does not interfere with the respirator face seal.

Full-face Respirator:

1. Make sure that the respirator is assembled correctly and is equipped with the proper

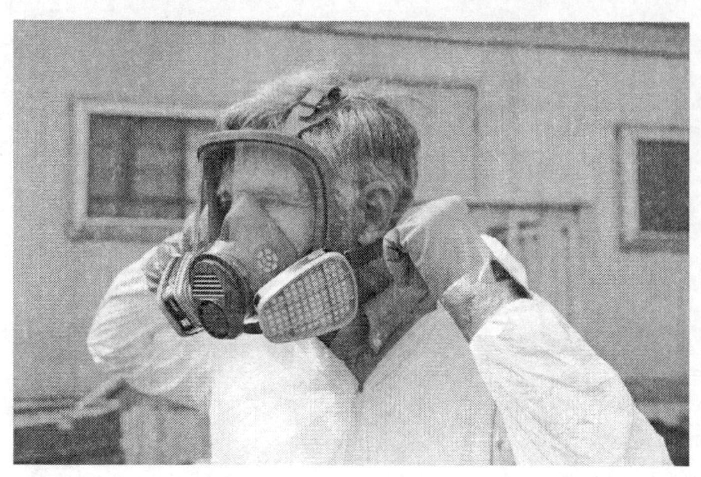

Figure 9-4 Tightening the lower straps of a full-face respirator

Figure 9-5 Negative pressure test for an air-purifying respirator

lens, lens clamp, valves, gaskets, headgear, filter and/or cartridge for the work assignment to be performed. OSHA regulations require that a respirator be inspected by the wearer before and after each use to ensure that it is in good working condition. Make certain that this is the same make and size of respirator used in the fit-tests.

2. Pull all six head straps out to the end tabs.
3. Clasp temple and side straps, as close to facepiece as possible, and stretch straps over the head, fitting mask against the chin first.
4. Pull the lower side straps as tightly as necessary to obtain a snug fit. At this stage, the mask may still hang lower than desired.
5. Pull temple straps next, tight enough to ensure a snug fit.
6. Pull top forehead straps last, completing airtight seal. To avoid forehead kinks, pull these two straps toward each other. Do not overtighten the mask since this can cause warping and create the potential for leaks.

9.5.4.2 Field Fit-Check Procedures

The seal of a respirator must be tested prior to entering a contaminated atmosphere by procedures recommended by the manufacturer or by one of the following qualitative fit checks:

Irritant or Odorous Chemical Agent Test— The wearer is exposed to an irritant smoke, isoamyl acetate vapor, or other suitable test agent easily detectable by irritation, odor, or taste. An APR must be equipped with the appropriate air-purifying element. If the wearer is unable to detect penetration of the test agent, the respirator is probably tight enough.

Negative Pressure Test—This test consists of closing off the inlet of the cartridges by covering with the palms, then inhaling gently so that the facepiece collapses slightly and holding the breath for 10 seconds. If the facepiece remains slightly collapsed and no inward leakage is detected, the respirator is probably tight enough.

Positive Pressure Test—This test is conducted by closing off the exhalation valve and exhaling gently into the facepiece. The fit is considered satisfactory if slight positive pressure can be built up inside the facepiece without any evidence of outward leakage. For some respirators, exhalation valve covers are of a shape or in a position where they cannot be blocked by the hand, and this fit-test cannot be conducted.

9.5.4.3 Respirator Fit-Testing

All users or potential users of negative-pressure type respiratory protection devices must be fit-tested to ensure proper facepiece-to-face seal of the respirator. The fit-test should be accomplished by use of an approved test aerosol and using standard OSHA methods found in 29 CFR 1910.134. Users are tested with a selection of brands of masks and allowed to choose the most comfortable from those that fit satisfactorily. Following respirator donning, a quantitative fit factor is determining by analyzing the air inside the mask and comparing it to the air analyzed outside the mask.

9.5.4.3.1 Quantitative Fit-testing; 29 CFR 1910.134 Appendix A. Quantitative fit-testing is the assessment of the adequacy of respirator fit that uses an instrument to numerically measure the presence of the test agent inside and outside of the respirator. In effect, it measures the amount of leakage of ambient air into the respirator. A nonhazardous aerosol such as corn oil, polyethylene glycol 400, or sodium chloride, is generated inside a test chamber. The subject, with full- or half-mask respirator, enters after adjusting the mask to a comfortable and secure fit. They are then tested in normal breathing, deep breathing, turning the head from side to side and moving it up and down, talking, grimacing (making a face), bending over, and finally, taking the respirator off, re-donning it and breathing normally. Throughout these exercises, the instruments monitor the "challenge" substance inside and outside the mask for evaluation of fit and suitability for purpose.

9.5.4.3.2 Qualitative Fit-testing; 29 CFR 1910.134 Appendix A. The qualitative fit-test is a pass/fail fit-test that relies on the subject's sensory response to detect a "challenge" agent. Using an irritant or odorous chemical agent test, the wearer is exposed to isoamyl acetate (banana oil) vapor, saccharin solution, Bitrex™ (denatonium benzoate), or irritant smoke (stannic chloride), any of which is easily detectable by irritation, odor or taste. An APR must be equipped with the appropriate air-purifying element for this test. If the wearer is unable to detect penetration of the test agent, the respirator may be deemed adequate. Once again, a variety of exercises, movements, readings, and breathing techniques are used to encourage unseating the facepiece and to determine the respirator suitability. Per OSHA, a qualitative fit test may only be used to test negative pressure APRs that must achieve a fit factor of 100 or less.

9.5.4.4 APR Maintenance

OSHA standards require an adequate maintenance program, but permit its tailoring to the type of plant, working conditions, and hazards involved. However, all programs are required to include an inspection for defects, leak checking, cleaning and disinfecting, repair, and storage. A proper maintenance program ensures that the worker's respirator remains as effective as when it was new. Common program components are as follows:

Inspection for Defects—Probably the most important part of a respirator maintenance program is frequent inspection. If conscientiously performed, inspections will identify damaged or malfunctioning respirators before they can be used. The OSHA requirements outline two primary types of inspection, one while the respirator is in use and the other while it is being cleaned. In a small operation where workers maintain their own respirators, the two types of inspection become essentially one and the same. In a large organization with a central respirator maintenance facility, the inspections differ.

Frequency of Inspection—OSHA requires that "all respirators be inspected before and after each use," and that those not used routinely, such as emergency escape and rescue devices, "shall be inspected after each use and at least monthly" This is important since it is highly unlikely that anyone needing a respirator in a hurry, as during an emergency, is going to perform a detailed inspection.

Inspection Procedures—Inspection procedures differ, depending on whether air-purifying or air-supplying devices are involved, and whether the inspection is to be conducted in the field during use or during routine cleaning. The OSHA standards require that respirator inspection include function tests and checks of the tightness of the connections, a check of the facepiece, valves, connecting tube, and canister, and a check of the regulator and warning devices on SCBA.

Field Inspection of APRs—Routinely used APRs should be checked before and after each use:

—Examine the facepiece for accumulated dirt, cracks, tears, holes, or distortion from improper storage, inflexibility (stretch and massage to restore flexibility), cracked or badly scratched lenses in full-facepieces, incorrectly mounted full-facepiece lens or broken or missing mounting clips, cracked or broken air-purifying element holders, badly worn threads, or missing gaskets (if required).

—Examine the head straps or head harness for breaks, loss of elasticity, broken or malfunctioning buckles and attachments, and excessively worn serrations on the head harness which might permit slippage.

—After removing its cover, examine the inhalation and exhalation valves for foreign material, such as detergent residue, dust particles, or human hair under the valve seat, cracks, tears, or distortion in the valve material, improper insertion of the valve body in the facepiece, cracks, breaks, or chips in the valve body particularly in the sealing surface, missing or defective valve covers, and improper installation of the valve in the valve body.

—Examine the air-purifying elements for incorrect cartridge, canister, or filter for the hazard; incorrect installation; loose connections; missing or worn gaskets; cross-threading in holder; expired shelf-life date on cartridge or canister; cracks or dents in the outside case of the filter, cartridge, or canister; and evidence of prior use of sorbent cartridge or canister indicated by absence of sealing material, tape, or foil, over the inlet.

Cleaning and Disinfecting—OSHA requirements in 29 CFR 1910.134 are not specific about cleaning and disinfecting procedures. The regulations state that "routinely used respirators shall be collected, cleaned, and disinfected as frequently as necessary to ensure that proper protection is provided" and that emergency use respirators "shall be cleaned and disinfected after each use." Manufacturers usually have recommendations for cleaning and disinfecting. During cleaning and rinsing, remember to keep water temperatures between 120° and 140°F to ensure that the components are not damaged.

9.6 Air-Supplying Respirators (ASR)

Air-supplying respirators do not purify the contaminated air but supply the wearer with "clean" air from some other source. The air may be either self-contained or airline-supplied. The two basic classes of air supplying respirators are the SAR and the SCBA. Unlike the SCUBA (self-contained *underwater* breathing apparatus), the SCBA cannot be used under water.

SCBAs can be purchased as closed circuit or open circuit. A closed-circuit breathing apparatus recycles the wearer's exhaled air. The carbon dioxide is chemically removed and fresh oxygen is introduced. This type of breathing apparatus is commonly referred to as a "rebreather." Due to weight and other factors, rebreathers are not commonly used on hazardous waste sites.

An open-circuit SCBA exhausts exhaled air directly into the environment. Air is supplied to the wearer by either a single-stage (conventional) or a two-stage regulator. Single-stage regulators reduce the air pressure from the air cylinder only once before reaching the wearer. This type of SCBA is usually characterized by a low pressure tube hanging from the facepiece and are not commonly used on hazardous waste sites. Two-stage regulators have an initial reduction of cylinder air pressure immediately exiting the air cylinder with a second regulator placed between the cylinder and at the facepiece.

9.6.1 TYPES OF AIR SUPPLYING RESPIRATORS

Positive-pressure respirators maintain a positive pressure in the facepiece during both inhalation and exhalation. The two main types of positive-pressure respirators are pressure-demand and continuous flow. In pressure-demand respirators, a pressure regulator and an exhalation valve on the mask maintain the mask's positive pressure. If a leak develops in a pressure-demand respirator, the regulator sends a continuous flow of clean air into the facepiece, preventing penetration into the facepiece by contaminated ambient air. Continuous-flow respirators send a continuous stream of air into the facepiece at all times. With SARs, the continuous flow of air prevents infiltration by ambient air, but uses the air supply much more rapidly than pressure-demand respirators. The air blowing over the face does tend to promote cooling on warm days. The continuous flow type is less protective than the pressure demand unit. Upon extreme exertion, the user may overbreathe the air supply in a continuous flow unit, thereby creating negative pressure in the facepiece. This can cause leakage of contaminated air into the facepiece. Airline respirators are not allowed in IDLH situations unless operated in pressure demand mode and equipped with an auxiliary self-contained air supply.

Negative-pressure respirators draw air into the facepiece via the negative pressure created by user inhalation. The main disadvantage of negative-pressure respirators is that if any leaks develop in the system (such as a crack in the hose or an ill-fitting mask or facepiece), the user also draws contaminated air into the facepiece during inhalation. Negative pressure respirators are not approved for use at uncharacterized hazardous waste sites.

9.6.2 PROPER USAGE OF ASR

9.6.2.1 SCBA

Key questions to ask when considering SCBA use:

Is the atmosphere at the hazardous waste site unknown?

Can an APR provide adequate protection?

Is the atmosphere IDLH or is it likely to become IDLH? If yes, a positive-pressure SCBA should be used. A positive-pressure SAR with an escape air bottle can also be used.

Is the duration of air supply sufficient for accomplishing the necessary tasks? If not, a larger cylinder should be used, a different respirator should be chosen (such as an SAR) or the work plan should be modified.

Will the bulk and weight of the SCBA interfere with task performance or cause unnecessary stress? If yes, use of an SAR may be more appropriate if conditions permit.

Figure 9-6 Self-Contained Breathing Apparatus with a 30-minute air tank

Will high temperature compromise respirator effectiveness or cause added stress to the worker? If yes, the work period should be shortened or scheduled during cooler hours of the day, or the mission postponed until favorable temperature occurs.

9.6.2.2 Supplied-Air Respirator

Key questions to ask when considering SAR use are:

Can an APR provide adequate protection?

Is the atmosphere IDLH or likely to become IDLH? If yes, an SAR/SCBA combination or SCBA should be used.

Will the hose significantly impair worker mobility? If yes, the work task should be modified or other respiratory protection should be used.

Is there a danger of the air line being damaged or obstructed (such as by heavy equipment, falling drums, rough terrain, or sharp objects) or permeated and/or degraded by chemicals (such as by pools of chemicals)? If yes, either the hazard should be removed or another form of respiratory protection should be used.

Can the work be completed within a 300-foot radius (the maximum distance from the compressor that a SAR may be used)?

If a compressor is the air source, is it possible for airborne contaminants to enter the air system? If yes, have the contaminants been identified, have appropriate alarms been installed, and are efficient filters and/or sorbents available that are capable of removing those contaminants? If not, either cylinders with certified breathing air should be used as the air source or the airborne contaminants removed.

Can other workers and vehicles that might interfere with the air line be kept away from the area? If not, another form of respiratory protection should be used.

9.6.3 AIR SUPPLYING RESPIRATOR OPERATIONS

9.6.3.1 Fit-Check Procedures

OSHA requires fit-testing for all employees using respirators. OSHA allows either a qualitative or quantitative fit-test. However, a quantitative fit-test must be used to validate fit factors of more then 100. A negative-pressure check may be performed in the field before connecting to the airsource by covering the bottom of the inhalation tube with the hand while inhaling.

9.6.3.2 Maintenance

Like the APR maintenance program, inspection for defects, cleaning and disinfecting, repair, and storage need to be included in the ASR maintenance program. Infrequently used or spare ASRs should be checked either weekly or monthly, but must be checked prior to any use.

Field inspection for a routinely used air supplying respirator should use the following procedures:

- If the device has a tight-fitting facepiece, use the procedures outlined previously for APR, with the exception of those pertaining to the air-purifying elements.
- If the device is a hood, helmet, blouse, or full suit, use the following procedures:
  - —examine the hood, blouse, or full suit for rips and tears, seam integrity, or other defects;
  - —examine the protective headgear (hard hat), if required, for general condition, with emphasis on the suspension inside the headgear; and
  - —examine the protective faceshield, if any, for cracks, breaks, or impaired vision due to particle abrasion or accumulation.
- Examine the air-supply system for:
  - —integrity and good condition of air-supply lines and hoses, including attachments and end fittings; and
  - —correct operation and condition of all regulators, valves, or other airflow regulators, and air-purifying and conditioning devices and contaminant alarms.
- For SCBA, determine if the high-pressure cylinder of compressed air is sufficiently charged for the intended use, preferably fully charged (mandatory for emergency devices).
- For cleaning and disinfecting, follow the cleaning and disinfecting procedures for the APR, except during disassembly. During disassembly, use the following procedures:
  - —remove the air tanks from the SCBA and send them to a charging station.
  - —check regulator for proper operation.

In addition to field maintenance, more detailed maintenance must be performed by trained personnel. They should be aware of their limitations and never try to replace components or make repairs and adjustments beyond the manufacturer's recommendations, unless they have been specially trained by the manufacturer. These restrictions apply primarily to maintenance of the more complicated components, especially SCBA regulators.

SCBA tanks must be hydrostatically tested at three to five year intervals, depending on the tank material. This test determines if the SCBA tank can safely hold the designed air pressure. After 15 years of service, tanks are considered obsolete and must be replaced or reconditioned by the manufacturer.

An important aspect of any maintenance program is having enough spare parts on hand. Only continual surveillance of replacement rates will determine what parts in what quantities should be kept in stock. It is desirable to have some sort of recordkeeping system to indicate spare parts usage and the inventory on hand.

9.7 Protective Clothing and Accessories

In this manual, personal protective clothing is considered to be any article offering skin and/or body protection. Each type of protective clothing has a specific purpose; many, but not all, are designed to protect against chemical exposure. Types include:

- fully-encapsulating suits
- non-encapsulating suits
- aprons, leggings, and sleeve protectors
- gloves
- firefighters' protective clothing
- proximity, or approach, garments
- blast and fragmentation suits
- cooling garments
- radiation-protective suits

9.7.1 SELECTION OF CHEMICAL PROTECTIVE CLOTHING

CPC is available in a variety of materials that offer a range of protection against different chemicals. The most appropriate clothing material will depend on the chemicals present and the task to be accomplished. Ideally, the chosen material resists permeation, degradation, and penetration (Figure 9-7). Permeation is the process by which a chemical dissolves in or moves through a protective clothing material on a molecular level. Degradation is the loss of or change in the fabric's chemical resistance or physical properties due to exposure to chemicals, use, or ambient conditions such as sunlight. Penetration is the movement of various modes of chemical entry through CPC, including chemicals through zippers, stitched seams or imperfections such as pinholes in a protective clothing material.

Selection of CPC is a complex task and should be performed by personnel with training and experience. Clothing is selected by evaluating the performance characteristics of the clothing against the requirements and limitations of the site and task-specific conditions. If possible, representative garments should be inspected before purchase and their use and performance discussed with someone who has experience with the clothing under consideration. In all cases, the employer is responsible for ensuring that all PPE necessary to protect employees from hazards at the work site is adequate and of safe design and construction for the work to be performed (see 29 CFR Part 1910.132-1910.137).

9.7.1.1 Permeation and Degradation

The selection of CPC depends greatly upon the type and physical state of the contaminants. This information is determined during site characterization. Once the chemicals have been identified, appropriate information sources should be consulted to identify CPC materials resistant to permeation and degradation by the known chemicals. One reference, *Guidelines for the Selection of Chemical-Protective Clothing*, provides a matrix of clothing material recommendations for approximately 300 chemicals based on an evaluation of permeation and degradation data from independent tests, vendor literature, and raw materials suppliers.

Charts indicating the resistance of various clothing materials to permeation and degradation are also available from manufacturers and other sources (Tables 9-4 and 9-5). It is important to note, however, that no material protects against all chemicals and combinations of chemicals. In reviewing vendor literature, it is important to be aware that the data provided are of limited value. For example, the quality of vendor test methods is inconsistent and may not relate well to field conditions. Vendors often rely on the raw material manufacturers for data rather than conducting their own tests and may not update data after initial tests. In addition, vendor data cannot address the wide vari-

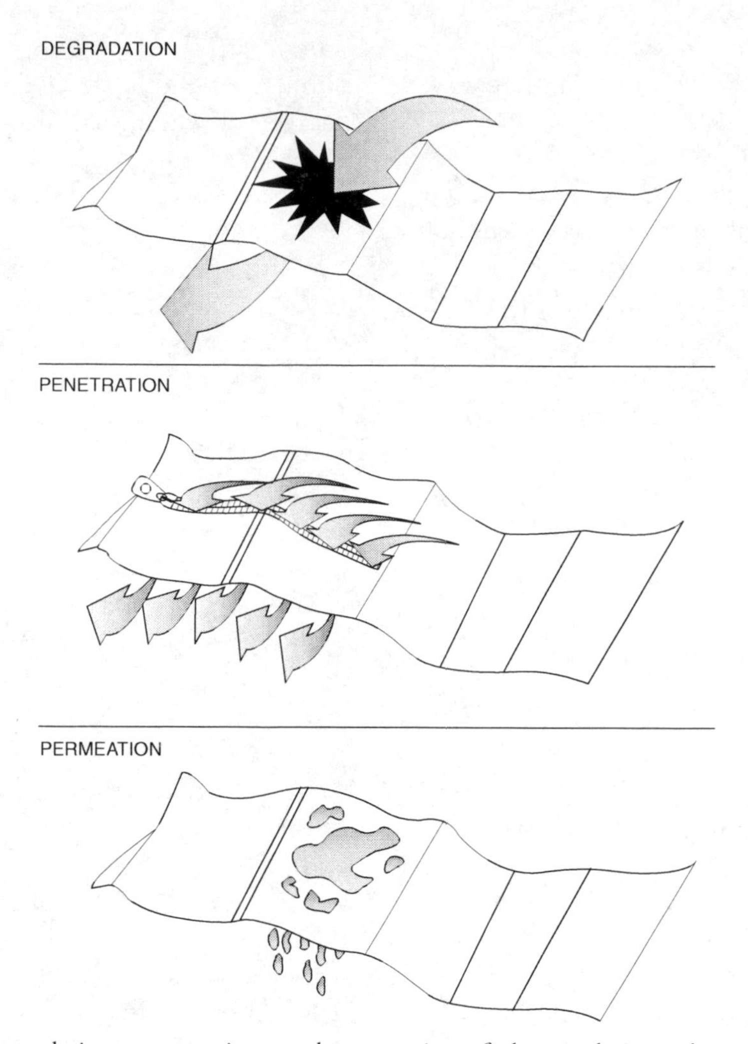

Figure 9-7 Examples of degradation, penetration, and permeation of chemicals into chemical protective clothing

ety of uses and challenges to which CPC may be subjected. Most vendors strongly emphasize this point in the descriptive text that accompanies their product. Also, permeation data typically do not provide information regarding the integrity of zipper closures and seams that can allow the penetration of a contaminant through the protective garment.

Another factor when selecting CPC is that the rate of permeation is a function of several factors, including clothing material type and thickness, manufacturing method, the concentration(s) of the hazardous substance(s), temperature, pressure, humidity, the solubility of the chemical in the clothing material, and the diffusion coefficient of the permeating chemical in the clothing material. Thus, permeation rates and breakthrough times (the time from initial exposure until a hazardous material is detectable on the inside of the CPC) may vary depending on these conditions. Permeation, which occurs on the molecular level, is most likely to cause an undetected exposure. In addition to chemicals, biological hazards, such as viruses may penetrate or permeate protective clothing. A trained professional such as a Certified Industrial Hygienist or Certified Safety Professional should be consulted when selecting chemical (or biological) protective clothing.

Permeation takes place in three distinct phases. Adsorption occurs when the chemical adheres to the surface of the CPC; diffusion is the movement of the chemical through the material, molecule by molecule; desorption takes place when the chemical breaks through the entire thickness of the fabric. In the laboratory, breakthrough times are generally measured by first weighing the garment, soaking it for a specific time in a chemical, drying it off, then re-weighing it a second time. The weight gain, if any, may indicate active adsorption and diffusion phases.

Table 9-4 **Chemical Permeation Rates For Selected Materials**

	Butyl Rubber	Natural Rubber	Neoprene Rubber	Nitrile Rubber	Polyethylene (PE)	Polyvinyl Alcohol (PVA)	Polyvinyl Chloride (PVC)	Teflon™
Common Uses	G, B, S	G	G, B, S	G, B	S	G	G, B, S	S
Acetone	>8 hrs.	NR	NR	NR	NR	NR	NR	>8 hrs.
Acetronitrile	>8 hrs.	NR	NR	NR	NR	C	NR	>8 hrs.
Benzene	NR	NR	NR	NR	NR	>8 hrs.	NR	>4 hrs.
1,3 Butadiene	C	NR	NR	NR	NR	Variable from 7 min. to 30 hrs.	NR	>4 hrs.
Carbon Disulfide	NR	NR	NR	NR	NR	>8 hrs.	>4 hrs.	>4 hrs.
Chlorine gas	>8 hrs.	>4 hrs.	>8 hrs.	>4 hrs.	NR	Unknown	NR	>8 hrs.
Diethylamine	NR	NR	NR	NR	NR	NR	NR	>8 hrs.
N,N-Dimethylformamide	>8 hrs.	NR	NR	NR	NR	NR	NR	>8 hrs.
Ethyl Acetate	C	NR	NR	NR	NR	>4 hrs.	NR	>4 hrs.
Ethylene Oxide	>4 hrs.	NR	NR	NR	NR	Unknown	NR	>4 hrs.
Formaldehyde	>8 hrs.	NR	C	>8 hrs.	NR	NR	C	>4 hrs.
Gasoline, unleaded	NR	NR	NR	>8 hrs.	C	>4 hrs.	NR	NR
n-Hexane	NR	NR	NR	>8 hrs.	NR	>8 hrs.	NR	>8 hrs.
Hydrogen Chloride	>8 hrs.	Unknown	>4 hrs.	Unknown	NR	Unknown	>4 hrs.	>8 hrs.
Methanol	>8 hrs.	NR	NR	NR	NR	NR	NR	>8 hrs.
Methylene Chloride	C	NR	NR	Unknown	NR	Unknown	NR	>4 hrs.
Nitrobenzene	>8 hrs.	NR	NR	NR	NR	>8 hrs.	NR	>8 hrs.
Sodium Hydroxide, 50%	>8 hrs.	>8 hrs.	>8 hrs.	>8 hrs.	>8 hrs.	NR	>8 hrs.	>8 hrs.
Sulfuric Acid, 70%	>8 hrs.	NR	C	NR	>8 hrs.	NR	C	>8 hrs.
Tetrachloroethylene	NR	NR	NR	NR	NR	>8 hrs.	NR	>4 hrs.
Tetrahydrofuran	NR	NR	NR	NR	NR	C	NR	>8 hrs.
Toluene	NR	NR	NR	NR	NR	>8 hrs.	NR	>8 hrs.

I = Immediate breakthrough, Not Recommended for working with this chemical
NR = Not Recommended; permeation in less than 1 hour
C = Caution; permeation between 1 and 4 hours (not an entire work shift)

G = Gloves
B = Boots
S = Suits

> = greater than
< = less than

(Continued)

Table 9-4 *Continued*

	Tyvek® QC	Viton®	Saranex™	4H™ & Silver Shield™	Barricade	CPF 3™	Responder™	Trellchem™	Tychem 10000™
Common Uses	S	G, S	S	G	S	S	S	S	S
Acetone	I	NR	NR	>8 hrs.	>8 hrs.	>8 hrs.	>8 hrs.	>8 hrs.	>8 hrs.
Acetonitrile	I	NR	NR	>8 hrs.	>8 hrs.	NR	>8 hrs.	>8 hrs.	>8 hrs.
Benzene	Unknown	>4 hrs.	NR	>8 hrs.	>8 hrs.	>8 hrs.	>8 hrs.	Unknown	>8 hrs.
1,3-Butadiene	I	>8 hrs.	>8 hrs.	Unknown	>8 hrs.	>8 hrs.	>8 hrs.	>8 hrs.	>8 hrs.
Carbon Disulfide	I	>8 hrs.	NR	>8 hrs.	>8 hrs.	NR	>8 hrs.	>8 hrs.	>8 hrs.
Chlorine Gas	I	>8 hrs.	>8 hrs.	>4 hrs.	>8 hrs.	>8 hrs.	>8 hrs.	>8 hrs.	>8 hrs.
Diethylamine	I	NR	NR	C	>8 hrs.	>8 hrs.	>8 hrs.	>8 hrs.	>8 hrs.
N,N-Dimethylformanide	I	NR	C	>8 hrs.	C	>8 hrs.	>8 hrs.	>8 hrs.	>8 hrs.
Ethyl Acetate	I	NR	NR	>8 hrs.	>8 hrs.	>8 hrs.	>8 hrs.	>8 hrs.	>8 hrs.
Ethylene Oxide	I	NR	NR	>4 hrs.	>8 hrs.	Unknown	>8 hrs.	>8 hrs.	>8 hrs.
Formaldehyde	Unknown	>8 hrs.	>8 hrs.	>4 hrs.	>8 hrs.	>8 hrs.	>4 hrs.	Unknown	Unknown
Gasoline, unleaded	I	>8 hrs.	NR	>4 hrs.	>8 hrs.	NR	>4 hrs.	Unknown	Unknown
n-Hexane	I	>8 hrs.	C	>4 hrs.	>4 hrs.	>8 hrs.	>4 hrs.	>8 hrs.	>8 hrs.
Hydrogen Chloride	I	>4 hrs.	>8 hrs.	Unknown	>8 hrs.	Unknown	>8 hrs.	>8 hrs.	>8 hrs.
Methanol	I	>8 hrs.	>8 hrs.	>8 hrs.	C	NR	>8 hrs.	>8 hrs.	>8 hrs.
Methylene Chloride	I	NR	NR	>8 hrs.	>4 hrs.	NR	>8 hrs.	>8 hrs.	>8 hrs.
Nitrobenzene	I	>8 hrs.	C	>8 hrs.	>8 hrs.	>8 hrs.	>8 hrs.	>8 hrs.	>8 hrs.
Sodium Hydroxide, 50%	>4 hrs.	>8 hrs.	>8 hrs.	>8 hrs.	>8 hrs.	>8 hrs.	>8 hrs.	>8 hrs.	>8 hrs.
Sulfuric Acid, 70%	>4 hrs.	>4 hrs.	>8 hrs.	>8 hrs.	>8 hrs.	>8 hrs.	>8 hrs.	>8 hrs.	>8 hrs.
Tetrachloroethylene	I	>8 hrs.	NR	>8 hrs.	>8 hrs.	>4 hrs.	>4 hrs.	>8 hrs.	>8 hrs.
Tetrahydrofuran	I	NR	NR	>8 hrs.	>8 hrs.	>8 hrs.	>8 hrs.	>8 hrs.	>8 hrs.
Toluene	I	>8 hrs.	NR	>8 hrs.	>8 hrs.	>8 hrs.	>8 hrs.	>8 hrs.	>8 hrs.

Sources: Forsberg, Krister, Keith, Lawrence H., *Chemical Protective Clothing Permeation and Degradation Compendium,* Lewis Publishers, Boca Raton, FL, 1995.
Forsberg, Krister, Mansdorf, S.Z., *Quick Selection Guide to Chemical Protective Clothing, 3rd. ed.,* Van Nostrand Reinhold, New York, NY, 1997.
Johnson, James S., Anderson, Kevin J. Editors, *Chemical Protective Clothing, Volumes I and II,* American Industrial Hygiene Association, Akron, OH, 1990.

Many hazardous wastes are mixtures, and permeation data for mixtures are usually not available. Due to a lack of testing, only limited permeation data for multicomponent liquids are currently available. Mixtures of chemicals can be significantly more aggressive towards CPC than can any single component alone. Even small amounts of a rapidly permeating chemical may provide a pathway that accelerates the permeation of other chemicals. To be on the safe side, the CPC should be selected for longer breakthrough times. However, if a range of times is given (such as from 30 minutes to three hours), the shorter time should be used to determine when to withdraw from the work area to decontaminate or to change into new clothing. NIOSH is currently developing methods for evaluating CPC materials against mixtures of chemicals and unknowns in the field. For hazardous waste site operations, CPC should be selected that offers the widest range of protection against the chemicals expected on site. Vendors are now providing CPC material—composed of two or even three different materials laminated together—that is capable of providing the best features of each material.

Remember, that while such materials increase protection, they may increase heat stress in the worker. In addition, these materials may be less stiff, and therefore cause the worker to expend more energy and increase the difficulty of performing tasks.

9.7.1.2 Heat Transfer Characteristics

The heat transfer characteristics of CPC may be an important selection factor. Since most CPC is virtually impermeable to moisture, evaporative cooling is limited. The “clo” value (thermal insulation value) of CPC is a measure of the capacity of CPC to dissipate heat through means other than evaporation. The larger the clo value, the greater the insulating properties of the garment and, the greater the likelihood of heat stress to the worker. Given other equivalent protective properties, clothing with the lowest clo value should be selected in hot environments or for high work rates. Unfortunately, clo values for clothing are rarely available at present.

Garments are available, worn beneath regular work clothing, that help reduce the body's core temperature. These generally work as a system of tubing attached to underwear through which cool water circulates from a heat exchanger. While heat stress can be reduced, an additional 20 pounds of equipment may be added so this type of cooling system would be used only in extreme conditions.

9.7.1.3 Other Considerations

In addition to permeation, degradation, penetration, and heat transfer, several other factors must be considered during clothing selection. These affect not only chemical resistance, but also the worker's ability to perform the required task. The following checklist summarizes these considerations.

Durability

Does the material have sufficient strength to withstand the physical stress of the task(s) at hand? Will the material resist tears, punctures, and abrasions? Will the material withstand repeated use after contamination or decontamination?

Flexibility

Will the CPC be supple enough to allow the workers to perform their assigned tasks? Will stress lines from flexing the CPC lead to break down or increased contamination penetration? This is particularly important to consider for gloves.

Temperature Effects

Will the material maintain its protective integrity and flexibility under hot and cold extremes. In other words, will temperature extremes cause it to crack, leading to increased contaminant penetration?

Decontamination

Are decontamination procedures available on site? Will the material pose any decontamination problems? Should disposable clothing be used? Disposable suits decrease the amount of decontamination solution needed, but can increase CPC costs as well as waste disposal costs.

Compatibility with Other Equipment

Does the clothing preclude the use of another, necessary piece of protective equipment; for example, suits that preclude hard hat use in hard hat areas?

Duration of Use

Can the required task be accomplished before contaminant breakthrough occurs, or degradation of the CPC becomes significant?

9.7.1.4 Closures

Chemical-protective ensembles leave gaps, especially between gloves and sleeves and between boots and pant legs and at the neck. Such gaps usually require closure. Standard duct tape is effective in providing a chemical resistant closure between PPE components. Velcro, elastic bands, and related closures are often ineffective in sealing PPE components. However, duct tape can usually provide a seal.

When closing gaps, tucking and folding of the PPE is necessary before applying the duct tape. Otherwise, small gaps will exist after taping and chemical entry is possible. Before applying tape, gloves are often interlocked into sleeves to provide a barrier to any liquids that might find a gap in the tape (Figure 9-8). After the tape is applied, the end of the tape should be folded back on itself to provide a tag. When removing the tape during decontamination, simply pull the tag.

9.7.1.5 Special Conditions

All site hazards must be considered when selecting protective clothing and equipment. In addition to chemical hazards, fire, explosion, heat, and radiation are special conditions requiring special protective equipment. The unique problems associated with radiation are beyond the scope of this manual. A qualified health physicist should be consulted if a radiation hazard exists.

Most CPC and accessories are not fireproof. Therefore, care must be taken when working in potentially flammable environments.

9.7.1.6 Eye and Face Protection

Eye and face protection is required by OSHA where there is reasonable probability of preventing injury when such equipment is used. Employers must provide a type of protection suitable for the work to be performed, and employees must use the protection devices. These stipulations also apply to supervisors and management personnel, and should apply to visitors while they are in hazardous areas.

A BLS study found that about 60 percent of workers who suffered eye injuries were not wearing eye protective equipment. When asked why they were not wearing face protection at the time of the accident, workers indicated that face protection was not normally used or practiced in their line of work, or it was not required

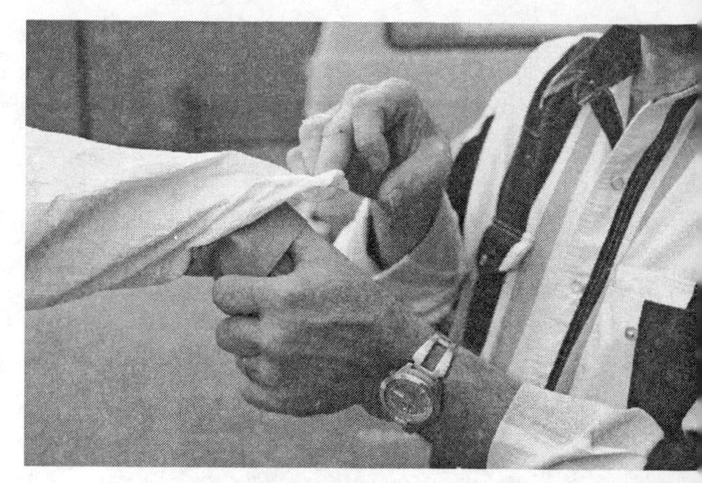

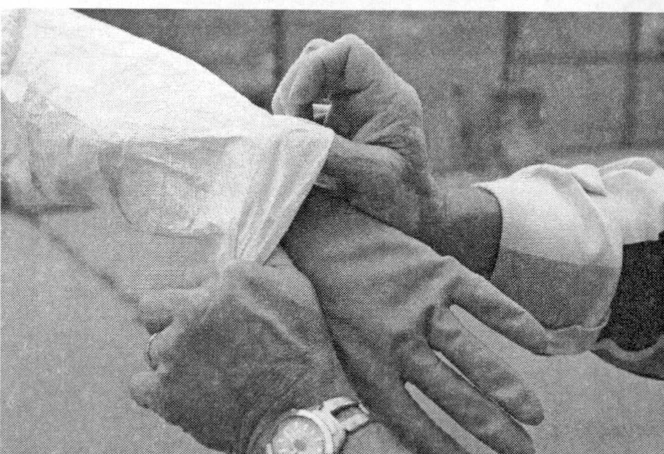

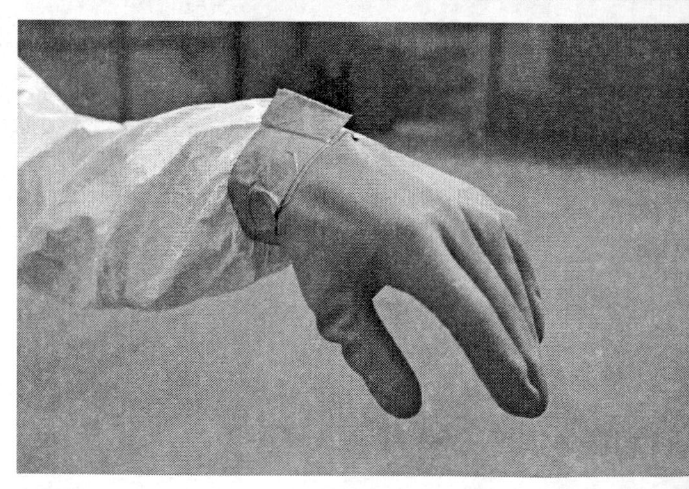

Figure 9-8 Folding, tucking, and taping chemical protective clothing

for the type of work performed at the time of the accident.

The design, construction, tests, and use of eye and face protection purchased prior to July 5, 1994, must be in accordance with ANSI Z87.1–1968 USA Standard Practice for Occupational and Educational Eye and Face Protection. Protective eye and face devices purchased after July 5, 1994, must comply with ANSI Z87.1–1989, American National Standard Practice for Occupational and Educational Eye and Face Protection.

Suitable eye protectors must be provided where there is a potential for injury to the eyes or face from flying particles, molten metal, liquid chemicals, acids, or caustic liquids, chemical gases or vapors, potentially injurious light radiation or a combination of these. These include, in order of increasing eye protection, safety glasses with side shields, safety goggles, and chemical splash goggles. In addition to eye protection, a face shield must also be worn to protect the face when a splash hazard exists. However, a face shield alone is not considered adequate eye protection. In other situations, employees must use equipment with filter lenses that have a shade number appropriate for the works being performed for protection from injurious light radiation.

OSHA and the National Society to Prevent Blindness recommend that emergency eyewashes be placed in hazardous locations. First-aid instructions should be posted close to potential danger spots since any delay to immediate aid or an early mistake in dealing with an eye injury can result in lasting damage.

9.7.1.6.1 Selection. Each eye or face-and-eye protector is designed for a particular hazard. In selecting the protector, consider the kind and degree of hazard, and choose the protective devices accordingly. When a choice of protectors is given, and the degree of protection required is not an important issue, worker comfort may be a deciding factor. The BLS survey showed that few workers ever complained about poor vision or discomfort with personal eye protection equipment.

The survey noted that typical injuries were caused by flying or falling blunt metal objects. Lacerations, fractures, broken teeth, and contusions were also common types of injuries reported.

Persons who use corrective spectacles and those who are required by OSHA to wear eye protection must wear face shields, goggles, or spectacles of one of the following types:

- spectacles with protective lenses providing optical correction;
- goggles or face shields worn over corrective spectacles without disturbing the adjustment of the spectacles; or
- goggles that incorporate corrective lenses mounted behind the protective lenses.

Goggles come in a number of different styles: eyecup, flexible or cushioned goggles, plastic eye shield goggles, and foundrymen's goggles. Goggles are manufactured in several styles for specific uses such as protecting against dusts and splashes, and in chipper's, welder's, and cutter's models.

Safety spectacles require special frames. Combinations of normal street-wear frames with safety lenses are not in compliance.

9.7.1.6.2 Fit. The fitting of goggles and safety spectacles should be done by someone skilled in the procedure. Prescription safety spectacles and goggles should be fitted only by qualified optical personnel.

9.7.1.6.3 Inspection and Maintenance. It is essential that the lenses of eye protectors be kept clean. Continuous vision through dirty lenses can cause eye strain—often used as an excuse for not wearing the eye protectors. Care should include the following:

- daily inspection and cleaning of the eye protector with soap and water, or with a cleaning solution and tissue;
- pitted or scratched lenses can be a source of reduced vision; they should be replaced since deeply scratched or excessively pitted lenses are apt to break more readily;
- slack, worn-out, sweat-soaked, or twisted headbands do not hold the eye protector in proper position and should be replaced;
- goggles should be kept in a case when not in use;
- spectacles should be given the same care as one's own glasses, since the frame, nose pads,

and temples can be damaged by rough usage; and

- PPE that has been previously used should be disinfected before being re-issued; also, when each employee is assigned protective equipment for extended periods, such equipment should be cleaned and disinfected regularly.

9.7.1.6.4 **Cleaning**. Several methods for disinfecting eye-protective equipment are acceptable. The most effective method is to disassemble the goggles or spectacles and thoroughly clean all parts with soap and warm water. Carefully rinse all traces of soap, and replace defective parts with new ones. Swab thoroughly and immerse all parts for 10 minutes in a solution of germicidal deodorant fungicide. Remove parts from solution and suspend in a clean place for air drying at room temperature or with heated air. Do not rinse after removing parts from the solution because this will remove the germicidal residue which retains its effectiveness after drying.

The dry parts should be placed in a clean, dust-proof container, such as a box, bag, or plastic envelope, to protect them until reissue.

9.7.2 SELECTION OF PROTECTIVE ENSEMBLES

The individual components of clothing and respiratory protection must be assembled into a full ensemble that both protects the worker from the site-specific hazards and minimizes the hazards and drawbacks of the PPE ensemble itself.

Personnel who work with hazardous waste or respond to emergencies involving hazardous materials and substances must wear protective clothing. Full-face respiratory protection provides the lungs, eyes, and GI tract with protection against airborne toxins. CPC protects the skin from direct contact with skin-destructive and skin-absorbable materials and substances.

Protective equipment has been divided into four categories, according to the degree of protection afforded against contact with known or anticipated hazardous materials, and are defined in 29 CFR Part 1910.120, Appendix B. These categories are as follows:

Level A should be worn when the highest level of respiratory, skin, and eye protection is required or needed;

Figure 9-9 Level A Protection

Level B should be worn when the highest level of respiratory protection is needed but a lesser level of skin protection is required;

Level C should be worn when the criteria for using air purifying respiratory protection is met; and

Level D is the minimum PPE ensemble for hazardous waste work and offers no respiratory protection and only minimal skin protection.

9.7.2.1 Level A Protection

Level A protection must be worn when the highest level of respiratory, skin, eye and mucous membrane protection is needed, such as in some emergency response situations where gas or vapor clouds are present.

The following constitute Level A equipment:

—positive-pressure, pressure-demand, full face-piece SCBA or positive pressure demand, SAR with escape SCBA approved by NIOSH;
—fully-encapsulating chemical resistant suit;
—gloves, inner, chemical resistant;
—gloves, outer, chemical resistant; and
—boots, chemical resistant, steel toe and shank, (depending on suit boot construction, worn over or under suit boot).

Optional PPE includes:

—underwear, cotton, long-john type;
—hard hat (under suit);

Figure 9-10 Level B Protection

Figure 9-11 Level C Protection

—coveralls (under suit); and
—two-way radio communicators (which are intrinsically safe).

9.7.2.2 Level B Protection

Level B protection should be selected when the situation requires the highest level of respiratory protection, but a lesser level of skin and eye protection. Level B protection is the minimum level recommended on initial entry to a site before the hazards have been fully identified and defined by monitoring, sampling, and other reliable methods of analysis, and PPE corresponding with those findings has been determined.

The following constitute Level B equipment:

—positive-pressure, pressure-demand, full-facepiece SCBA or pressure-demand, SAR with escape SCBA that is NIOSH approved;
—hooded, chemical-resistant clothing (overalls and long-sleeved jacket, coveralls, two-piece chemical splash suit, or disposable chemical-resistant overalls);
—gloves, outer, chemical-resistant;
—gloves, inner chemical-resistant; and
—boots, outer, chemical-resistant, steel toe and shank.

Optional PPE includes:

—coveralls (under splash suit);
—boots, outer, chemical resistant (disposable);
—two-way radio communicators (which are intrinsically safe);
—hard hat; and
—face shield.

9.7.2.3 Level C Protection

Level C protection should be selected when the type of airborne substance is known, its concentration has been measured and the criteria for using APRs have been met. Periodic monitoring of the air must still be performed when this level of protection is being used.

The following constitute Level C equipment:

—full-face or half-face APR (NIOSH approved); typically used in conjunction with safety glasses or safety goggles and based on the degree of eye protection needed;
—chemical-resistant clothing (one-piece coverall, hooded two-piece chemical splash suit, chemical resistant hood and apron, or disposable chemical resistant coveralls);
—gloves, outer, chemical-resistant; and
—gloves, inner, chemical-resistant.

Optional PPE includes:

—boots, outer, chemical-resistant (disposable);
—cloth overalls (inside CPC);
—two-way radio communicators (which are intrinsically safe);
—hard hat;
—goggles;
—escape mask; and
—face shield.

Figure 9-12 Level D Protection

9.7.2.4 Level D Protection

Level D is primarily a work uniform. It should not be worn on any site where respiratory or skin hazards are a potential problem for on-site workers. Level D equipment provides protection from only physical site hazards. Steel-toed shoes, eye protection, hard hat, coveralls, safety glasses or chemical splash goggles, escape mask, face shield, and gloves are common components of Level D protection.

In summary, levels A through D ensembles can be used as a starting point for worker protective equipment. However, each ensemble must be tailored to the specific situation in order to provide the most appropriate level of protection. For example, if work is being conducted at a highly contaminated site or if the potential for contamination is high, it may be advisable to wear a disposable covering, such as Tyvek® coveralls or PVC splash suits over the cloth coveralls.

"Modifying" a given level of protection means including some components of the next higher level of protection. For example, modified Level D may include all CPC specified for Level C, but exclude the APR. Modified Level B may include a fully-encapsulating suit, but one offering less vapor protection than required for Level A.

The type of equipment used and the overall level of protection should be reevaluated periodically as information about the site increases and as workers are required to perform different tasks. Personnel should be able to upgrade or downgrade their level of protection with concurrence of the Site Safety Officer and approval of the Field Team Leader.

Reasons to upgrade can include:

- known or suspected presence of dermal hazards;
- occurrence or likely occurrence of gas or vapor emission;
- change in work task that will increase contact or potential contact with hazardous materials; and/or
- request of the individual performing the task.

Reasons to downgrade can include:

- new information indicating that the situation is less hazardous than was originally thought;
- change in site conditions that decreases the hazard; and/or
- change in work task that will reduce contact with hazardous materials.

Disposable protective garments are commonly used in many work areas such as those dealing with fiberglass, medicine, pesticide application, laboratories, radioactive dust, lead particulates, and asbestos abatement. One of the most often cited trade names is DuPont's Tyvek which, because of its wide usage has become a common synonym for "disposable." Tyvek and related products are nonwoven fabrics with the qualities of film, woven fabric and paper to provide protection against dry particulates. They are widely acceptable to workers because of their light weight, thinness and flexibility as well as being tough, abrasion resistant, and chemically inert, providing a barrier to dry particles while allowing limited air and water to vapor to pass through. This practical ability to "breathe" helps worker comfort in warm or humid climates. Such garments come in a variety of styles and sizes, including built-in hoods and booties, collars, zippers, elastic wrists and/or ankles, coveralls, pants, shirts, separate booties, and arm covers. Colors range from standard white to high visibility yellows and oranges.

In addition, Tyvek can be laminated with Saranex 23-P® film for protection against a broad range of chemicals. These suits are still considered to be light duty and offers increased splash protection. Coating Tyvek with polyethylene provides splash protection for some acids, bases and other liquid chemicals and can be made strong enough to classify as Level B protection. Before any PPE is selected, it must be determined by permeation, breakthrough charts and/or consultation with the supplier, manufacturer or a Certified Safety Professional or Certified Industrial Hygienist, whether or not the material is proper for the type of protection needed. Sizes of most garments range from small through extra-extra large.

9.7.3 PPE MAINTENANCE

The technical details of maintenance procedures for reusable PPE varies. Manufacturers frequently restrict the sale of certain PPE parts to individuals or groups who are specially trained, equipped, and "authorized" by the manufacturer to purchase them. Explicit procedures should be adopted to ensure that the appropriate level of maintenance is performed only by individuals having this specialized training and equipment. This is generally carried out by the company health and safety office.

9.7.4 INSPECTION

An effective PPE inspection program features five different inspections:

1. inspection and operational testing of equipment received from the factory or distributor;
2. inspection of equipment as it is issued to workers;
3. inspection after use or training and prior to maintenance;
4. periodic inspection of stored equipment; and
5. periodic inspection when a question arises about the appropriateness of the selected equipment or when problems with similar equipment arise.

Each inspection will cover somewhat different areas in varying degrees of depth. Detailed inspection procedures, where appropriate, are usually available from the manufacturer. The inspection checklists provided in Table 9-5 may also help. Inspection checklists are available from OSHA, NIOSH, industrial hygiene professional associations, and some equipment manufacturers.

Records must be kept of all inspection procedures and individual identification numbers assigned to all reusable pieces of equipment. Records of each inspec-

Table 9-5 **Sample PPE Inspection Checklists**

Before use:

- Determine that the clothing material is correct for the specific task at hand.
- Visually inspect for:
  - —imperfect seams
  - —nonuniform coatings
  - —tears
  - —malfunctioning closures
- Hold up to light and check for pinholes.
- Flex product:
  - —observe for cracks
  - —observe for other signs of shelf deterioration
- If the product has been used previously, inspect inside and out for signs of chemical attack:
  - —discoloration
  - —swelling
  - —stiffness

During the work task, periodically inspect for:

- Evidence of chemical attack such as discoloration, swelling, stiffening and softening. Keep in mind, however, that chemical permeation can occur without any visible effects.
- Closure failure.
- Tears.
- Punctures.
- Seam discontinuities.

GLOVES

Before use:

Pressurize glove to check for pinholes. Either blow into glove, then roll gauntlet toward fingers or inflate glove and hold under water. In either case, no air should escape.

FULLY-ENCAPSULATING SUITS

Before use:

- Check the operation of pressure relief valves.
- Inspect the fitting of wrists, ankles and neck.
- Check faceshield, if so equipped, for:
  - —cracks
  - —fogginess

RESPIRATORS

All respirators:

- Inspect all respirators:
  - —before and after each use
  - —at least monthly when in storage
  - —every time they are cleaned

(Continued)

Table 9-5 *Continued*

RESPIRATORS *(Con't.)*

All respirators:

- Check all connections for tightness.
- Check material conditions for:
  —signs of pliability
  —signs of deterioration
  —signs of distortion
  —also perform inspections for respirators as previously described in this unit
- Check for proper setting and operation of regulators and valves (according to manufacturers' recommendations).
- Inspect SCBAs.
- Check operations of alarms at least monthly when in storage.
- Check the polymer facepiece for:
  —excessive dirt
  —cracks, tears, or holes
  —distortion
  —cracked, scratched, or loose fitting lenses
- Check headstraps for:
  —breaks and tears
  —loss of elasticity
  —broken or malfunctioning buckles or attachments
  —slippage at the faceplate
- Check inhalation valve and exhalation valve for:
  —detergent residue, dust particles, or dirt on valve or valve seat
  —cracks, tears, or distortion in the valve material or valve seat
  —missing or defective valve cover
- Check cartridge/canister/filter element(s) for:
  —proper filter for the hazard
  —approval designation
  —missing or worn gaskets
  —worn filter threads and faceplate threads
  —cracks or dents in filter housing

Additional inspection elements for atmosphere supplying respirators include the following:

- Check the air supply system for:
  —breathing tubes integrity
  —breathing air quality
  —breaks or kinks in air suply hoses and end fitting attachments
  —tightness of connections
  —proper setting of regulators and valves
  —correct operation of air-purifying elements and carbon monoxide or high-temperature alarms
  —air cylinders (recharge when ≤90% of manufacturer's recommended pressure level)

tion should include the ID number, date, inspector, and any unusual findings or conditions. These should be retained with other site records so that cross-checks can be done correlating workers with specific pieces of equipment. Damaged equipment should immediately be tagged and removed from service.

9.7.5 DEVELOPMENTS

Much of the literature dealing with PPE, both clothing and respiratory protection, is technically complex and difficult to understand. Whenever there is any doubt as to what constitutes appropriate protection, a Certified Industrial Hygienist or Safety Professional should be consulted. Technical support for permeation information is also available through phone service, FAX and Internet web sites from commercial suppliers and manufacturers such as DuPont, DOW Chemical, Siebe North, Kappler Company and Trelleborg Company. For those individuals putting together site safety/health and PPE plans, computer index systems are available with information about generic protective clothing materials that have been tested against certain specific chemicals or classes of chemicals. One program, Chemical-Protective Clothing Permeation and Degradation Database, is set up for industrial hygienists and environmental professionals. It is detailed enough to allow an initial selection of protective clothing based on materials available from manufacturers. A final selection must still be made, since manufacturers are not required to use uniform warnings and instructions for the direct comparison of one brand with another.

9.7.6 HAZARD DETERMINATION ASSESSMENT FOR PPE

Prior to any final selection, OSHA (revised 29 CFR 1910, Subpart I) requires that an employer select PPE for their employees based on an assessment of hazards in the workplace as well as provide written proof of this compliance. This hazard evaluation covers hazards that may be found in occupational operations and processes. Normally, for hazardous waste sites, this is accomplished through the site-specific health and safety plan. Based upon the determinations made during the assessment, appropriate protective devices for particular hazards must be selected. The walk-through survey of the area must take into consideration the basic categories of hazards such as:

- impact from falling or dropping objects;
- sharp objects that can pierce or penetrate;
- compression hazards from rolling or pinching objects;
- sources of chemical hazards;
- sources of dust and particulates;
- sources of light (optical) radiation from welding, brazing, cutting, furnaces, heat treating;
- sources of high temperatures;
- sources of motion—movement of tools, machinery, personnel; and/or
- electrical hazards.

The information obtained from such a survey is then used by the employer in selecting the proper PPE. If questions arise, they may be addressed to OSHA offices or to Certified Industrial Hygienists or Certified Safety Professionals.

In the future, manufacturers of protective clothing are expected to continue improving their products, including more durable seams; lighter, stronger, more breathable fabrics; new fabric combinations; "smart" or instrumented garments; antilacerative, better-fitting, and longer-wearing gloves; ensembles in which gloves, boots and suits are one solid piece; lower cost disposable units; laminated Tyveks and other materials with increased permeation-resistance; flame-resistant fabrics; and a broader range of sizes and shapes for more body types.

9.8 Sources of Information

American Conference of Governmental Industrial Hygienists: *Guidelines for the Selection of Chemical Protective Clothing.* 3rd ed., Vols. I and II. Cincinnati, OH: American Conference of Governmental Industrial Hygienists, 1987.

American Industrial Hygiene Association, *Chemical Protective Clothing* edited by J.S. Johnson and K.J. Anderson. Vols. I and II. Akron, OH: 1990.

Balden-Anslyn, Roxanna: Hazard Assessment and PPE Selection. *Occupational Health and Safety 65, Number 10*:154–158 (October 1996).

"Benzene," *Code of Federal Regulations* Title 29, Part 1910.1028. As amended 8 January 1998.

Cherniak, Michael J.: Proper PPE Use Requires Training. *Environmental Protection 3, Number 2*:34+ (March 1992).

Clambacher, Charles W.: Index Speeds Chore of PPE Selection. *Environmental Protection 3, Number 2*:44+ (March 1992).

Clambacher, Charles W.: PPE Selection Includes Replacing Gear, Adapting to Different Work Hazards. *Occupational Health and Safety*, p. 94+ (May 1993).

Colton, Craig E.: Respiratory Program Develops Training, Fit-Testing Routines. *Occupational Health and Safety*, p. 54+ (May 1993).

Currance, Phillip L.: Personal Hazmat Protection. *Rescue*, p. 33+ (March/April 1990).

Forsberg, Kirster: *Chemical Protective Clothing Permeation and Degradation Database*. Lewis Publishers, Inc./CRC Press, 1992.

Forsberg, Kirster and L.H. Keith: *Chemical Protective Clothing Permeation and Degradation Compendium*. Boca Raton, FL: Lewis Publishers, 1995.

Forsberg, Kirster and S.Z. Mansdorf: *Quick Selection Guide to Chemical Protective Clothing*. 3rd ed. New York: Van Nostrand Reinhold, 1997.

Goldstein, Lynn: Eliminate Misuse of Protective Clothing. *Occupational Hazards*, p. 59 (October 1993).

Johnson, Linda F.: A PPE Checklist. *Occupational Health and Safety 67, Number 2*:38 (February 1998).

National Institute for Occupational Safety and Health: *Occupational Safety and Health Guidance Manual for Hazardous Waste Site Activities*. Washington, DC: U.S. Government Printing Office, 1985.

National Institute for Occupational Safety and Health: *NIOSH Guide to the Selection and Use of Particulate Respirators Certified Under 42 CFR 84* (Publication No. 96–101). Washington, DC: U.S. Government Printing Office, 1996.

National Institute for Occupational Safety and Health: *NIOSH 42 CFR 84: Procedures for Testing and Certifying Air-Purifying/Particulate Respirators*. Washington, DC: National Institute for Occupational Safety and Health, 1996.

Newcomb, William: Fit Testing Under the New 42 CFR Part 84: What It Means To Respirator Users. *Industrial Hygiene News 21, Number 6*:4 (September 1998).

Occupational Safety and Health Administration: *Inspection Procedures for the Respiratory Protection Standard* (Directive Number CPL 2–0.120). In www.osha.gov/ [Occupational Safety and Health Administration (OSHA) World Wide Web homepage]. Washington, DC: U.S. Department of Labor, September 1998. pp. 1–37.

Occupational Safety and Health Administration: *Personal Protective Equipment*. Washington, DC: U.S. Government Printing Office, 1995.

Occupational Safety and Health Administration Standard Interpretation and Compliance Letters. "Selection of air purifying respirators for gases and vapors with poor warning properties (diisocyanates)," July 18, 2000, www.osha.gov/oshdoc/interp_data.

"Respiratory Protection: Final Rule," *Code of Federal Regulations* Title 29, Parts 1910.134 and 1926.103. 1998.

"Respiratory Protection: Final Rule," *Federal Register 63*, Number 5 (8 January 1998). pp. 1270–1300.

"Respiratory Protection for Asbestos Fibers," *Code of Federal Regulations* Title 29, Part 1910.1001. As amended 8 January 1998.

Roychowdhury, Mahendra: OSHA's Revised Respiratory Protection Standard. *Professional Safety, Journal of the American Society of Safety Engineers*, pp. 10, 48–49 (August 1998).

Sheppard, Rob: Figuring Out Fit-Testing. *Occupational Health and Safety 67, Number 11*:34–36 (November 1998).

Smith, S.L.: The Last Line of Defense. *Occupational Hazards*, pp. 48–51 (March 1992).

TEN

Decontamination

10 | CONTENTS

Decontamination—the process of removing or neutralizing contaminants that have accumulated on personnel and equipment—is critical to health and safety at hazardous waste sites. Decontamination protocols, covered in 29 CFR 1910.120 (k), have been formulated to protect workers from hazardous substances that may contaminate protective clothing, respiratory equipment, tools, vehicles, and other equipment used on site. They protect site personnel by minimizing the transfer of harmful materials into clean areas and help prevent the mixing of incompatible chemicals. They also protect the community by preventing the uncontrolled transportation of contaminants from the site.

Decontamination is required when any personal protective clothing or equipment is suspected of being contaminated. It is necessary to decontaminate before personnel move from "dirty" to "clean" work areas; prior to eating, drinking, smoking, or using restroom facilities; and before transport trucks or other equipment leave the site.

This unit describes the types of contamination that workers may encounter at a waste site, the factors that influence the extent of contamination, and methods for preventing or reducing contamination. It also provides general guidelines for designing and selecting decontamination procedures at a site, and presents decision aids for evaluating the health and safety aspects of decontamination methods. This unit does not cover decontamination of radioactively contaminated personnel or equipment. A health physicist should be consulted if this situation arises.

10.1 Decontamination Plan

Decontamination plans must be site specific. Individual sites will have particular needs for decontamination based on facility locations, cleaning solutions, processes, and waste containment. The decontamination plan should be developed as part of the Site Health and Safety Plan, and should be implemented before any personnel or equipment may enter areas where the potential for exposure to hazardous substances exists. The decontamination plan should:

- determine the number and layout of decontamination stations;
- determine the decontamination equipment needed;
- determine appropriate decontamination methods and procedures;
- establish procedures to prevent the contamination of clean areas;
- establish methods and procedures to minimize worker contact with contaminants during

removal of personal protective clothing and equipment; and
- establish methods for disposing of CPC, site tools and equipment that is not completely decontaminated.

The plan should be revised whenever the type of personal protective clothing or equipment changes, the site conditions change, or the site hazards are reassessed based on new information.

The Site Health and Safety Plan should also anticipate the types of emergencies that could arise on the site, evaluate the toxicity and characteristics of contaminants, and establish decontamination protocols for both physical injuries (non life-threatening) and medical emergencies (life-threatening).

In an emergency, the primary concern is to prevent the loss of life or severe injury to site personnel. Decontamination usually precedes medical treatment. Without adequate decontamination, medical treatment can expose both the victim and the rescuers to toxic chemicals. During an emergency, provisions must also be made for protecting medical personnel and for disposing of contaminated clothing and equipment. Unless they are specifically equipped, hospitals are poor choices for use as decontamination sites.

10.2 Prevention of Contamination

More desirable than good decontamination facilities is the non-contamination of personnel and equipment in the first place. The first step in decontamination is to establish standard operating procedures that minimize contact with waste, thus lessening the potential for contamination. For example:

- use remote sampling, handling, and container opening techniques such as drum grapplers and pneumatic wrenches;
- emphasize work practices that minimize contact with hazardous substances (for example, do not walk through areas of obvious contamination, do not directly touch potentially hazardous substances);
- protect monitoring and sampling instruments by placing them in clear plastic bags, making openings in the bags for sample ports and sensors that must analyze site materials;
- wear disposable outer garments and use disposable equipment where appropriate;
- cover equipment and tools with a strippable coating that can be removed during decontamination; and
- encase the source of contaminants (for example, place plastic sheeting over the contamination or use overpacks for leaking drums).

In addition, standard operating procedures that maximize worker protection and facilitate decontamination should be established. Proper procedures for dressing prior to entering the Exclusion Zone (the contaminated or potentially contaminated part of the site) will minimize the potential for contaminants to bypass the protective clothing and escape decontamination. All fasteners should be used (zippers fully closed, all buttons used, all snaps closed, etc.) and all junctures at sleeves and boots should be interlocked and taped to prevent contaminants from getting inside. Frequently, workers operate with a "buddy," helping each other with protective equipment.

Prior to each use, PPE should be inspected for cuts or punctures that could expose workers to hazardous substances. Minor injuries, such as cuts and scratches, can become serious when chemicals or infectious agents penetrate into the body. Therefore, particular care should be taken to protect these areas. Workers with severe skin abrasions should be kept from working on-site until the skin heals.

Common sense procedures and training should not be ignored. Simple, logical precautions should be part of worker training.

- know the limitations of all protective equipment being used;
- do not enter a contaminated area unless it is necessary to carry out a specific function;
- walk upwind of contamination, if possible;
- do not sit or lean against anything in a contaminated area. If kneeling is necessary, use a plastic ground sheet. Likewise, do not set instruments on the ground;

- before sampling any hazardous waste, read the label and manifest (if available) for all containers to determine the identity of the substance, the potential contamination hazard, and the possible incompatibility with PPE in use; and
- check drums and waste containers for evidence of chemical incompatibility or instability such as bulging drums, blistered paint, exploded drums, vapors, dead vegetation, melted plastic, etc. These may signal a condition to be avoided.

10.3 Types of Contamination

Contaminants can cling to the surface of PPE by adsorption, or permeate into the PPE material through absorption. Surface contaminants may be easy to detect and remove. However, contaminants that have permeated a material are difficult or impossible to detect and dislodge. If such contaminants are not removed by decontamination, they may continue to permeate, and in time may be found on the inside surface of the material where they can cause unexpected exposure.

Five factors of greatest concern which affect the extent of permeation are:

1. **Concentration**—This encompasses molecules flowing from areas of high concentration to areas of low concentration. As the concentrations of hazardous substances increase, the potential for permeation of personal protective clothing increases.
2. **Contact time**—The longer a contaminant is in contact with an object, the greater the probability of permeation and the greater the concentration. For this reason, minimizing contact time is one of the most important objectives of a decontamination program.
3. **Temperature**—An increase in temperature generally increases the permeation rate of contaminants.
4. **Size of contaminant molecules and pore space**—Permeation increases as the contaminant molecule becomes smaller and as the pore space of the material permeated becomes larger.
5. **Physical state of wastes**—As a rule, gases, vapors, and low-viscosity liquids tend to permeate more readily than high-viscosity liquids or solids.

10.4 Decontamination Methods

All personnel, clothing, equipment, and samples leaving the Exclusion Zone must be decontaminated to remove harmful chemicals or infectious organisms. Alternately, materials may be discarded when leaving the Exclusion Zone, followed by containerization and disposal as contaminated waste. Decontamination methods (1) physically remove contaminants; (2) deactivate contaminants by chemical detoxification, disinfection or sterilization; or (3) remove contaminants by a combination of both physical and chemical means. Methods for contaminate removal from equipment are often more abrasive and use stronger chemicals than contaminate removal from PPE. Decontamination methods must therefore be evaluated for compatibility with PPE.

10.4.1 PHYSICAL REMOVAL

In many cases, gross contamination (immediately noticeable) can be removed by physical means involving dislodging, displacement, rinsing, wiping, or evaporation. Physical methods involving pressure or heat, such as high pressure wash or steam cleaning, should be used only as necessary and with caution since they can spread contamination and can cause burns or abrasions. Contaminants that can be removed by physical means may be categorized as follows:

Loose contaminants—Dusts and vapors that cling to equipment and workers or become trapped in small openings, such as the weave of clothing fabrics, can sometimes be removed with water or other liquids. Removal of electrostatically attached materials can be enhanced by coating the clothing or equipment with anti-static solutions. These are available commercially as wash additives or anti-static sprays. A vacuum (usually equipped with a HEPA filter) can be used to remove loose contamination. However, compressed air is not used (except under specified, controlled conditions) because of the high likelihood of contaminant dispersal and spread.

Adhering contaminants—Some contaminants adhere by forces other than electrostatic attraction. Adhesive qualities vary greatly with specific contaminants. Glues, cements, resins, and mud have adhesive properties that make them difficult to remove by physical means. Methods for removing gross contaminants include scraping, brushing, and wiping, although some of these methods may be too abrasive for the material being treated. Tears may occur in PPE, increasing rather than decreasing the likelihood of contamination. In addition, these methods may produce air-borne contaminants, creating a respiratory hazard during decontamination. Removal of adhesive contaminants can be enhanced through special methods such as solidifying, freezing (for example, using dry ice or ice water), adsorption (kitty litter) or absorption (powdered lime), or by melting; however, these methods are rarely used.

Volatile liquids—Volatile liquid contaminants can be removed from protective clothing or equipment by evaporation followed by a water rinse. Evaporation of volatile liquids can be enhanced by using steam jets (usually restricted to equipment). With any evaporation or vaporization process, care must be taken to prevent worker inhalation of the vaporized chemicals. Total quantities of such released volatiles may require permitting, monitoring or filtering of the air. Additionally, chemicals that evaporate may leave a residue that must also be removed.

10.4.2 CHEMICAL REMOVAL

Following the physical removal of contaminants, chemical techniques may be necessary to complete the decontamination process. Examples of chemical decontamination techniques follow.

Dissolving contaminants—Chemical removal of surface contaminants can be accomplished by dissolving them in a solvent that is chemically compatible with the equipment being cleaned. Compatibility is particularly important when decontaminating personal protective clothing constructed of organic materials that could be damaged or dissolved by organic solvents. In addition, care must be taken in selecting, using, and disposing of any organic solvents that may be flammable or potentially toxic. Organic solvents include alcohols, ethers, ketone, aromatics, straight-chain alkanes, and common petroleum products.

Chlorinated solvents are generally incompatible with PPE and are also toxic. They should only be used for decontamination in extreme cases when other cleaning agents will not remove the contaminant.

Table 10-1 provides a general guide to the solubility of several categories of contaminants in four types of solvents: water, dilute acids, dilute bases, and organic solvents. Due to the potential hazards, decontamination using chemicals should be done only if approved by a Certified Industrial Hygienist or Certified Safety Professional.

Surfactants—Surfactants augment physical cleaning methods by reducing the adhesion forces between contaminants and the surface being cleaned, and by preventing redepositing of the contaminants. Household detergents are among the most common surfactants. Some detergents can be used with organic solvents to enhance dissolving capabilities and the dispersal of contaminants into the solvent.

Solidification—Solidifying liquid or gel-like contaminants can enhance their physical removal. The mechanisms of solidification are: (1) removal through the use of absorbents such as ground clay or powdered lime; (2) chemical reactions via polymerization catalysts and chemical reagents; and (3) freezing using ice water or dry ice.

Rinsing—Rinsing removes contaminants by physical attraction and by solubilization. Multiple

Table 10-1 **General Guides To Solubility of Contaminants in Four Solvent Types**

Solvent	Soluble Contaminants
Water (most desirable)	Low-chain hydrocarbons Inorganic compounds Salts Some organic acids and other polar compounds
Dilute Acids	Basic (caustic) compounds Amines Hydrazines Metals
Dilute Bases For example: —detergent —soap	Acidic compounds Phenols Thiols Some nitro and sulfonic compounds
Organic Solvents[a] For example: —citric solvents —alcohols —ethers —ketones —aromatics —straight-chain alkanes (such as hexane) —common petroleum products (such as fuel oil, kerosene)	Nonpolar compounds (such as fuels and oils, benzene, chloroform, and others)

Adopted from: *Occupational Safety and Health Guidelines Manual for Hazardous Waste Activities*, OSHA, 1985.

[a]WARNING: Some organic solvents can permeate and/or degrade protective clothing and most create chemical hazards.

rinses with clean solutions remove more contaminants than a single rinse with the same total volume of solution. Continuous rinsing with large volumes of solution remove even more contaminants than multiple rinsing with a lesser total volume. However, rinsing with large volumes of liquids potentially creates a large volume of hazardous waste. Pressured rinses are often used, combining solubilization with physical abrasion to remove contaminants.

Disinfecting/Sterilization—Chemical disinfectants are a practical means of inactivating infectious agents. Unfortunately, standard sterilization techniques are generally impractical for large equipment and for personal protective clothing and equipment. For this reason, disposable PPE is recommended for protection against infectious agents.

Disposal—Industry has been increasing its emphasis on "dry decontamination" practices in which equipment and clothing are thrown away after a single use. This reduces the volume of waste generated during decontamination. However, dry decontamination is not entirely dry. Most decon procedures rinse the PPE to remove loose contaminants that may fall off during PPE removal.

Many factors, such as cost, availability, ease of implementation, and handling of additional wastes can influence the selection of a decontamination method. From a health and safety standpoint, two key questions must be addressed:

- is the decontamination method effective for the specific substances present; and
- does the method itself pose any health, safety or environmental hazards?

10.5 Testing the Effectiveness of Decontamination

Decontamination methods vary in their effectiveness at removing different contaminants. The effectiveness of any decontamination method should be assessed at the beginning of a program, and monitored and reevaluated periodically through the lifetime of the program. If contaminated materials are not being removed, or if they are penetrating protective clothing, the decontamination program must be revised. The following methods may be useful in assessing the effectiveness of decontamination.

10.5.1 VISUAL OBSERVATION

There are rarely any reliable tests to immediately determine the effectiveness of decontamination. In some cases, effectiveness can be estimated by visual observation.

Natural light—Discoloration, stains, corrosive effects, visible dirt, or alterations in clothing fabric may indicate that contaminants have not been removed. However, not all contaminants leave visible traces; many contaminants can permeate clothing without being easily observed.

UV light—Certain contaminants, such as polycyclic aromatic hydrocarbons (common in many refined oils and tar wastes) fluoresce and can be visually detected when exposed to UV light. UV light can be used to observe contamination on skin, clothing, and equipment. However, certain areas of the skin, such as scar tissue, may fluoresce naturally, thereby introducing an uncertainty into the test. Caution: Never use UV-C light for these purposes. In addition, the use of some types of UV light can increase the risk of skin cancer and eye damage. Therefore, a Certified Industrial Hygienist or Certified Health Physicist should assess the benefits and risks associated with UV light prior to its use at a waste site.

10.5.2 WIPE SAMPLING

Wipe sampling provides after-the-fact information about the effectiveness of decontamination. In this procedure, a dry or wet cloth, glass fiber filter paper, or swab is wiped over the surface of the potentially contaminated object and then analyzed in a laboratory. Both the inner and outer surfaces of protective clothing should be tested for permeation. Skin may also be tested using wipe samples.

10.5.3 CLEANING SOLUTION ANALYSIS

Another test for the effectiveness of decontamination procedures is to analyze for contaminants left in the cleaning solutions. Elevated levels of contaminants in the final rinse solution may suggest that additional cleaning and rinsing are needed. In order to obtain the best analysis of the final rinse and establish the effectiveness of the procedure, distilled or laboratory grade water should be used.

10.5.4 TESTING FOR PERMEATION

Testing for the presence of permeated chemical contaminants requires that wipe samples or pieces of the protective garments be sent to a laboratory for analysis.

10.5.5 MONITORING

If volatile chemicals are used or produced during decontamination, air monitoring may be necessary. Direct reading instruments (Unit 8) are commonly used when monitoring is necessary.

10.5.6 HEALTH AND SAFETY HAZARDS

While decontamination is performed to protect health and safety, it may also pose hazards under certain circumstances. These can include the following:

- The decontamination method may be incompatible with the hazardous substances being removed. For example, a decontaminant such as a citrus-based solvent may react with strong oxidizing agents to produce an explosion, heat, and toxic decomposition products such as carbon monoxide.
- The clothing or equipment being decontaminated can be permeated or degraded by deteriorating solvents such as organic solvents. For example, a commercial water-based inorganic zinc primer must be used with solvent-impermeable gloves such as butyl rubber.
- Vapors of arsenic and sodium oxide are produced from a specific household detergent reacting with active metals, strong oxidizers, or strong mineral acids.

The chemical and physical compatibility of the decontamination solutions or other decontamination materials must be determined before they are used. Any decontamination method that permeates, degrades, damages, or otherwise impairs the safe functioning of the PPE is incompatible and should not be used. If a decontamination method does pose a direct health hazard, measures must be taken to protect both decontamination personnel and workers.

10.6 Decontamination Facility Design

At a hazardous waste site, decontamination facilities should be located in the Contamination Reduction Zone or CRZ (the zone between contaminated and uncontaminated zones). In relation to other operations, it should be the first facility set up and the last taken down.

The level and types of decontamination procedures required depend on several site-specific factors including:

- the chemical, physical, and toxicological properties of the wastes;
- the pathogenicity of infectious wastes;
- the potential for, and location of, exposure based on assigned worker duties, activities, and functions;
- the potential for wastes to permeate, degrade, or penetrate materials used for personal protective clothing and equipment, vehicles, tools, buildings, and structures;
- the proximity of incompatible wastes;
- the movement of personnel and/or equipment between different zones;
- emergencies;
- the methods available for protecting workers during decontamination; and
- the impact of the decontamination compounds and processes on workers' safety and health.

Decontamination procedures must provide an organized process by which levels of contamination are reduced. These processes should consist of a series of procedures performed in a specific sequence. For example, outer, more heavily contaminated items, particularly outer boots and gloves, are decontaminated and removed first, followed by decontamination and removal of inner, less contaminated items, such as inner gloves. Since decontamination mobilizes contaminates, respiratory protection equipment is among the last item to be removed. Each procedure should be performed at a separate station in order to prevent cross-contamination. The sequence of such stations is called a decontamination line.

Stations should be physically separated to prevent cross-contamination and should be arranged in order of decreasing contamination, preferably in a straight line (Figures 10-1 and 10-2). Flow patterns and stations should isolate workers from contamination zones that may contain incompatible wastes. Entry and exit points to and from the contaminated areas into contamination reduction areas and to clean areas should be separate. They should be clearly marked, and any dressing stations (changing clothing to protective or to nonwork gear) should also be separated. This helps isolate wastes as well as prevent cross-contamination. Personnel who wish to enter clean areas of the decontamination facility, such as locker rooms, should be completely decontaminated.

Additionally, certain precautions must be taken to ensure the safety of workers undergoing decontamination, as well as to prevent further contamination. These include having adequate personnel to help in the line, providing hand-holds where boots are washed and removed; being aware of slippery surfaces such as plastic sheeting, and providing nonwooden stools for personnel to sit when suits and boots are being removed. All shower and change rooms which are outside a contaminated area must meet the requirements of OSHA under 29 CFR 1910.141(d)(e). Finally, unauthorized individuals must not remove protective clothing or equipment from these change rooms.

10.7 Decontamination Equipment Selection

Table 10-2 lists recommended equipment for decontamination of personnel, personal protective clothing, and equipment. In selecting decontamination equipment, consider whether the equipment can be easily disposed of or can be decontaminated for reuse. Table 10-3 lists recommended equipment for the decontamination of large equipment and vehicles. Other types of equipment not listed in Tables 10-2 and 10-3 may be appropriate in specific situations.

10.8 Disposal Methods

All equipment used for decontamination must be decontaminated or disposed of properly. Buckets, brushes, clothing, tools, and other contaminated equipment should be collected, placed in containers,

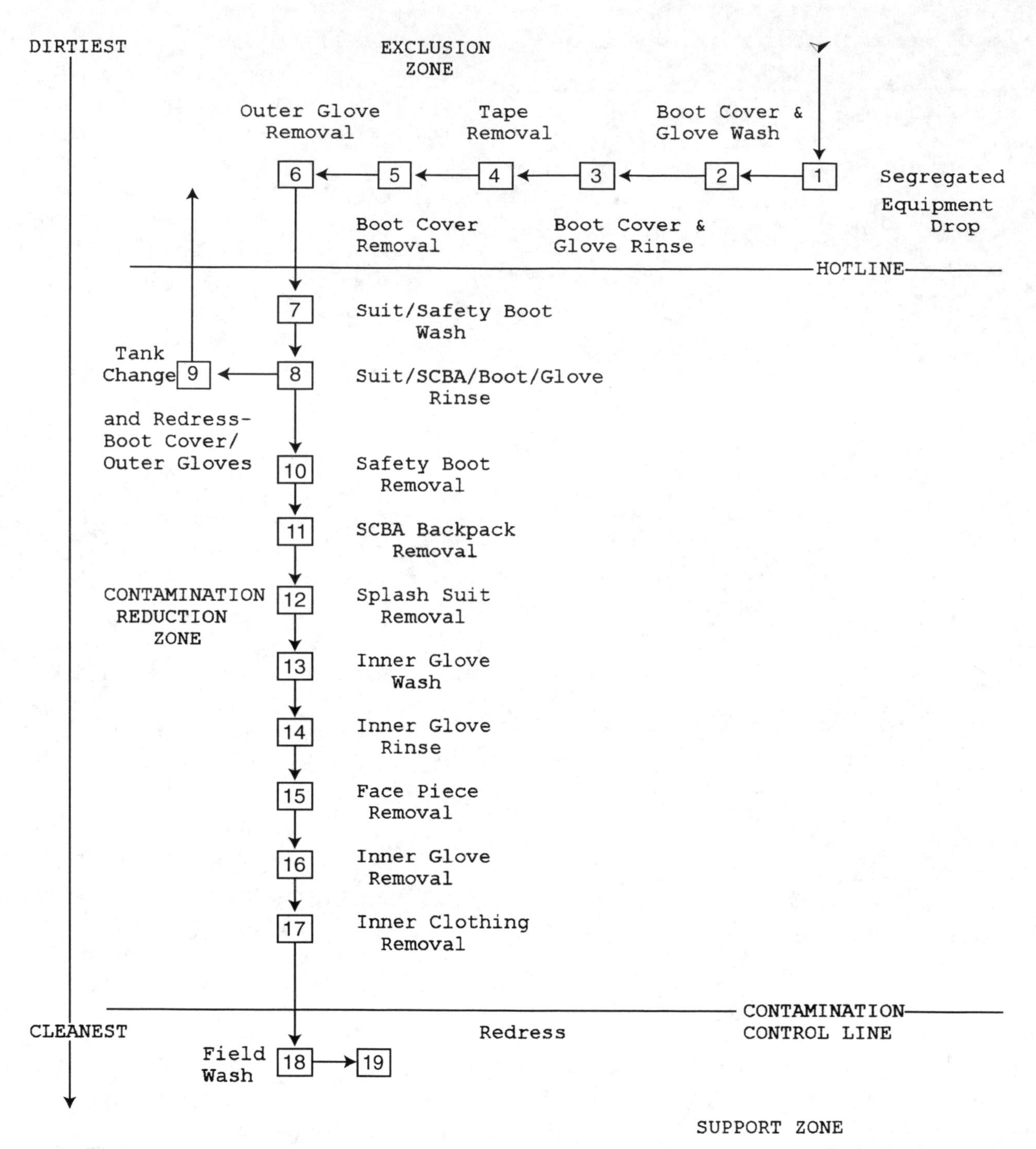

Figure 10-1 Suggested maximum decontamination layout for Level B protection. Adopted from *Occupational Safety and Health Guidelines Manual for Hazardous Waste Activites,* OSHA, 1985.

and labeled. All used solutions and wash water must be collected and disposed of properly according to local, state, and federal regulations. Clothing that is not completely decontaminated should be placed in plastic bags pending further decontamination or disposal.

10.9 Personal Protection

Decontamination workers who initially come in contact with personnel and equipment leaving the Exclusion Zone will require more protection from contaminants than decontamination workers who are assigned

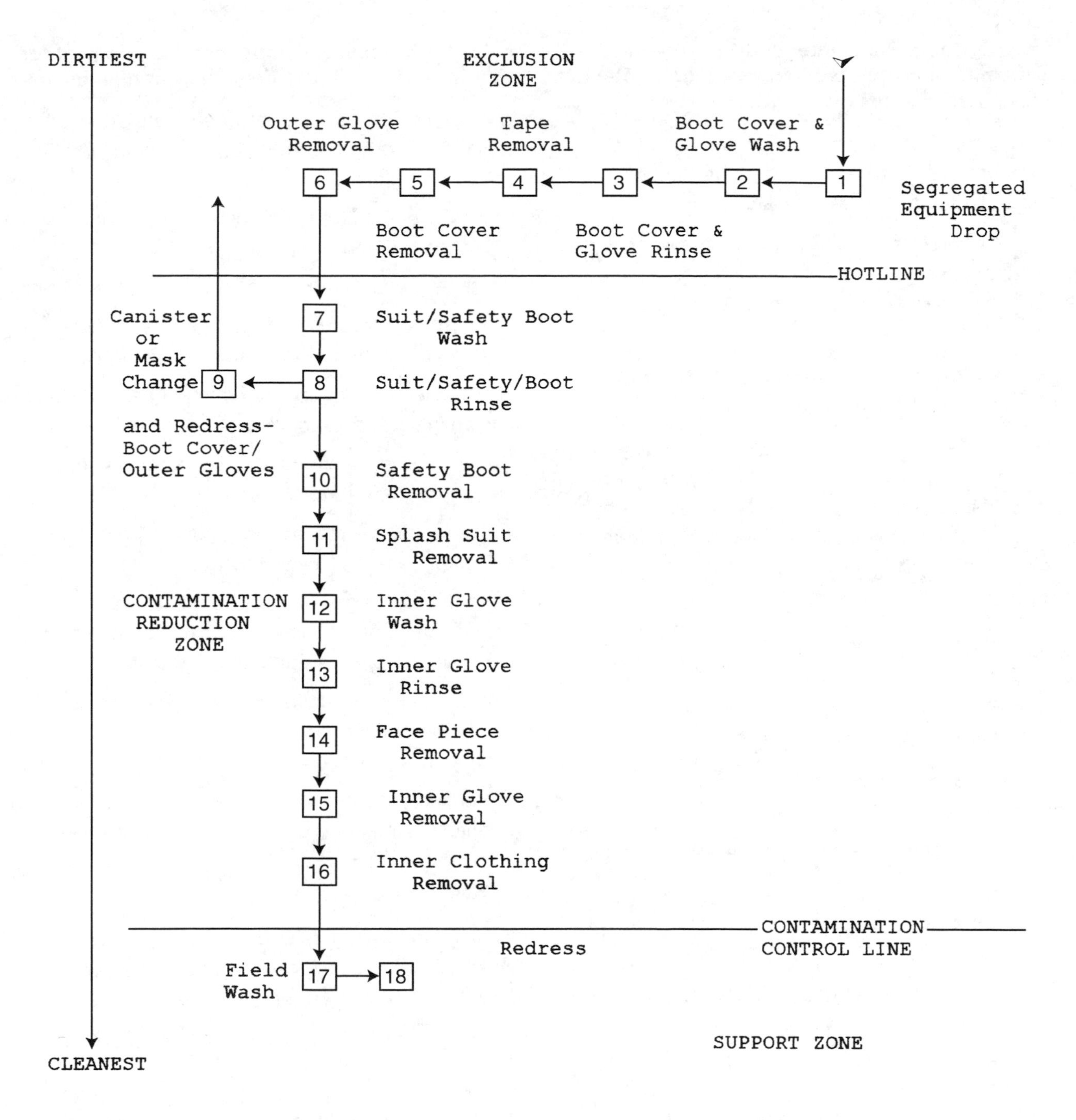

Figure 10-2 Suggested maximum decontamination layout for Level C protection. Adopted from *Occupational Safety and Health Guidelines Manual for Hazardous Waste Activites,* OSHA, 1985.

to the last station in the decontamination line. In some cases, decontamination personnel should wear the same levels of protection as workers in the Exclusion Zone. In other cases, decontamination personnel may be sufficiently protected by wearing one level lower protection, for example, wearing Level C protection while decontaminating workers wearing Level B.

The level of protection required will vary with the type of decontamination equipment used. For example, workers using a steam jet may need a different type of respiratory protection than other decontamination personnel because of the high moisture levels produced and the potential volatilization of contaminants. In some situations, the cleaning solutions used and

Table 10-2 **Some Recommended Equipment for Decontamination of Personnel, Personal Protective Clothing, and Equipment**

Dropcloths of plastic or other suitable materials on which heavily contaminated equipment and other protective clothing may be deposited;
Collection containers, such as drums or suitably lined trash cans, for storing disposable clothing, heavily contaminated personal protective clothing or equipment that must be discarded;
Lined box with absorbents for wiping or rinsing off gross contaminants and liquid contaminants;
Large galvanized tubs, stock tanks, or children's wading pools to hold wash and rinse solutions; (these should be at least large enough for a worker to place a booted foot in, and should have either no drain or a drain connected to a collection tank or appropriate treatment system;
Wash solutions selected to remove and reduce the hazards associated with the contaminants;
Rinse solutions selected to remove contaminants and contaminated wash solutions;
Long-handled, soft-bristled brushes to help wash and rinse off contaminants;
Paper or cloth towels for drying protective clothing and equipment;
Metal or plastic cans or drums for contaminated wash and rinse solutions;
Plastic sheeting, sealed pads with drains, or other appropriate methods for containing and collecting contaminated wash and rinse solutions spilled during decontamination;
Lockers or closets for clean clothing and personal linen storage;
Shower facilities for full body wash or, at a minimum, personal wash sinks (with drains connected to a collection tank or appropriate treatment system); and/or
Soap or wash solution, wash cloths, and towels for personal use.

Adopted from: *Occupational Safety and Health Guidelines Manual for Hazardous Waste Activities*, OSHA, 1985.

Table 10-3 **Some Recommended Equipment for Heavy Equipment and Vehicle Decontamination**

Storage tanks or appropriate treatment systems for temporary storage or treatment of contaminated wash and rinse solutions;
Drains or pumps for collection of contaminated wash and rinse solutions;
Long-handled brushes for general exterior cleaning;
Wash solutions selected to remove and reduce the hazards associated with site contamination;
Rinse solutions selected to remove contaminants and contaminated wash solutions;
Pressurized sprayers for washing and rinsing particularly hard-to-reach areas;
Curtains, enclosures, or spray booths to contain splashes from pressurized sprays;
Long-handled brushes, rods, and shovels for dislodging contaminants and contaminated soil caught in tires and the undersides of vehicles and equipment;
Containers to hold contaminants and contaminated soil removed from tires and the undersides of vehicles and equipment;
Wash and rinse buckets for use in the decontamination of operator areas inside vehicles and equipment;
Brooms and brushes for cleaning operator areas inside vehicles and equipment; and/or
Containers for storage and disposal of contaminated wash and rinse solutions, damaged, or heavily contaminated parts and equipment to be discarded.

Adopted from: *Occupational Safety and Health Guidelines Manual for Hazardous Waste Activities,* OSHA, 1985.

wastes removed during decontamination may generate harmful vapors. Appropriate equipment and clothing used to protect personnel in contact with these vapors should be selected by a qualified expert such as a certified industrial hygienist or safety professional.

All decontamination workers are considered to be in a contaminated area and must be decontaminated before entering the clean Support Zone. The extent of decontamination should be determined by the types of contaminants they may have contacted and the type of work they have performed.

10.10 Sources of Information

Currance, Phillip L.: Staging Decon Operations. *Rescue* (July/August 1989).

Hildebrand, Michael S.: The Nuts and Bolts of Hazardous Materials Emergencies, Part XI: Standard Operating Procedures for Decontamination. In *HAZMAT Response Team Leak and Spill Guide*. Stillwater, OK: Fire Protection Publicists/Oklahoma State University, 1984.

U.S. Department of Health and Human Services, National Institute for Occupational Safety and Health, Occupational Safety and Health Administration, U.S. Coast Guard, and Environmental Protection Agency: *Occupational Safety and Health Guidance Manual for Hazardous Waste Site Activities*. Washington, DC: U.S. Government Printing Office, 1985.

ELEVEN

Planning and Organization

11 | CONTENTS

Adequate planning is a critical element of hazardous waste site activities. By anticipating and taking steps to prevent potential hazards to health and safety, work at a waste site can proceed with a minimum risk to workers and to the public.

Three aspects of planning are discussed in this unit: (1) developing an overall organizational structure for site operations; (2) establishing a comprehensive Work Plan that considers each specific phase of the operation; and (3) developing and implementing a Site Health and Safety Plan (HASP).

The organizational structure should identify all personnel needed for the overall operation, establish the chain-of-command, and specify the responsibilities of each employee. The Work Plan should establish the objectives of site operations and should detail the logistics and resources required to achieve those objectives. The site HASP should address the concerns for each phase of the operation and define the requirements and procedures for worker and public protection.

A fourth important aspect of planning, though not discussed in this unit, is coordination with the established response community. The National Response Center (NRC), established by a congressionally-mandated NCP, implements procedures for coordinating responses to releases of hazardous substances into the environment. Local community officials, units, or bureaus oversee such plans. Finally, an important contact for hazardous waste site activities is the EPA-designated official responsible for coordinating federal activities related to cleanup at hazardous waste sites.

All plans must be site-specific, and planning must be an ongoing process. Cleanup activities and the site HASP must continuously adapt to site conditions and new information. Changes may occur daily, seasonally, and from one site sector to another. The site HASP is a dynamic document that must be designed to adjust quickly to changing conditions. In general, planning is easy for physical hazards since they can usually be seen. Chemical hazards are more nebulous, so planning for their occurrence may require a more detailed site analysis. This unit is intended to serve as a starting point for planning the response activities at hazardous waste sites.

11.1 Organizational Structure

An organizational structure that supports the overall objectives of the project should be developed in the first stage of planning. This structure should:

- identify a leader who has the authority to direct all activities;
- identify a Site Safety and Health Officer (SSHO);
- identify other personnel needed for the project, and assign their general functions and responsibilities; and
- identify the interface with the response community.

As the project progresses, it may be necessary to modify some organizational aspects, such as personnel responsibilities and authority, so that individual tasks can be performed as efficiently and safely as possible. For example, specialists may be called upon for specific tasks, either off-site or on-site, and these specialists should be shown in the organizational structure. Any changes to the overall organizational structure must be recorded in the appropriate parts of the Work Plan or HASP developed for individual phases or tasks and must be communicated to all parties involved.

Figure 11-1 presents one example of the organizational framework for a hazardous waste site response team. It shows the lines of authority for 12 categories of off-site and on-site personnel. The responsibilities and functions of each category are described in Tables 11-1 and 11-2. On-site categories are divided into essential personnel who are necessary for a safe and efficient response, and optional personnel who may be desirable in a large operation where responsibilities can be delegated to a greater number of people.

This example is intended to illustrate the scope of responsibilities and functions that may be required of site personnel. The personnel categories can be used as a starting point for designing an organizational structure appropriate to a particular situation. For smaller investigative and response efforts, single individuals often perform several functions.

All response teams should include an SSHO who is responsible for implementing health and safety requirements. The SSHO should have ready access to other occupational health and safety professionals, particularly an industrial hygienist. Once an organizational system has been developed, all individuals responsible for establishing and enforcing health and safety requirements should be identified and their respective authorities clearly explained to all members of the response team.

One of the most critical elements in worker safety is the attitude of all levels of project management. A strong, visible, and viable commitment to worker safety must be present from the beginning of a project. Such an attitude sets the tone for the entire operation. The SSHO and the Project Team Leaders must have the clear support of senior-level management for establishing, implementing, and enforcing safety programs from the outset of the project. The importance of management's attitude toward safety throughout the project cannot be overemphasized. Site personnel are more likely to participate in the safety program if they sense a genuine concern on the part of management.

Several indicators of successful worker safety programs include:

- strong management commitment to safety, as defined by actions reflecting management's support and involvement in safety activities;
- close contact and interaction among workers, supervisors, and management, enabling open communication on safety as well as other job-related matters;
- a high level of good housekeeping, orderly workplace conditions, and effective environmental quality control (QC);
- well-developed worker selection, job placement, and advancement procedures, plus other employee support services;
- training practices emphasizing early indoctrination, site specific and task-specific training, and follow-up instruction in job safety procedures;
- features added to conventional safety practices that enhance the effectiveness of those practices; and
- an effective disciplinary plan that discourages employees from using unsafe work practices.

Overall, the most effective industrial safety programs are successful in dealing with "people" variables. Open communication among workers, supervisors, and management concerning work site safety is essential.

The effective management of response actions at hazardous waste sites requires a commitment to the health and safety of the general public as well as to the on-site personnel. Prevention and containment of contaminant release into the surrounding community should be addressed in the planning stages of a project. Not only must the public be protected, they must also be made aware of the HASP and must have confidence in those who implement it. To accomplish these goals, the Project Team Leader or Public Information Officer under the supervision of the Project Team Leader

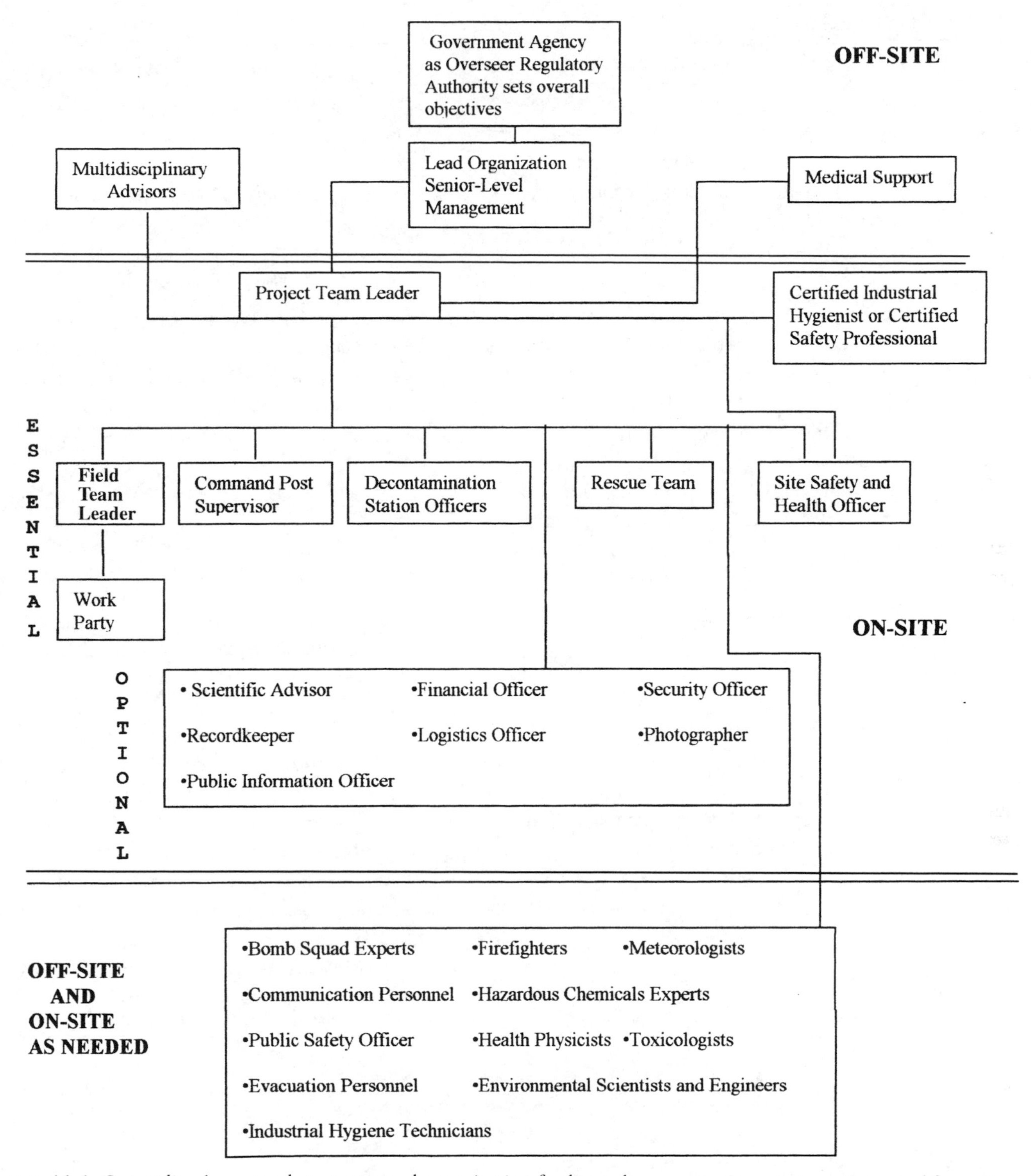

Figure 11-1 Generalized approach to personnel organization for hazardous waste site activities. Extracted from *Occupational Safety and Health Guidelines Manual for Hazardous Waste Activites,* OSHA, update, 1998.

Table 11-1 **Off-Site Personnel**

Title	General Description	Specific Responsibilities
Senior-Level Management	Responsible for defining project objectives, allocating resources, determining the chain-of-command, and evaluating program outcome (Generally the contracted company sets policies.)	• Provide the necessary facilities, equipment, funding, personnel and time resources to conduct activities safely; • Support the efforts of on-site management; and • Provide appropriate disciplinary action when unsafe acts or practices occur;
Multi-Disciplinary Advisors	Includes representatives from upper-level and on-site management, a field team member, and experts in fields such as: Chemistry, Engineering, Industrial hygiene, Information/public relations, Law, Medicine, Pharmacology, Physiology, Radiation health physics, Toxicology	• Provide advice on the design of the Work Plan and the HASP (The number of advisors and the work performed depend on the type of site involved.)
Medical Support	Consulting and examining physicians	• Become familiar with the types of materials on-site, the potential for worker exposures, and recommend the medical program for the site; • Conduct initial, annual, and termination medical exams as appropriate for the work tasks and potential exposures;
	Medical personnel at local hospitals and ambulance personnel	• Provide emergency treatment and decontamination procedures for the specific types of exposures that may occur at the site; • Obtain special drugs, equipment, or supplies necessary to treat such exposures; and • Provide emergency treatment procedures appropriate to the hazards on-site.

Extracted with modifications, from *Occupational Safety and Health Guidelines Manual for Hazardous Waste Activities,* OSHA, updated 1998.

Table 11-2 **On-Site Personnel**

Title	General Description	Specific Responsibilities
Project Team Leader	Reports to upper-level management; has authority to direct response operations; assumes total control over site activities	• Ensures preparation of the Work Plan, the Site Health and Safety Plan and adequate resources and personnel for the Field team; • Obtains permission for site access and coordinates activities with appropriate officials; • Ensures that the Work Plan is completed and on schedule; • Ensures that the HASP is completed and on schedule; • Briefs the field teams on their specific assignments; • Directs the Site Safety and Health Officer to ensure that safety and health requirements are met; • Prepares the final report and support files on the response activities; • Serves as the liaison with public officials; • Gives briefings, daily or otherwise as needed;
Site Safety and Health Officer or Site Safety Officer (SSO)	Advises the Project Team Leader on all aspects of health and safety on-site; recommends stopping work if any operation threatens worker or public health or safety	• Selects protective clothing and equipment in accordance with the HASP; • Periodically inspects protective clothing and equipment; • Ensures that protective clothing and equipment are stored and maintained; • Controls entry and exit at the Access Control Points; • Coordinates safety and health program activities with the Scientific Advisor, and CIH or CSP; • Confirms each team member's suitability for work based on a physician's recommendation; • Monitors the work parties for signs of stress, such as cold exposure, heat stress, and fatigue; • Monitors on-site hazards and conditions; • Participates in the preparation of and implements the Site Health and Safety Plan; • Conducts periodic inspections of sitework tasks to determine if the HASP is being followed or if it needs updating; • Enforces the "buddy" system; • Knows emergency procedures, evacuation routes, and the telephone numbers of the ambulance, local hospital, poison control center, fire department, and police department; • Notifies, when necessary, local public emergency officials; • Coordinates emergency medical care; and • Ensures maintenance of required training and medical records for each individual on-site.

(Continued)

Table 11-2 *Continued*

Title	General Description	Specific Responsibilities
Field Team Leader	May be the same person as the Project Team Leader and may be a member of the work party. Responsible for field team operations and safety. (More than one field team leader may be necessary depending on the number of teams involved.)	• Manages field operations; • Executes the Work Plan and schedule; • Enforces safety procedures; • Coordinates with the Site Safety Officer in determining protection level; • Enforces site control; • Documents field activities and sample collection; and • Serves as a liaison with public officials.
Command Post Supervisor	May be the same person as the Field Team Leader. Responsible for communications and emergency assistance. (Generally separate individual at large site.)	• Notifies emergency response personnel by telephone or radio in the event of an emergency; • Assists the Site Safety Officer in a rescue, if necessary; • Maintains a log of communication and site activities; • Assists other field team members in the clean areas, as needed; and • Maintains communication contact with the work parties via walkie-talkies, signal horns, or other means.
Decontamination Station Officer(s)	Responsible for decontamination procedures, equipment, and supplies. (May be the same as the SSO.)	• Sets up decontamination lines and the decontamination solutions appropriate for the type of chemical contamination on-site; • Controls the decontamination of all equipment, personnel, and samples from the contaminated areas; • Assists in the disposal of contaminated clothing and materials; • Ensures that all required equipment is available; and • Advises medical personnel of potential exposures and consequences.
Rescue Team	Used primarily on large sites with multiple work parties in the contaminated area.	• Stands by, partially dressed in protective gear, near hazardous work areas; and • Rescues any worker whose health or safety is in danger.
Work Party	Depending on the size of the field team, any or all of the field team may be in the Work Party, but the Work Party should consist of at least two people (buddy system).	• Safely completes the on-site tasks required to fulfill the Work Plan; • Complies with Site Health and Safety Plan; and • Notifies Site Safety Officer or supervisor of unsafe conditions.

Extracted, with modifications, from *Occupational Safety and Health Guidelines Manual for Hazardous Waste Activities,* OSHA, updated 1998.

should establish community liaison well before any response action is begun.

11.2 Work Plan

To ensure a safe response, a Work Plan describing anticipated cleanup activities must be developed before beginning on-site response actions. The Work Plan should be periodically reexamined and updated as new information about site conditions is obtained.

The following steps should be taken in formulating a comprehensive Work Plan:

- review available information, including:
  - —site records
  - —waste inventories
  - —generator and transporter manifests
  - —previous sampling and monitoring data
  - —site photos
  - —site owner records
  - —state and local environmental and health agency records
  - —possible interviews with former employees and/or people living near the site;
- define work objectives;
- determine methods for accomplishing the objectives, such as a sampling plan, an inventory, or disposal techniques;
- determine personnel requirements;
- determine the need for additional training of personnel by evaluating their current knowledge/skill level against the tasks they will perform and situations they may encounter;
- determine equipment requirements by evaluating the need for special equipment or services, such as drilling equipment or heavy equipment and operators; and
- determine the data quality objectives (DQOs) and quality assurance (QA) procedures.

DQOs are qualitative and quantitative statements established prior to data collection that specify the quality of the data required to support regulatory decisions for site decisions and remedial response activities. The DQOs for a particular site vary according to the end use of the data. For example, DQOs in support of a preliminary assessment will differ from those used to support a remedial design. The DQOs are used to develop a conceptual model of the site that describes suspected sources, contaminant pathways, and potential receptors. They also determine sampling approaches and the analytical options to allow more timely or cost-effective data collection and evaluation. The DQO methods used to obtain data of acceptable quality are specified in such documents as the SAP (Sampling and Analysis Plan) or the Work Plan.

QA procedures are written instructions that address such activities as:

- chain of custody requirements;
- sample preservation and transportation;
- data and records management;
- equipment use, maintenance, and calibration;
- field documentation;
- waste management;
- internal and field QC;
- data reduction, validation, and reporting;
- audits and assessments;
- corrective action procedures; and
- QA/QC reporting procedures.

Preparation of the Work Plan requires a multidisciplinary approach and, therefore, requires input from all levels of on-site and off-site management. Consultants may also be useful in developing sections of the Work Plan. For example, chemists, laboratory technicians, geologists, occupational health and safety professionals, and statisticians may be needed to develop the sampling plan.

11.3 Site Health and Safety Plan

A site HASP establishes policies and procedures to protect workers and the public from the potential hazards posed by a specific hazardous waste site. Therefore, this plan must be developed before site activities proceed. The HASP must anticipate potential accident scenarios and provide measures to minimize accidents or injuries that may occur during normal daily activities or during adverse conditions such as hot or cold weather. It must address each and every work task to be performed at a hazardous waste site. As such, it is much more detailed than the Work Plan.

The following section describes a planning process that might be developed for normal site operations. It does not include emergency situations. Development of a written site HASP helps ensure that all safety aspects of site operations are thoroughly examined prior to commencing field work, and it should be modified as needed for all stages of site activity.

Planning requires information, so planning and site characterization (refer to Unit 12) should be coordinated. An initial HASP should be developed so the preliminary site assessment can proceed in a safe manner. The information from this assessment can then be used to refine and expand the site HASP.

Development of a site HASP should involve both off-site and on-site management, and should be reviewed by occupational and industrial health and safety experts or other appropriate personnel.

Briefly, the plan should:

- name key personnel responsible for site safety;
- describe the risks associated with each operation in the Work Plan;
- specify training required for personnel to perform their jobs and to handle specific hazardous situations that may be encountered;
- describe various protective clothing and equipment necessary for various operations;
- describe site-specific medical surveillance and services that may be necessary, including the location of and a map to the nearest medical service facility;
- describe programs for air monitoring, personal monitoring, and environmental sampling, if needed;
- describe how the work environment might be made less hazardous by correcting or eliminating existing hazards;
- define site control measures;
- establish decontamination procedures for personnel and equipment;
- establish a Contingency Plan for safe and effective response to emergencies, including the names and phone numbers of emergency personnel;
- establish a confined space entry plan;
- establish a spill containment plan; and
- list Standard Operating Procedures (SOPs) for those activities that can be standardized, such as decontamination and respirator fit testing. Care must be taken in defining SOPs so they cover appropriate topics and are thoroughly understood by all personnel.

This information is generic and should be used as a guide, not as the standard, for designing a site HASP.

The establishment of an HASP is not enough. The plan must be read, understood, and used by everyone on the site. The major limitation of such a plan is its non-use. As site workers become complacent toward the hazards with which they work every day, they may tend to follow the safety plan less carefully, putting themselves at risk.

11.4 Safety Meetings and Inspections

To ensure that the site HASP is being followed, the SSHO or designee should conduct a safety meeting prior to initiating any site activity and before beginning each work shift. Sometimes the SSHO may delegate these duties to help emphasize the importance of involving everyone in taking responsibility for safety. The purpose of these safety meetings is to:

- describe the assigned tasks and their potential hazards;
- coordinate activities;
- identify methods and precautions to prevent injuries;
- plan for emergencies;
- describe any changes in the site HASP; and
- evaluate worker feedback on conditions affecting safety and health to determine the effectiveness of the HASP.

The SSHO should also conduct frequent inspections of the site, including its conditions, facilities, equipment, and activities, to determine whether the HASP is adequate and being followed.

The importance of a clear, user-friendly safety plan

cannot be overstated. At a hazardous waste site, risks to workers can change rapidly and dramatically when there are changes in:

- work and other site activities (such as uncovering buried drums);
- state of degradation of containers and containment structures (bottoms may be rusted and unstable);
- state of equipment maintenance (general oversight, incomplete decontamination); and
- weather conditions (heat/cold stress, wind, rain, lightning, and ice).

In order to make safety inspections effective, the following guidelines should be observed:

- develop an inspection checklist for each site, listing the items that should be inspected;
- review the results of these inspections with supervisors and workers;
- reinspect any identified problems to ensure they have been corrected; and
- document all inspections and subsequent follow-up actions. Retain these records until site activities are completed and/or as long as required by a regulatory agency, project management, the Work Plan, or quality procedures.

The frequency of inspections should be described in the HASP, and will vary according to the characteristics of the site and the equipment used on-site. Factors that need to be considered are:

- severity of risk;
- regulatory requirements;
- operation and maintenance requirements;
- expected effective lifetime of clothing, equipment, vehicles, and other items; and
- recommendations based on professional judgment, laboratory test results, and field experience.

11.5 Sources of Information

Briggum, Sue M., et al: *Hazardous Waste Regulation Handbook–A Practical Guide to RCRA and Superfund*. Revised ed. New York: Executive Enterprises Publications Co., Inc., 1985.

Dennison, Mark S.: *Hazardous Waste Regulation Handbook: A Practical Guide To RCRA & Superfund*. New York: John Wiley & Sons, 1994.

Einolf, David M.: Complying with HAZWOPER & Understanding Your Training and Planning Requirements. *ECON* (January 1993).

Roughton, James E.: A Guide to Safety and Health Plan Development for Hazardous Waste Operations. *Hazmat World* (January 1993).

Smith, R. Blake: Hazmat Handling Field Faces Risks Above and Beyond Chemical Exposure. *Occupational Health and Safety* (November 1992).

Smith, R. Blake: "Real Life" HAZWOPER Exercises Help Students Master Skills, Teamwork. *Occupational Health and Safety* (February 1993).

Solinger, Tim (ed.): *Hazardous Waste Regulatory Guide*, Neenah, WI: J.J. Keller and Associated, Inc. Staff, 1998.

Sullivan, Thomas S. (ed.): *Environmental Law Handbook*. 14th ed. Rockville, MD: Government Institutes, Inc., 1997.

U.S. Department of Health and Human Services: *NIOSH/OSHA/USCG/EPA Occupational Safety and Health Guidance Manual for Hazardous Waste Site Activities*. Washington, DC: U.S. Government Printing Office, 1998.

U.S. Environmental Protection Agency: *Risk Assessment Guidance for Superfund Volume I, Human Health Evaluation Manual (Part A)*. Washington, DC: Office of Emergency and Remedial Response, 1989.

Wheeler, John R.: *Toxic Substance Control Act Compliance Guide and Service, 3 Vol.* Plano, TX: Environmental Books, 1997.

Worobec, Mary: *Toxic Substances Control Guide*. 2nd ed. Washington, DC: The Bureau of National Affairs, Inc., 1992.

TWELVE

Site Characterization

12 | CONTENTS

Site characterization provides information needed to identify site hazards, select worker protection methods, and plan the details of site activities, especially sampling and remediation. The more accurate, detailed, and comprehensive the information is about a site, the more precisely protective measures can be tailored to the actual hazards workers may encounter.

The person with primary responsibility for site characterization and assessment is the Project Team Leader. In addition, other as-needed personnel and outside experts, such as chemists, health physicists, industrial hygienists, and toxicologists, may be required to accurately and fully interpret the available information on site conditions.

Site characterization generally proceeds in three phases:

- prior to site entry, conduct off-site characterization through interviews, records research, and perimeter reconnaissance;
- next, conduct on-site surveys, restricting site entry to reconnaissance personnel; then
- once the site has been determined safe for commencement of other activities, perform ongoing monitoring to provide a continuous source of information about site conditions.

It is important to recognize that site characterization is a continuous process. Each phase should obtain and evaluate information to define the hazards the site may pose. This assessment can then be used to develop a HASP for the next phase of work. In addition, all site personnel should constantly be aware of new information concerning site conditions.

The following sections detail the three phases of site characterization and provide a general guide which should be adapted to meet specific conditions and contingencies. Within each phase, the most appropriate procedures should be determined, particularly if time or budget considerations limit the scope of the work. Wherever possible, all information sources should be pursued.

As much information as possible should be obtained before site entry so that hazards can be evaluated and preliminary controls instituted to protect initial entry personnel.

12.1 Off-Site Characterization

Off-site information can be obtained through interviews/records research and perimeter reconnaissance.

12.1.1 INTERVIEWS/RECORDS RESEARCH

As much data as possible should be collected before anyone enters a site. Where possible, the following information should be obtained:

- exact location of the site;
- history of the site, using maps, aerial photographs, archive records, historical insurance maps, etc.;
- detailed description of the activities that occurred at the site;
- duration of these activities;
- meteorological data, such as prevailing wind direction, precipitation levels, temperature profiles;
- site terrain, using U.S. Geological Survey topographic quadrangle maps, land use maps, and land cover maps;
- geologic and hydrologic data;
- habitation (or population at risk);
- accessibility, including roadways;
- regulatory compliance history;
- hazardous substances records; and
- possible pathways of contaminant dispersion (the air, soil, ground water, and surface water should be evaluated using a variety of records that often include topographic maps, prevalent wind charts, ground water surveys, housing drains, culverts, pipes, and buried utilities).

Information sources useful for evaluating the possible presence of hazardous wastes may include:

- company records, particularly those that relate to chemical purchase and disposal;
- records from federal, state, and local environmental regulatory and enforcement agencies, state occupational safety and health agencies, and state and local emergency response organizations;
- interviews with site employees (if any), site retirees, previous owners, and nearby residents that may have information on the use and disposal of hazardous substances (verify all information received from interviews);
- water department and sewage district records;
- blueprints;
- environmental permits and compliance issues; and
- media reports (verify all information received from the media).

This list is a clear indication that no single source of data will provide all necessary information; therefore, numerous potential information sources must be examined for pertinent data. Commercial companies that specialize in environmental record searches are often employed to obtain most of the records mentioned above. This data collection process should follow the assessment procedures outlined in the most recent version of standard E 1527 from the American Society for Testing and Materials (ASTM).

12.1.2 PERIMETER RECONNAISSANCE

At a site where hazards are largely unknown or where there is no need to go on-site immediately, a perimeter reconnaissance is recommended. The purpose of this reconnaissance is to examine the site without actual entry. Visual observations, air monitoring, and perhaps limited sampling are part of the reconnaissance process. (Never assume the perimeter is safe since the adjacent property may be more polluted than the site.) Perimeter reconnaissance of a site should involve the following actions:

- develop a preliminary site map with the locations of buildings, containers, impoundments, pits, ponds, wells, sewage and drainage pipes, ditches, and tanks;
- make visual observations (that may be aided by use of binoculars), including:
  - —the appearance of unnatural depressions, quarries, pits, lagoons, etc.
  - —variation in revegetation or disturbed areas
  - —mounding or uplift in disturbed areas or paved surfaces
  - —modifications in grade
  - —changes in traffic patterns at the site
  - —noticeable physical hazards
- record any labels, markings, or placards on containers or vehicles;
- check the amount of deterioration or damage to containers or vehicles;
- note biologic indicators such as dead animals or plants;
- look for unusual conditions, such as vapor clouds, discolored liquids, oil slicks, or other evidence of chemicals being actively released;

- monitor the ambient air at the site perimeter for:
  —toxic substances
  —combustible and flammable gases
  —ionizing radiation
  —specific materials, if known
- using appropriate personal protection methods, monitor soils around the perimeter as well as drainage points coming from the site; and
- note any unusual odors.

As the level of housekeeping decreases on a property, the probability of chemical contamination increases. At this point, it is possible to prepare a preliminary site HASP, allowing for unobserved or unreported contingencies. With this plan as a guide, it is now permissible to enter the work site. Changes to the HASP may take place as soon as individuals enter the actual site and begin to assess the situation.

12.2 On-Site Survey

The purpose of an on-site survey is to verify and supplement information from the off-site characterization. The off-site characterization should be used to develop a site HASP for initial site entry that addresses the work to be accomplished and prescribes procedures necessary to protect the health and safety of the entry team.

Figure 12-1 Air monitoring instruments and binoculars can be useful during a perimeter reconnaissance

12.2.1 PROTECTION OF ENTRY PERSONNEL

Information from interviews/records research and perimeter reconnaissance is used as the basis for selecting protective equipment for the initial site survey. The proposed work to be accomplished must also be considered. For example, if the purpose of the survey is to inspect on-site conditions, count containers, and measure the ambient air, the level of protection may be less stringent than if containers are to be opened and samples taken.

The ensemble of clothing and equipment referred to as Level B protection is generally the minimum level recommended for an initial entry if site hazards are unknown or poorly documented. A decontamination plan should be established and a station set up to prepare for egress and emergencies.

12.2.2 THE ENTRY TEAM(S)

The composition of the entry team depends on the site characteristics, but should consist of at least four persons—two workers who will enter the site and two outside support persons, who are also suited in PPE and prepared to enter the site in case of emergency. Upon initial entry of the site, entry personnel should:

- monitor the air for IDLH and other conditions that may cause death or serious harm (combustible or explosive atmospheres, oxygen deficiency, and toxic substances). Confined spaces are considered to be IDLH and are not to be entered without special precautions and documented training; and
- monitor for ionizing radiation if prior information indicates a suspected radiological hazard.

Table 12-1 **Visible Indicators of Potential IDLH and Other Dangerous Conditions**

- Physical hazards such as stacked drums, corroded or rotten building structures or tank racks, slip and trip hazards, excavations
- Large containers or tanks that must be entered
- Enclosed spaces such as boarded-up buildings or trenches that must be entered
- Potentially explosive or flammable situations (indicated by bulging drums, effervescence, gas generation, or instrument readings)
- Extremely hazardous materials noted from labels or records (such as cyanide, phosgene, or radiation sources)
- Visible vapor clouds

Figure 12-2 Storage containers come in all sizes and shapes

IDLH conditions are of particular concern during initial entry. Visible indicators of such conditions are summarized in Table 12-1.

A video camera may be useful in recording site conditions. The video camera should be covered with a plastic bag to aid in decontamination. If IDLH or other dangerous conditions are not noted, the survey should continue by performing the following actions:

- conduct further air monitoring as necessary;
- note the types of containers, impoundments, or other storage systems such as:
  —paper or wood packages

Figure 12-3 Deteriorating and bulging container represents a potential IDLH situation

  —metal or plastic barrels or drums
  —underground tanks (from visible vents, piping, etc.)
  —above-ground tanks
  —compressed gas cylinders
  —pits, ponds, or lagoons
- determine the condition of waste containers and storage systems in terms of:
  —intactness (undamaged)
  —visibly rusted or corroded
  —leaking
  —bulging
  —types and quantities of material in containers (quantity may not be possible to ascertain since moving containers to determine weight may be dangerous)
  —labels on containers indicating corrosive, explosive, flammable, radioactive, or toxic materials (remember that labels may not be indicative of what the container actually holds)
- observe the physical state of the material (if possible without actually handling or contacting the containers), making note of the following:
  —gas, liquid, or solid
  —color and turbidity
  —behavior (e.g., corroding, foaming, or vaporizing)
  —conditions conducive to splash or contact

- identify natural wind barriers, such as buildings, hills, tanks, etc., which are potential areas for wind-blown contaminants to lodge;
- determine the potential pathways of dispersion, such as:
  - —air
  - —biologic routes, such as animals and food chains
  - —groundwater
  - —drainage ditches, culverts, creeks, or rivers
  - —land surface
  - —surface water
  - —soil
- note any indicators of potential exposure to hazardous substances, such as:
  - —dead fish, animals, or vegetation
  - —dust or spray in the air
  - —fissures or cracks in solid surfaces that expose deep waste layers
  - —pools of liquid
  - —foams or oils on liquid surfaces
  - —gas generation or effervescence
  - —deteriorating containers
  - —cleared land areas or possible landfilled areas
- look for any physical hazards, including:
  - —conditions of site structures
  - —obstacles to entry and exit
  - —terrain homogeneity
  - —terrain stability
  - —stability of stacked material
- identify any reactive, incompatible, flammable, or highly corrosive wastes from labels or signs on the containers without moving them;
- record land features;
- look for the presence of any potential naturally occurring skin irritants or dermatitis-inducing agents, such as poison ivy, poison oak, stinging nettle or poison sumac, and other biological hazards such as bee nests or rodent excreta;
- record any tags, labels, markings, or other identifying indicators; and
- assess the need for the collection of samples to augment the HASP.

Figure 12-4 Stacked drums can represent both a physical and chemical hazard

12.3 Information Documentation

Proper documentation and document control are important for ensuring accurate communication and the quality of the data collected, providing the rationale for safety decisions, and substantiating possible legal actions. Documentation can be accomplished by recording information pertinent to field activities, sample analysis, and site conditions in several ways, including:

- logbooks;
- field data records;
- graphs;
- photographs/video taping;
- sample labels;
- chain-of-custody forms;
- analytical records; and
- QA performance records

These documents should be controlled to ensure that they are all accounted for when the project is completed. The task of document control should be assigned to one individual on the project team whose responsibilities should include:

- numbering each document (including sample labels) with a unique number;
- listing each document in a document inventory;

Table 12-2 **Example of Field Logbook Entries Describing Sampling**

- Date and time of entry
- Purpose of sampling
- Name, address, and affiliation of personnel performing sampling
- Type of material being sampled (sludge, wastewater, air, etc.)
- Description of material container
- Physical description of sample
- Chemical components and concentrations, if known
- Number and size of samples taken
- Description and location of the sampling point
- Date and time of sample collection
- Difficulties experienced in obtaining sample (e.g., Is it representative of the bulk material?)
- Visual references such as maps or photographs of the sampling site
- Field observations such as weather conditions during sampling periods (for example, strong winds, rain, and temperatures)
- Field measurements of the materials (for example, explosiveness, flammability, or pH)
- Status of chain-of-custody forms for any samples

- recording the location of each document in a separate document register so that any document can be readily accessed. In particular, the name and location of site personnel having documents in their possession should be recorded;
- collecting all documents at the end of each work period and keeping them in a secured area until needed;
- making certain that all document entries are made in waterproof ink; and
- filing all documents in a central file at the completion of the site response.

Field personnel should record all on-site activities and observations in a field logbook (such as a bound book with consecutively numbered pages). Entries should be made during or just after completing a task to ensure thoroughness and accuracy. Table 12-2 shows the level of detail that should be recorded during sampling.

Photographs may be needed to supplement observations. Digital cameras are particularly useful in this respect. For each photograph, the following information should be recorded in the field logbook:

- date, time, name of site;
- name of photographer;
- general compass direction of the orientation of the photograph;
- general description of the subject;
- sequential number of the photograph and the film roll number; and
- camera, lens, and film type used for photography.

If using a video camera for additional documentation, a unit that automatically records the date, time, etc. is preferable. A unit that allows real-time narration is valuable, but only if a viable and safe method of voice recording can be used through a PPE.

12.4 Monitoring

Site activities and weather conditions change. Therefore, an ongoing air monitoring program should be implemented after characterization has initially determined that the site is safe for entry.

The ongoing monitoring of atmospheric chemical hazards should be conducted using a combination of stationary sampling equipment, personal monitoring devices, and periodic area monitoring with direct-reading instruments. Data obtained during off-site and on-site surveys can be used to develop a plan that details procedures for monitoring ambient conditions during cleanup operations. Where necessary, routes of exposure in addition to inhalation should be monitored. For example, skin swipe tests may be used to determine the effectiveness of personal protective clothing. Depending on the physical properties and toxicity of the on-site materials, community exposures resulting from hazardous waste site operations may need to be assessed.

Monitoring also includes continuous evaluation for any changes in site conditions or work activities that could affect worker safety. Some indicators of the need for reassessment are:

- commencement of a new work phase, such as the start of drum sampling;

- change in job tasks during a work phase;
- change in season;
- change in weather; and
- discovery of new hazards.

Whenever changes take place, the site HASP must be modified and all workers and the community must be informed, as appropriate.

12.5 Exposure Record Retention

Results of monitoring must be recorded in the field notebook or other appropriate documents. Monitoring records, including lists of MSDSs, become part of the exposure records for the workforce. OSHA regulations require exposure records to be maintained for thirty years under 29 CFR 1910.1020 (d)(ii)(A).

12.6 Sources of Information

American Society for Testing and Materials, *Standard E 1527*, ASTM, 100 Barr Harbor Drive, West Conshohocken, PA, 2001.

Guidance Manual for Hazardous Waste Site Activities. Washington, DC: U.S. Government Printing Office, 1985.

Koren, Herman: *Handbook of Environmental Health and Safety Principles and Practices*. 2nd ed. Vols. I and II. Chelsea, MI: Lewis Publishers, 1991.

Koren, Herman: *Illustrated Dictionary of Environmental Health and Occupational Safety*. Chelsea, MI: Lewis Publishers, 1995.

Koren, Herman and M.S. Bisesi: *Handbook of Environmental Health and Safety: Principles and Practice*. 3rd ed. Vols. I and II. Chelsea, MI: Lewis Publishers, 1995.

National Institute for Occupational Safety and Health, Occupational Safety and Health Administration, U.S. Coast Guard, and Environmental Protection Agency: *Occupational Safety and Health Guidance Manual for Hazardous Waste Site Activities*. Washington, DC: U.S. Government Printing Office, 1985.

U.S. Department of Labor: *OSHA, 29 CFR 1910.1020*. Washington, DC: U.S. Government Printing Office, 2001.

THIRTEEN

Site Control

13 | CONTENTS

Once the site has been characterized and a work plan established, procedures should also be set in place to manage materials, workers, and equipment on and off the locale. This process is known as site control.

The purpose of site control is to minimize potential contamination of workers, direct the orderly movement of materials and personnel in and out of the site, protect the public from the site's hazards, and prevent vandalism. In addition, site control must function during an emergency. This unit describes the basic components of a program aimed at controlling the activities and movements of people and equipment at a hazardous waste site.

Several site control procedures can be implemented to reduce worker and public exposure to chemical, physical, biological, and safety hazards. They include:

- compiling an accurate site map (required by OSHA);
- preparing the site for subsequent activities;
- establishing work zones;
- using the buddy system, when necessary;
- establishing and strictly enforcing decontamination procedures for both personnel and equipment;
- establishing site security measures;
- setting up communication networks;
- enforcing safe work practices (SOPs); and
- planning for emergencies.

This unit discusses the general aspects of seven of these control measures, which are the minimum components of a site control program according to OSHA 29 CFR 1910.120(d)(3). These seven are the site map, site preparation, site work zones, the buddy system, site security, communication systems, and safe work practices. Discussion of decontamination was included in Unit 10. Emergency planning is included in Unit 14.

The degree of site control depends on site characteristics, site hazards, work activity, site size, public relations, and the demographics of the surrounding community. The site control program should be established in the planning stages of a project and modified based on new information and site activity. The appropriate sequence for implementing these measures should be determined on a site-specific basis. In some cases, it may be necessary to implement several measures simultaneously.

13.1 Site Map

A site map showing topographic features, prevailing wind direction, drainage, and the location of buildings, containers, impoundments, pits, ponds, and tanks is helpful for:

- planning activities;
- identifying access routes, evacuation routes, and problem areas;
- identifying areas of the site requiring the use of PPE; and for
- supplementing the daily safety and health briefings of the field teams.

The map should be prepared prior to site entry and updated throughout the course of the operations to reflect:

- changes in site activities;
- hazards not previously identified;
- new materials introduced on site;
- vandalism and theft (for example, contamination may have been spread and equipment disabled); and
- weather-caused hazards (for example, standing water, mud, or snow).

Overlays can be used to help convey information without cluttering the map. In addition, a map showing specific routes from the site to emergency medical care facilities is necessary.

13.2 Site Preparation

Site preparation can be as hazardous as site cleanup, requiring safety measures of the same quality as those used during actual cleanup. Specific issues addressed during site preparation may include the movement of equipment and people, hearing protection, load restrictions, traffic safety, location of underground utilities, and general construction safety. Table 13-1 presents the major steps necessary to prepare a site for remediation.

Table 13-1 **Examples of Site Preparation Activities**

- Construct roadways to provide ease of access and a sound roadbed for heavy equipment and vehicles.
- Arrange traffic flow patterns to ensure safe and efficient operations.
- Eliminate physical hazards from the work area as much as possible, including:
  - —ignition sources in flammable hazard areas;
  - —exposed or ungrounded electrical wiring and low overhead wiring that may entangle or contact equipment;
  - —sharp or protruding edges, such as glass, nails, and torn metal, which can puncture protective clothing and equipment and inflict puncture wounds;
  - —debris, holes, loose steps or flooring, protruding objects, slippery surfaces, or unsecured railings, which can cause falls, slips, and trips;
  - —unsecured objects, such as bricks and gas cylinders near the edges of elevated surfaces, catwalks, roof tops, and scaffolding, which may dislodge and fall on workers;
  - —debris and weeds that obstruct visibility; and
  - —unstable piles or stacks of materials or drums.
- Install skid-resistant strips and other anti-skid devices on slippery surfaces.
- Construct temporary but solid foundations for mobile facilities and temporary structures.
- Construct loading docks, processing and staging areas, and decontamination pads.
- Provide adequate illumination for work activities, and equip temporary lights with guards to prevent accidental contact.
- Install all wiring and electrical equipment in accordance with the applicable code such as a ground fault circuit interrupter (GFCI).

13.3 Site Work Zones

Preventing the accidental spread of hazardous substances by workers from the contaminated area to the clean area is a high priority. Establishing clear work zones will help reduce the chance of accidentally spreading chemicals from a contaminated area to a clean area. Such deliniations will also reduce the number of personnel authorized in high-risk areas, establish required levels of PPE, and help plan emergency evacuation routes.

Hazardous waste sites should be divided into as many different zones as necessary to meet operational and safety objectives. This manual describes three frequently used zones.

- The Exclusion Zone (EZ) is the area where contamination is suspected or known to be present.

- The Contamination Reduction Zone (CRZ) is the area where decontamination takes place, and is also the main entrance and egress zone.
- The Support Zone is the uncontaminated area where workers should not be exposed to hazardous substances.

These three zones should be distinct based on sampling and monitoring results, and on an evaluation of potential routes for and the extent of contaminant dispersion in the event of a release. Movement of personnel and equipment through these zones should be minimized and restricted to specific Access Control Points (ACPs) to prevent cross-contamination from contaminated areas to clean areas. Sign-in/sign-out logs should be used to document entry, the time spent in each zone, and the exit of personnel working at the site. A schematic representation of the layout of work zones is given in Figure 13-1.

13.3.1 EXCLUSION ZONE

The EZ is the area where contamination does or could occur. The primary activities performed in the EZ are:

- site characterization (for example, photographing, mapping, and sampling);
- installation of monitoring equipment (for example, groundwater monitoring wells and air monitoring stations); and
- cleanup work (for example, drum movement, staging, materials bulking, soil excavation, building decontamination and demolition, soil treatment, and groundwater clean-up).

The outer boundary of the EZ, called the "Hotline," should be established according to the criteria listed in Table 13-2. It should be clearly marked by lines, placards, hazard tape, and/or signs, or enclosed by physical barriers, such as chains, fences, or ropes. ACPs should be established at the periphery of the EZ to regulate the flow of personnel and equipment into and out of the zone, and to help verify that proper procedures for entering and exiting are followed. If possible, separate entrances and exits should be established for personnel and equipment.

The EZ can be subdivided into a variety of areas based on the known or expected type and degree of hazard, or on the incompatibility of waste streams. This allows more flexibility in safety requirements, operations, decontamination procedures, and use of resources. Generalized requirements for hazard control may not apply to all areas, and specialized requirements for one hazard may not apply to others. Therefore, specific hazard zones within the EZ are often designated.

The personnel working in the EZ may include the Field Team Leader, work parties, and specialized personnel such as heavy equipment operators. All personnel within the EZ must wear the level of protection required by the site HASP. Within the zone, different levels of protection may be justified based on the degree of the hazard. The level of personal protection required in each sub-area should be specified and marked. Signs or notices clearly indicating each of these special PPE levels should be posted within each sub-area. However, worker changes in respiratory equipment are allowed only at the CRZ, and not within the EZ.

The required level of protection in the EZ varies according to job assignment. For instance, a worker who collects samples from open containers might require Level B protection, while one who performs walk-through ambient air monitoring might only need Level C protection. When appropriate, different levels of protection within the EZ should be assigned to promote a more flexible, effective, and less costly operation while still maintaining a high degree of safety.

13.3.2 CONTAMINATION REDUCTION ZONE

The CRZ is the transition area between the contaminated area and the clean area. This zone is designed to prevent the clean Support Zone from becoming contaminated or affected by other site hazards. Routing of personnel and equipment leaving the EZ through the CRZ for decontamination limits the physical transfer of hazardous substances into clean areas. The boundary between the CRZ and the EZ is the Hotline. The degree of contamination in the CRZ decreases as one moves from the Hotline to the Contamination Control Line.

Decontamination procedures take place in a designated area within the CRZ called the Contamination Reduction Corridor (CRC). Decontamination starts at the Hotline and proceeds from dirty to clean through

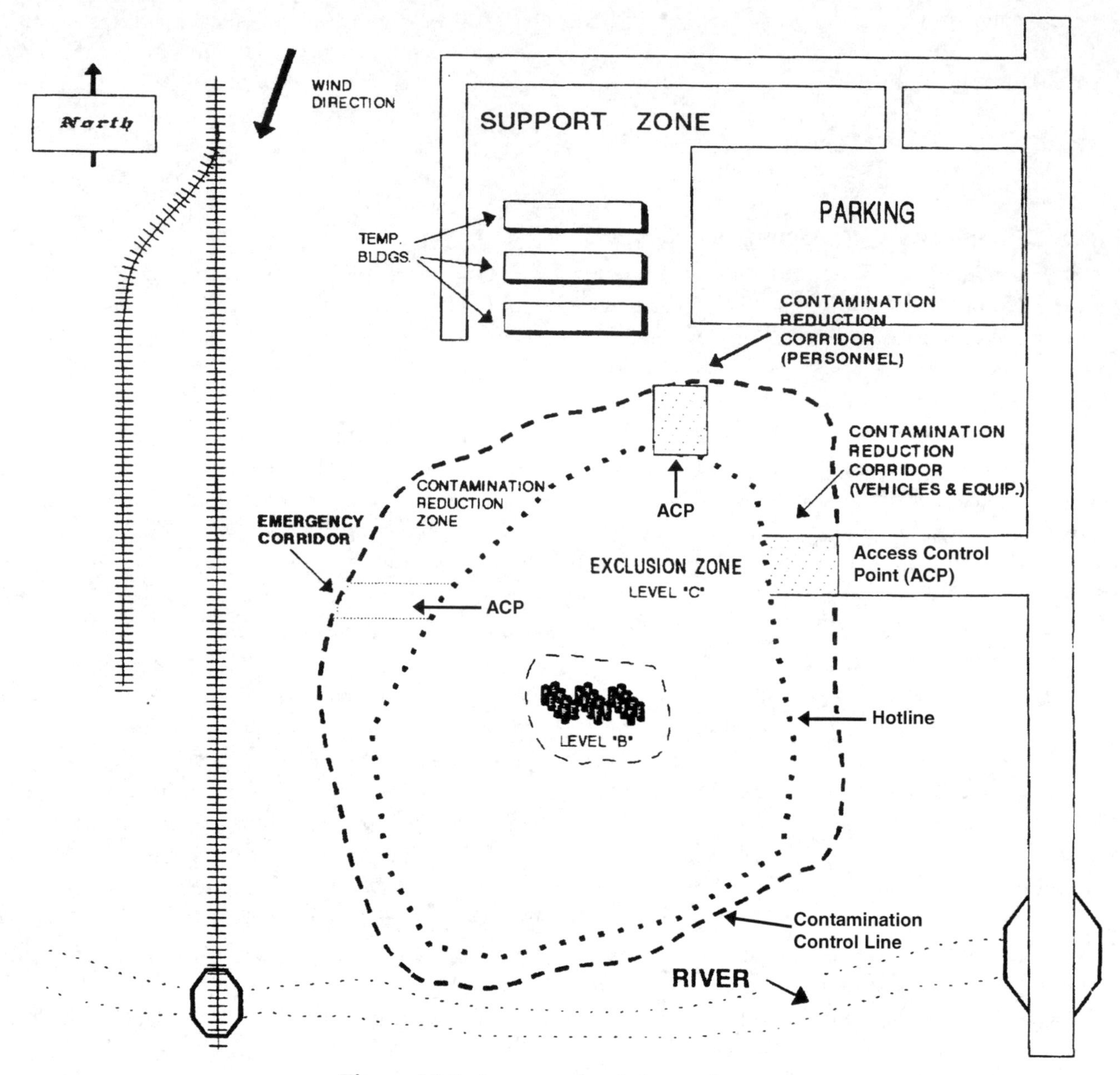

Figure 13-1 An example of site work zones

a series of planned steps as previously described in Unit 10. At least two lines of decontamination stations should be set up within the CRC, one for personnel and one for heavy equipment. It is important not to mix people and equipment during decontamination because each requires different decontamination procedures. A large operation may require more than two lines. Access into and out of the CRZ from the EZ is through the ACPs. If feasible, separate entry and exit ACPs for personnel and equipment should be used.

The boundary between the Support Zone and the CRZ is the Contamination Control Line. It separates the lower contamination area from the clean Support Zone. Access to the CRZ from the Support Zone is through ACPs. Personnel entering the CRZ must wear the personal protective clothing and equipment prescribed for working in that zone. To reenter the Support Zone, workers should remove any protective clothing and equipment worn in the CRZ, perform any decontamination that may be required, and leave through the personnel ACP exit.

Personnel stationed in the CRZ may include the Project Team Leader and a Personnel Decontamination Station (PDS) Operator. Additional personnel may

Table 13-2 **Criteria Often Used to Establish the Hotline**

- Visually survey the immediate area.
- Determine the locations of:
  —hazardous substances;
  —drainage, leachate and spilled materials; and
  —visible discolorations, drums, and other potential hazards.
- Evaluate data from the initial site survey indicating the presence of:
  —combustible gases;
  —organic and inorganic gases, particulates or vapors; and
  —ionizing radiation.
- Evaluate the results of soil and water sampling.
- Consider the distances needed to prevent an explosion or fire from affecting personnel outside the EZ.
- Consider the distances personnel must travel within the EZ. If the Hotline is too wide, unnecessary physical exertion and breathing air may be used by personnel working in the EZ.
- Consider the physical area necessary for site operations such as turning or maneuvering equipment and moving equipment into and out of the site.
- Consider meteorological conditions and the potential for contaminants to be blown from the area.
- Once established, secure or mark the Hotline and modify its location (if necessary) as more information becomes available.

assist the PDS operator such as those individuals needed to conduct abbreviated decontamination procedures for sample containers.

The CRZ must be well designed to facilitate:

- decontamination of equipment, PDS operators, EZ personnel, and samples;
- emergency response such as transport for injured personnel (safety harness, stretcher), first-aid equipment (bandages, blankets, eye wash, splints, and water), and containment equipment (absorbents, diking materials);
- equipment resupply, such as air tank changes, personal protective clothing and equipment (booties and gloves), sampling equipment (bottles and glass rods), and tools;
- sample packaging and preparation for on-site or off-site laboratories;
- capture of water and other liquids that are used during decontamination; and
- temporary rest areas for workers, including toilet facilities, benches, chairs, liquids, and shade. Water and other potable liquids should be clearly marked and stored properly to ensure that all drinking vessels are clean. Wash facilities should be located near drinking facilities to allow employees to wash before drinking. Drinking, washing, and toilet facilities should be located in a safe area where protective clothing can be partially removed. Facilities should be cleaned and inspected regularly. This level of detail is generally only established in the case of very large incidents or clean-ups where an extensive CRZ has been established. In other instances, these tasks will be performed in the Support Zone.

Personnel within the CRZ should be required to maintain internal communications, line-of-sight contact with work parties in the EZ, work party monitoring (for example, monitoring workers for remaining air time, fatigue, heat stress, hypothermia), and site security.

13.3.3 SUPPORT ZONE

The Support Zone is the location of administrative and other support functions needed to keep operations in the EZ and the CRZ running smoothly. The Command Post Supervisor should be present in the Support Zone. Other personnel present will depend on the functions being performed, and can include the Project Team Leader and field team members who are preparing to enter or who have returned from the EZ.

Personnel may wear normal work clothes within the Support Zone. Any potentially contaminated clothing, equipment, and samples must remain in the CRZ until they are decontaminated or disposed.

Support Zone personnel are responsible for alerting the proper agency in the event of an emergency. All emergency telephone numbers, evacuation route maps, and vehicle keys should be kept in the Support Zone.

Necessary support facilities (listed in Table 13-3)

Table 13-3 **Examples of Support Zone Facilities**

Facility	Function
Command Post	Supervision of all field operations and field teams; Maintenance of communications, including emergency lines of communication and recordkeeping such as: —accident reports —daily logbooks —medical records —site inventories —up-to-date site HASPs —chain-of-custody records —manifest directories and orders —personnel training records —site safety map Provides access to current safety and health manuals and other reference materials; Liaison with the public, including government agencies, local politicians, medical personnel, the media, and other interested parties; Monitors work schedules and weather changes; Maintains site security; Sanitary facilities.
Medical Station	First-aid administration; Medical emergency response; Medical monitoring activities; Sanitary facilities.
Equipment and Maintenance Center	Maintenance and repair of communications, respiratory, and sampling equipment; Maintenance and repair of vehicles; Replacement of expendable supplies; Storage of monitoring equipment and supplies (storage may be here or in an on-site field laboratory).
Administration	Sample shipment; Communications with home office; Emergency telephone numbers, evacuation route maps, and vehicle keys; Coordination with transporters, disposal sites, and appropriate federal, state, and local regulatory agencies.
Field Laboratory	Coordination and processing of environmental and hazardous waste samples (copies of sampling plans and procedures should be available for quick reference in the laboratory); Packaging materials for analysis following the decontamination of the outsides of the sample containers performed in the CRZ (packaging can also be performed in a designated location in the CRZ) (shipping papers and chain-of-custody files should be kept in the Command Post); Maintenance and storage of laboratory notebooks (these should be kept in designated locations in the Command Post when not in use).

are located in the Support Zone. When placing these facilities within the Support Zone, consider factors such as:

- accessibility (topography, available open space, location of highways and railroad tracks, ease of access for emergency vehicles);
- resources (adequate roads, power lines, telephones, shelter, and water);
- visibility (line-of-sight to all activities in the EZ);
- wind direction (upwind of the EZ, if possible); and
- distance (as far from the EZ as practicable).

Always remember that the Support Zone should contain no chemical hazards; however, physical hazards, such as uneven terrain and slippery ground, may still exist and may cause injury to unwary personnel.

13.4 The Buddy System

Most activities in contaminated or otherwise hazardous areas should be conducted with a buddy who is able to:

- provide their partner with assistance;
- observe their partner for signs of chemical exposure, heat stress, and cold stress;
- periodically check the integrity of their partner's protective clothing; and
- notify the Command Post Supervisor or others if emergency help is needed.

The ACP for personnel to the EZ is a convenient location for enforcing the buddy system.

The buddy system alone may not be sufficient to ensure that help will be provided in an emergency. At all times, workers in the EZ should be in line-of-sight or in communication with the Command Post Supervisor or a backup person outside the EZ.

13.5 Site Security

Site security is necessary to:

- prevent the exposure of unauthorized, unprotected people to site hazards;
- avoid increased hazards from vandals or persons seeking to abandon other wastes on the site;
- prevent theft; and
- avoid interference with safe working procedures.

To maintain site security during working hours:

- provide security in the Support Zone and at all ACPs to the site;
- establish an identification system to recognize authorized persons and any limitations to their approved activities;
- assign responsibility for enforcing authority over entry and exit requirements;
- erect a fence or other physical barrier around the site;
- post signs around the perimeter if the site is not fenced, and use guards to patrol the perimeter (guards must be fully apprised of the hazards involved and trained in emergency procedures); and
- have the Project Team Leader approve all visitors to the site making certain they have a valid purpose for entering (have trained site personnel accompany visitors at all times and provide them all with the appropriate protective equipment).

To maintain site security during off-duty hours, one or more of the following techniques might be appropriate:

- assign trained, in-house technicians for site surveillance who are familiar with the site, the nature of the work, the site's hazards, and respiratory protection techniques;
- use security guards to patrol the site boundary;
- enlist public enforcement agencies, such as the local police department, if the site presents a significant risk to the health and safety of the community or if elements within the community present a risk to workers; and
- lock or otherwise secure all equipment, buildings, etc.

Table 13-4 **Internal Communication Devices**

Radios, including: • citizen's band • FM Noisemakers, including: • bells • compressed air horns • megaphones • sirens • whistles	Visual signals, including: • flags • hand signals • lights • signal boards • whole body movements

13.6 Communication Systems

Two types of communication systems—internal communication among personnel on-site and external communication between on-site and off-site personnel—should be established.

Internal communication is used to:

- alert team members to emergencies;
- pass along safety information, such as weather changes, rest period notifications, air changes, heat stress checks, etc.;
- communicate changes in tasks or work methods; and
- maintain site control.

Verbal communication at a site can be impeded by on-site background noise and by the use of PPE. Speech transmission through a respirator can be poor, and hearing can be impaired by protective hoods and respirator air flow. For the most effective communication, it is important to remember that all types of signals (audio and/or visual) can help convey messages. To be most effective, the symbols must be agreed to in advance and be made part of the worker's protection devices through periodic practice.

Table 13-4 lists common internal communication devices. Both a primary and backup system are necessary. A set of unmistakable signals should be established for use only during emergencies.

Effective internal communication also requires the identification of individual workers so that commands can be addressed to a specific person. The worker's name should be marked on the suit and, for long-distance identification, color coding, numbers, or symbols can be added. Flags can also be used to help locate personnel in areas where visibility is poor due to obstructions such as accumulated drums, equipment, and waste piles.

All communication devices used in a potentially explosive atmosphere must be intrinsically safe and should be checked daily to ensure they are functioning properly.

An external communication system between on-site and off-site personnel is necessary to coordinate emergency response, report to management, and to maintain contact with essential off-site personnel.

The primary means of external communication are telephone and radio. If telephone lines are not installed at a site, all team members should know the location of the nearest telephone. Correct change and necessary telephone numbers should be readily available in the Support Zone. Cellular phones and even satellite communication and location devices are excellent additions to work sites.

13.7 Safe Work Practices

To maintain strong safety awareness and enforce safe procedures at a site, a list of standing orders or SOPs should be developed. These should state the practices that must be followed and those that must never occur in any area of the work site. SOPs are required as part of work plans and the site HASP. Everyone who enters the site should be aware of and familiar with the SOPs; therefore, these procedures should be:

- posted as appropriate warnings, in specific work areas;
- posted conspicuously at the Command Post;
- communicated in detail at safety meetings or special training meetings (worker training and competence in using and following these procedures must be initially documented and periodically verified through site QA and safety surveillance); and
- periodically reviewed by the Field Team Leader or Project Team Leader.

In addition to SOPs, a hazardous substance information form (similar to an MSDS) listing the names and properties of chemicals present on-site should be prepared and conspicuously posted. Employees should be briefed on the chemical information at the beginning of the project or whenever they first join the work team. Additionally, daily safety meetings should be held for all employees, stressing any new identifications as they occur.

Tools and heavy equipment can also be major hazards. Injuries result from equipment hitting or running over personnel, impacts from falling and flying objects, burns from hot exhaust manifolds, and damage to protective equipment such as SAR systems. The following precautions can help lessen injuries due to such hazards:

- train personnel in proper operating procedures;
- install adequate on-site roads, signs, and lights;
- install appropriate equipment guards and engineering controls on tools and equipment (such as rollover protective structures, seat belts, emergency shutoff in case of rollover, backup warning lights and signals);
- at the start of each work day, inspect brakes, hydraulic lines, light signals, fire extinguishers, fluid levels, steering, and spark protection;
- provide equipment such a cranes, derricks, and power shovel with signs saying "Unlawful to operate this equipment within 10 feet of all power lines;"
- use equipment and tools that are safe (not capable of sparking) as well as pneumatically and hydraulically driven equipment;
- where portable electric tools and appliances must be used (where there is no potential for flammable or explosive conditions), three-wire grounded extension cords or double insulated tools and GFCI are necessary to prevent electric shocks;
- when using hydraulic power tools, choose fire-resistant fluid that is capable of retaining its operating characteristics at the most extreme temperatures;
- keep all non-essential personnel out of the work area;
- prohibit loose-fitting clothing or loose, long hair around moving machinery;
- keep cabs free of all non-essential items and secure all loose items;
- do not exceed the rated load capacity of any vehicle;
- instruct equipment operators to report to their supervisor(s) any abnormalities such as equipment failure, oozing liquids, unusual odors, etc.;
- when an equipment operator must negotiate in tight quarters, provide a second person to direct them for adequate clearance around obstacles;
- have a signal man direct backing as necessary;
- all on-site internal combustion engines should have spark arrestors that meet the requirements for hazardous atmospheres;
- only refuel in safe areas; do not fuel engines while vehicle is running; prohibit ignition sources near a fuel area;
- lower all blades and buckets to the ground; set parking brakes before shutting off the vehicle;
- implement an ongoing maintenance program for all tools and equipment. Inspect all tools and moving equipment regularly to ensure that parts are secured and intact with no evidence of cracks or areas of weakness, that the equipment turns smoothly with no evidence of wobble, and that it is operating according to manufacturers' specifications;
- promptly repair or replace any defective items;
- keep maintenance and repair logs;
- store tools in clean, secure areas to prevent damage or loss; and
- keep all heavy equipment that is used in the EZ in that zone until the job is done; and
- completely decontaminate equipment before moving it to the clean zone.

13.8 Sources of Information

Koren, Herman: *Handbook of Environmental Health and Safety Principles and Practices.* 2nd ed. Vol. II. Chelsea, MI: Lewis Publishers, 1991.

Koren, Herman, and M.S. Bisesi: *Handbook of Environmental Health and Safety: Principles and Practices.* 3rd ed. Vols. 1 and 2. Chelsea, MI: Lewis Publishers, 1995.

Koren, Herman: *Illustrated Dictionary of Environmental Health & Occupational Safety.* Chelsea, MI: Lewis Publishers, 1995.

U.S. Department of Health and Human Services, National Institute for Occupational Safety and Health, Occupational Safety and Health Administration, U.S. Coast Guard, and Environmental Protection Agency: *Occupational Safety and Health Guidance Manual for Hazardous Waste Site Activities.* Washington, DC: U.S. Government Printing Office, 1985.

FOURTEEN

Emergency Planning and Spill Control

14 | CONTENTS

Before workers begin working at a hazardous waste site, and in accordance with OSHA 1910.138(a) and 1910.120(l), employers must develop and implement an emergency response plan. This written plan is a required section of the site HASP and must be available to employees and their representatives as well as to personnel from OSHA and other regulatory agencies. This plan must be compatible with and integrated into local, state, and federal fire and/or disaster response plans.

The nature of work at hazardous waste sites makes emergencies a constant possibility (Figure 14-1). When they actually occur, emergencies happen quickly and unexpectedly, and require immediate response. At a hazardous waste site, an emergency may be as confined as a worker experiencing heat stress, or as widely encompassing as an explosion that spreads toxic fumes throughout a community. Any on-site hazard can precipitate an emergency. Chemicals, biological agents, radiation, or physical hazards may act alone or together to create explosions, fires, spills, toxic atmospheres, or other dangerous and harmful situations. Table 14-1 lists common causes of site emergencies.

Table 14-1 **Causes of Emergencies at Hazardous Waste Sites**

Worker-Related:

- Minor accidents (slips, trips, falls, cuts, abrasions)
- Chemical exposure
- Medical problems (heat and cold stress, lacerations, personal medical conditions)
- Personal protective equipment failure (air source, torn clothing, etc.)
- Physical injury (injuries from hot or falling objects, tangling loose clothing in machinery, serious falls, vehicle accidents)
- Electrical (burns, shock, electrocution)

Waste-Related:

- Fire
- Explosion
- Leak
- Discovery of radioactive materials
- Release of toxic vapors
- Reaction of incompatible chemicals
- Collapse of containers

Site emergencies have the potential for complexity. Uncontrolled toxic chemicals may be numerous and unidentified, and their effects, if combined, may be synergistic. Rescue personnel attempting to remove injured workers may themselves become victims. Advance planning is essential in protecting workers and the health and safety of the community. Anticipat-

Figure 14-1 Example of a hazardous waste site emergency

ing a variety of emergency scenarios and thoroughly preparing for such events are integral parts of this advance planning.

Because of the Emergency Planning and Community Right-to-Know Act (also known as SARA Title III), the roles of local government in hazardous materials (HAZMAT) planning and response have greatly increased. Title III provides a strong statutory mandate to plan for hazardous chemical incidents and requires industry to provide information that may be needed by local authorities to meet that responsibility.

This unit outlines the basic factors to be considered when planning for and responding to emergencies. It defines the nature of site emergencies and lists the types of emergencies that may occur. It also outlines a Contingency Plan and its components, including personnel roles, lines of authority, training, communication systems, site mapping, site security and control, refuges, evacuation routes, decontamination, medical program, step-by-step emergency response procedure, documentation, and reporting to outside agencies. This unit, however, does not qualify trainees for work on an emergency response team.

14.1 Planning

The emergency response plan is a written document that sets forth policies and procedures for responding to site emergencies. It should incorporate the following:

- pre-emergency planning with outside parties
- personnel
  - —roles
  - —lines of authority
  - —training for emergency recognition
  - —emergency warning communications
- site information
  - —mapping
  - —security and control
  - —refuges
  - —evacuation routes
  - —decontamination stations
- medical/first aid
- equipment, including PPE
- emergency procedures
- documentation
- reporting of the event and follow-up reports

Overall, an emergency response plan should be:

- designed as a section of the site HASP;
- compatible and integrated with the pollution, disaster, fire, and emergency response plans of local, state, and federal agencies;
- rehearsed regularly using drills and mock situations; and
- reviewed periodically in response to new or changing site conditions or information.

14.2 Personnel

This component of the plan addresses on-site and off-site personnel with specific emergency response roles, plus other individuals who may be on-site, such as contractors, agency representatives, and visitors.

Emergency personnel may be deployed in a variety of ways. Depending on the nature and scope of the emergency, the size of the site, and the number of available personnel, the emergency response staff may include individuals, small or large teams, or several interacting teams. Although deployment is determined on a site-by-site basis, general guidelines and recommendations are discussed below. In all cases, the organizational structure should show a clear chain-of-command with every individual knowing his or her position and authority. The chain-of-command must also be flexible enough to handle multiple emergencies, such as a rescue and a spill response, multiple rescues, or a fire and spill response.

14.2.1 ON-SITE PERSONNEL

The emergency response plan must identify all individuals and teams who will participate in emergency response and must define their roles. All personnel, whether directly involved in emergency response or not, must know their own responsibilities in an emergency. They must also know the names of those in authority and the extent of their authority.

14.2.1.1 Incident Commander

In an emergency situation, one person must assume total responsibility for decision-making on-site. This person may be the Project Team Leader, the SSHO, a Field Team Leader, or another individual who is familiar with the site. This person must be identified by name in the emergency response plan. The incident commander must also:

- have an alternate who is specified by name in the emergency response plan;
- have the authority to resolve all disputes about health, safety, and environmental requirements and precautions;
- determine who assumes control of the incident when outside responders, such as police and fire departments, enter the scene; and
- have management support.

14.2.1.2 Teams

Although individuals may perform certain tasks in emergencies, response teams provide greater efficiency and safety in most cases. Teams composed of on-site personnel may be created for specific emergency purposes, such as decontamination or rescue. Rescue teams can be used during a particularly dangerous operation or at large sites having multiple work parties in the EZ. Other teams may be formed to respond to containment emergencies and fires until off-site assistance arrives.

14.2.2 OFF-SITE PERSONNEL

Off-site personnel involved in emergency response may include individual experts, such as meteorologists or toxicologists, and representatives or groups from local, state, and federal organizations who offer assistance in rescue or spill response operations. As part of advance planning, site personnel should:

- make arrangements with individual experts to provide guidance as needed;
- make arrangements with appropriate organizations (such as the local fire department, state environmental agency, EPA regional office), and provide them with specific information on the types of emergencies that may arise; and
- establish a contact person and means of notification with each organization.

14.2.3 FEDERAL RESPONSE ORGANIZATIONS

Site emergencies that involve significant chemical releases should be coordinated with federal response organizations. The federal government has established a NCP to promote the coordination and direction of federal and state response systems, and to encourage the development of local government and private capabilities to handle emergencies involving chemical releases.

If a chemical release occurs that exceeds the reportable quantity listed in federal regulations, the NRC in Washington, D.C., must be contacted at 1-800-424-8802. In addition, EPA's regulations concerning reportable quantities must be followed. The reporting requirements for the NRC are found in CERCLA, parts 102 and 103, and 40 CFR 302. The NRC may activate federal response under the NCP. When contacting the NRC, it is essential that they obtain on-site information about the emergency.

14.3 Training

Since an immediate, informed response is essential in an emergency, all site personnel as well as others entering the site (visitors, contractors, off-site emergency response groups, other agency representatives) must have some level of emergency training. According to 29 CFR 1910.120(q), five levels of emergency training are recognized. Individuals may be trained in any or all of the five levels of response. The five levels include, in increasing order of complexity and responsibility:

- **Level 1**—awareness of a release and the ability to notify the proper authorities;
- **Level 2**—defensive response to contain a release to keep it from spreading and to prevent exposure;

- **Level 3**—stopping the release of hazardous substances (such as plugging or patching);
- **Level 4**—site-specific response measures for the chemicals and processes used in a particular industry or location; and
- **Level 5**—assuming control of an incident beyond the initial response effort.

All personnel must be aware of the hazards present on-site and the site-specific safety rules. They must also know what to do in case of an emergency. Even visitors should be briefed on basic emergency procedures such as decontamination, emergency signals, and evacuation routes.

Personnel without defined emergency response roles (such as contractors and federal agency representatives) must still receive a minimum level of training, which includes:

- hazard recognition;
- signaling an emergency (for example, which alarm to use, how to summon help, what information to give and to whom it should be given);
- evacuation routes and refuges; and
- the person or station to report to when an alarm sounds.

On-site emergency personnel who have emergency roles in addition to their ordinary duties must have a thorough understanding of emergency response procedures as specified in 29 CFR 1910.120(q). In addition, training should be directly related to their specific emergency roles and should:

- relate directly to site-specific, potential situations;
- be brief and repeated often;
- be realistic and practical;
- provide an opportunity for special skills to be practiced regularly; and
- feature frequent drills (site-specific mock rescue operations).

Off-site emergency personnel such as local firefighters and emergency medical responders are often first responders and run a risk of acute hazard exposure equal to that of any on-site worker. These personnel should be informed of ways to recognize and deal effectively with on-site hazards. Lack of information may inadvertently worsen an emergency, such as spraying water on a water-reactive chemical and causing an explosion. Inadequate knowledge of the on-site emergency chain-of-command may cause confusion and delays. Site management should, at a minimum, provide off-site emergency personnel with information about site-specific hazards, appropriate response techniques, site emergency procedures, and decontamination procedures.

14.4 Emergency Response Actions

Hazardous waste workers may be trained in one or more response techniques or a separate team may be formed and trained specifically for responding to hazardous emergencies. This unit provides only basic guidance for emergency response operations. The extent of involvement for any individual will depend on their level of training and responsibility. The following sections present only general information on the initial actions that might be performed during a site emergency, as well as more specific methods of identifying, categorizing, and acting upon emergency situations.

14.4.1 METHODS OF RESPONSE

Strategic decisions must be made when dealing with hazardous substance emergencies, the first of which is whether to respond in a defensive or an offensive mode. Offensive operations are those that actively attempt to stop or mitigate a release, and require that the proper PPE is available and personnel have been properly trained. Defensive operations are those that focus on stopping the flow of the substance from a safe distance in which no direct contact is made with the substance in question.

The objectives that must be accomplished as soon as possible during an emergency include, but are not limited to:

- isolation of the site;
- rescue of people inside the isolation area;
- protection from exposures;

- fire extinguishing;
- confinement of the substance; and
- recovery of the area to working conditions.

It is important to realize that it may not be possible to accomplish all of these objectives. For example, without proper PPE, only site isolation might be accomplished. Worker training must emphasize that they cannot exceed the on-site capabilities and resources when responding to an emergency, thereby exacerbating an already dangerous situation.

When too much risk is involved for the rescue personnel, rescue may not be possible. There are many examples where rescuers that were inadequately outfitted and/or trained died while attempting to rescue a victim. Emergencies involving hazardous substances necessitate an evaluation of the situation prior to attempting rescue. This evaluation should include:

- probability of victim survival;
- risks to rescuers;
- difficulty of rescue;
- capabilities of rescuers;
- possibility of explosions or sudden product releases;
- available escape routes and safe refuges;
- constraints of time and distance; and
- availability of proper PPE.

14.4.2 HAZARD AND RISK ANALYSIS

Risk analysis is an important factor when responding to a HAZMAT emergency. The ability to anticipate the course of HAZMAT incidents will allow responders to think ahead of the problem and reduce the risk to response personnel. For example, certain physical evidence may indicate that a HAZMAT incident is escalating. Such signs should be sought out when responding to a HAZMAT emergency and heeded. These signs may include:

- frost forming near a leak;
- deformed containers;
- operation of pressure-relief valves;
- "pinging" of heat-exposed vessels;
- extraordinary fire conditions;
- peeling of container finish;
- boiling of unheated materials;
- colored vapor clouds;
- smoking or self-igniting materials;
- peculiar smells;
- deterioration of equipment; and/or
- unexplained site changes of any type.

Any of these signs may indicate a rapid degeneration in site conditions. If any are noted, a complete site evaluation should be completed immediately.

Emergency responders must be able to analyze the risks and do what is necessary to mitigate the situation. They should:

- always consider personal safety;
- positively identify the chemicals involved;
- obtain accurate information on commercial chemicals from the shipper or the manufacturer (use of chemical manuals can be beneficial);
- record, if possible, all pertinent information received from first responders and pass this on to the appropriate parties;
- regard abnormal containers since fire and heat can create pressure within containers;
- maintain control of the scene. Use the minimum number of people needed to safely handle an incident, keeping others away to avoid added problems. Law enforcement personnel are a help in isolating an area.
- consider weather conditions since changes in the weather may completely alter the chosen course of action; and
- use instruments to detect contaminants in the atmosphere.

14.4.3 D.E.C.I.D.E.

During an emergency, certain basic decisions must be made for any intervention to be successful. Emergency response personnel should:

- **Detect Hazardous Material Presence**
  This is critical to any emergency. If project planners are not aware that a hazardous material problem exists, it is much more difficult for them to protect personnel against a nasty surprise.

- **Estimate Likely Harm Without Intervention**
  This is a difficult but indispensable step. Prior to defining operational objectives, problems and any possible harmful effects must be determined. An evaluation of potential problems, probable outcomes, and necessary protective actions will help determine the success of emergency operations.
- **Choose Response Objectives**
  Once problem areas are determined, it is necessary to choose the appropriate prevention or protective measures to guard against possible harmful effects of the problem.
- **Identify Action Options**
  Once objectives are defined, the options available to accomplish them must be examined. All practical options should be considered, with the outcome of each action being evaluated for its direct ability to provide an acceptable solution to the problem, while minimizing or preventing harm to people, property, or the environment.
- **Do Best Option**
  When multiple options are available, one should be selected that provides the greatest gain and the least loss.
- **Evaluate Progress**
  Once a course of action is in place, evaluate the progress of that decision. Is what was expected to happen actually happening? If not, the problem must be reviewed and another course of action must be selected.

14.4.4 MAKING DECISIONS IN EMERGENCIES

Estimating likely harm without intervention is an important element in planning for chemical emergencies. This concept includes predicting what is likely to happen in an emergency; and describing the likely outcome based on that assessment.

The first step in estimating likely harm without intervention is "events analysis," defined as the process of breaking down complex actions into smaller, more easily understandable parts. The process of events analysis provides a means to track and predict the sequence of past and current events to determine the time and type of actions necessary to change that sequence. Events analysis helps to define problems in a sequential and logical way, thus minimizing confusion, guesswork, mistakes, delays, unnecessary harm, and loss of control at emergencies. The steps in events analysis are:

- identify all components of an emergency;
- identify the effects of each component; and
- identify all possible outcomes.

For example, a steel drum with flammable contents is involved in a fire caused by leaking fuel. What is likely to happen in this situation? Events analysis can be used to systematically evaluate this incident.

The first step in events analysis is to identify the components causing the emergency. In this case, the components include a drum with flammable contents that have been exposed to a fire.

Secondly, identify the effects each component contributes to the emergency. The burning fuel is heating the drum, the drum is absorbing heat from the burning fuel and heating the contents, and the contents are absorbing heat from the drum.

The fuel spill will keep burning until all the fuel is consumed. Thus, if nothing is done, the burning fuel will continue to surround and heat the drum. As the drum absorbs heat, the contents will expand and its physical state may change from liquid to gas. Internal pressure inside the drum will increase causing increased stress on drum parts.

This discussion of events breaks up complex actions in the emergency into smaller components. In this case, the components are the fire, the drum, its contents and the effect of the fire on the released contents.

Possible consequences as a result of the increased pressure are:

- the flat drum head will begin to round out as the internal pressure continues to rise;
- the chime between the drum head and the drum wall will begin to yield; and
- the drum head will separate from the drum wall.

When the drum head breaks away from the side wall, what will happen with the contents?

- As the pressure is relieved through the breach in the drum, the heated contents will start to expand and flow through the breach.
- Contents will flash to vapor as they reach the unpressurized atmosphere outside the drum, creating a new component—vaporized flammable contents.
- Escaping contents can produce a propulsive effect on the drum, propelling it like a rocket.
- If the drum is still surrounded by the burning fuel, the vaporized contents will ignite, forming a fireball and escalating the problem.

When the contents are released, what will be the outcome?

- The drum, propelled by the escaping contents, flies along a trajectory, which is dependent upon where the drum was heated. Obstructions can cause a change in direction or limit travel.
- The released contents may fall along the flight path of the drum, leaving a trail of burning material along the ground.

The third step in events analysis visualizes the sequential interaction. Each event is broken down and placed in logical sequence so that points of intervention become readily apparent.

Events analysis is the first step in estimating likely harm without intervention in the D.E.C.I.D.E. process. The process of visualizing likely events in sequence helps identify gaps in understanding the problem.

If what is likely to happen cannot be visualized from the information available, additional information should be gathered from a reputable source. If additional information is not available, experience during past emergencies can also help. Past experience is critical to understanding and visualizing events in their proper sequence.

14.4.4.1 Methodology For Predicting Outcome

As emergency events are visualized, the type of harm likely to occur with and without intervention should be estimated. Estimates should indicate the outcome in terms of:

- fatalities
- injuries
- property damage
- critical system disruption
- environmental damage

Outcome estimates may be modified as additional information becomes available.

Part of the role of the emergency Incident Commander is to collate and analyze outcome estimates, coordinate community preparations for "worst case" scenarios, and coordinate information flow from the on-site manager to appropriate policy and operational elements within the community.

14.4.4.2 Hazardous Materials Behavior

Factors affecting the behavior of HAZMAT in an unexpected or catastrophic release are:

- inherent properties of the hazardous material;
- quantity of the hazardous material;
- built-in characteristics of the container;
- natural laws of physics and chemistry; and
- environmental factors, including the physical surroundings (buildings and structures), atmospheric conditions (weather), and topography (water flow).

Determining the interrelationship among these factors can help considerably when visualizing what is likely to happen in an emergency. Experience has shown that hazardous material behavior in a sudden or catastrophic release follows a basic sequence of events. The Hazardous Materials General Behavior Model defines the sequence of events which characterizes a HAZMAT emergency. The events are:

- stress
- breach
- release
- engulf
- impinge
- harm

14.4.4.2.1 Stress Event. Stress is applied force or system of forces that tends to strain or deform a container and trigger a change in the condition of the container.

There are three basic forms of stress: thermal, mechanical, and chemical. Thermal stress is caused by fire, sparks, friction, electricity, ambient temperature changes, and extreme or intense cold. Mechanical stress is caused by an object physically contacting the container. The object may puncture, gouge, bend, break, tear, or split the container. Chemical stress is caused by a chemical action, such as acids corroding the container or pressure being generated by decomposition, polymerization, or corrosion.

14.4.4.2.2 Breach Event. If the container is stressed beyond its recoverable limits, it will open or breach. Different containers breach in different ways. There are five basic types of breach, including:

- disintegration, which is the total loss of integrity (visualize a glass jar shattering);
- runaway cracking (like the shell of an egg), which is associated with closed containers. A BLEVE is an example;
- open attachments, such as a pressure relief device;
- punctures, such as a drum puncture; or
- a split or tear, such as torn bags or boxes, or a split in the seam of plastic drums.

14.4.4.2.3 Release Event. Once the container is breached, the material can escape. There are four types of releases, including:

- detonation, which is an explosive chemical reaction that occurs in less than 1/100th of a second. Examples are military munitions, dynamite, and organic peroxides.
- violent ruptures, such as runaway cracking, resulting in an instantaneous release and ignition of vapor. A BLEVE occurs in less than one second.
- rapid release, which occurs through pressure relief devices, damaged valves, punctures, or broken piping. This type of release occurs over several seconds to several minutes.
- a spill or leak through openings in fittings, splits or tears, or punctures. This non-violent release of materials may take minutes to days or even longer, such as in leaking underground storage tanks (LUST).

The release event shows how hazardous materials escape from containers during an emergency. The escape may be in the form of matter, energy, or a combination of both.

14.4.4.2.4 The Engulfing Event. Once contents are released, they are free to travel, and the emergency is likely to escalate. Factors that may affect the movement of material include the form of material (gas, liquid, or solid), thermal differential, self-propelled forces, wind, gravity, and diffusion. The movement path may be linear, radial, random, or could follow the topographic contour. The dispersion pattern may be in the form of a cloud, cone, plume, stream, and/or irregular deposition.

The dispersion pattern is also dependent on the chemical state (gas, liquid, or solid) of the hazardous material. Gases often escape under pressure, such as leaks from a cylinder, forming a cloud or plume. If enclosed, the cloud will fill available spaces. Gases may also be carried by the wind as a plume. If the vapor density is greater than 1.0, the material is heavier than air and may settle in low places or travel along at ground level as a plume. Liquids may flow along the ground as a stream simultaneously vaporizing and acting as a gas (stream with plume) or being absorbed into the ground or clothing worn at the scene. Solids may scatter (irregular deposits), form dust clouds that are carried by the wind (plume), stick to surfaces, or be carried away from the scene (irregular deposits).

14.4.4.2.5 Harm Event. Factors which affect or determine harm from a release include:

- time of the release (for example, speed of escape and travel, length of exposure);
- size of the area covered; and
- lethality in terms of dosage or concentration received from individual exposure.

Types of harm that may occur in this and other release events include:

- thermal (heat or cold)
- radiation
- asphyxiation
- toxic
- corrosive
- etiologic
- biological
- mechanical

14.5 Confinement and Containment

A common type of chemical emergency involves a spill where the basic response is to confine and contain the spilled product. Spills can occur at chemical use facilities, either during the processing of the material or during the loading/off loading of the material. Spills also occur in field operations such as when moving drums or when unknown buried tanks and drums are ruptured.

During emergency releases in the field, potential environmental damage must be evaluated before confinement or containment procedures can begin. Therefore, the emergency response plan should address:

- volume of hazardous materials released to the environment and the rate of leakage;
- dangers to on-site and off-site personnel;
- nature of the damage to the container and the repairs attempted to mitigate the problem;
- transfer of material to an alternate container;
- dike construction around the spill area;
- characteristics of the spill area and how the material has spread over the area;
- proximity to waterways or sewers;
- danger of fire or explosion,
- effect of rain and wind on the spilled material; and
- availability of equipment and supplies necessary to confine the material.

14.5.1 TYPES OF RELEASES

14.5.1.1 Air Releases

If the material released has a vapor density less than 1.0, the only effective containment is to plug the leak or evacuate the area. Evacuation is dependent on the size of the vapor cloud created. Restriction of the air space over the incident should also be considered.

If the material released has a vapor density greater than 1.0, the material may hug the ground at the level of release or lower. This can be greatly affected, however, by the temperature of the material and by meteorological conditions, including thermal inversions and wind patterns. In some cases, the only effective containment is to evacuate the area downwind and downhill of the material.

14.5.1.2 Land Spills

When material escapes from a container and penetrates into the ground, it may be considered a land spill. Land spills also include contamination from byproducts of combustion (for example, a fire at a chemical storage site occurs, but was only hot enough to change chemicals from one form to another). The easiest method of confinement of a land spill is an earthen dike. The materials and manpower to construct this type of dike are usually available and inexpensive. However, over time, both vertical and horizontal seepage through and around the dike will occur. This process can be slowed by the use of plastic sheets, absorbents, or adsorbents.

An alternative is to transfer as much of the material as possible to another container (if any remains in the leaking container). It still may be necessary to dike around the original spill while waiting for the second container to arrive.

14.5.1.3 Surface Water Discharges

This form of contamination has been researched longest; therefore, several options exist to contain spills in, on, and under the water. For insoluble chemicals that have a specific gravity of greater than 1.0, the best method of confinement in moving waterways is an overflow dam. The released chemical will settle at the base of the dam, thus allowing for product recovery. This system works best on slow moving waterways.

If the spilled material has a specific gravity of less than 1.0 and is insoluble or only slightly soluble in water, the best method of containment is a floating boom. There are several different types of booms on the market. Also, sorbent booms are available that adsorb the spill rather than confining it. Neither product works well in rough water or fast currents.

If the material spill is soluble in water, there is very little that the responder can do. If the waterway is small, dam the waterway and recover or filter the water. The other option is to neutralize the chemical, rendering the chemical inert. This may require the resources of the EPA and state health department for technical assistance before it is attempted.

14.5.1.4 Groundwater Contamination

Groundwater contamination is not normally handled by first responders. However, the potential for ground water contamination during a response must be evaluated and can be reduced by minimizing the volume of any liquids used during spill response. In addition, when spills are retained by dikes, make sure that the retained spill does not back up into an area that might allow for rapid movement of liquids into the groundwater. For example, retaining a spilled chemical in a graveled area may promote product movement into groundwater where product recovery will be difficult.

14.5.2 RESPONSE OPTIONS

Options for confining and containing chemical releases include diversion, diking, retention, and plugging and patching. These techniques are discussed below.

14.5.2.1 Diversion Techniques

A flowing land-based spill can be quickly diverted to another location by placing a barrier (normally dirt) in advance of the spill. To be successful, the barrier must be built well in advance of the spill and responders should be prepared to sacrifice some territory to protect sensitive areas. The effective and quick construction of a diversion barrier requires teamwork.

14.5.2.2 Diking Techniques

Dikes are effective if they can be constructed quickly and contain most of the release at the desired location. They can be constructed from practically any available materials. Several readily available items include dirt, boards, roof ladders, plastic tubes filled with water, and plastic tarps or sheeting.

Construction of dikes by hand will usually be done by first responders using brute force. The decision to build a containment dike should be based on a quick evaluation of resources versus the quantity of material on the ground. Responders make the common errors of underestimating the number of people and time required to build a dike, and underestimating the amount of spilled materials.

Slow moving or heavy materials in a surface land spill should be confined by building a circle dike. Faster moving products can be confined by constructing a V-shaped dike in a low area.

All dike construction should begin with bulky foundation materials such as concrete, rocks, boards, roof ladders, etc. Whenever possible, this material should be surrounded (top and bottom) with plastic sheeting to prevent the foundation material from becoming contaminated. A seam of dirt on the inside or outside is used to keep the plastic in place and to minimize seepage beneath the plastic. Keep in mind, if the dirt is placed on the inside, it will be contaminated.

There are several limiting factors to dike construction. They include the following:

- If the surrounding area is concrete or asphalt and there is no easily accessible soil, the paved area may have to be sacrificed or appropriate materials will have to be brought to the site;
- If the ground is frozen, snow may be used as a partial barrier in conjunction with other materials like ladders and plastic. Otherwise, materials such as sand may have to be brought to the site;
- If essential equipment is unavailable, dike construction time will increase and options will decrease;
- If chemical hazards exist, appropriate PPE must be worn. This will increase response time.

14.5.2.3 Retention Techniques

In some cases, retention techniques can be implemented independently of diversion or diking to achieve the same results or act as a backup. Basic techniques include the following:

- Place salvage covers or plastic sheeting over storm drains;
- Shovel dirt onto the cover, burying the storm drain;

- If a curb is adjacent to a storm drain, lay a plastic sheet across the drain and secure it with shovels full of dirt;
- If the spill is primarily liquid or slurry, has a specific gravity less than water, and is not water-reactive or soluble in water, it may be possible to flood the retention area with water. The spill will then float. Any leakage into the storm drain system should be water. This technique, however, requires careful and consistent management and a permit for a successful resolution;
- When bulk quantities of hazardous materials are leaking, the situation may warrant bringing in heavy equipment to construct large retention pits. Other options include portable chemical tanks, drafting basins, and inflatable swimming pools;
- Overpacking may be appropriate if a drum is leaking. This is accomplished by the use of an oversized container. Overpack containers should meet the needs of the hazards involved. If the material is to be shipped, DOT-specification overpack containers must be used.

Figure 14-2 Overpacking a leaking drum

14.5.2.4 Plugging and Patching Techniques

The 55-gallon drum is the most common container for storing and shipping hazardous materials. Because of their widespread use, these drums are involved in over 30 percent of the HAZMAT incidents reported to CHEMTREC.

Of the 55-gallon drums, those made out of steel are most common. Such drums are manufactured by welding a sheet of steel to form a cylinder and clamping or welding a heavier gauge of steel on the top and bottom. The top or bottom perimeter rim is usually called the "chime" and the screw cap on the top is the "bung." Plastic drums are also common.

Although this unit focuses on the 55-gallon drum, plugging and patching techniques are basically the same, regardless of drum or container size. In addition, all plugging operations follow the same basic safety rules. Prior to initiating plugging activities, identify the product. Based on the chemical hazards, select appropriate PPE and plugging materials that are non-reactive to the hazardous material. If the event is beyond the equipment, knowledge, or ability of the initial responders, call for help before taking any defensive actions.

14.5.2.4.1 Causes of Leaks in Drums Puncture leaks can be caused by penetration from fork lifts, by placing the drum on a sharp object, or can occur during accidents and falls when the drum is forced into a sharp object. Pressure can also cause leaks. Pressure can be caused by an external heat source, including exposure to sunlight, which raises the temperature of the contained liquid. A chemical reaction within the drum can also cause a pressure buildup. As the liquid expands, a leak or rupture may develop. Leaks can also occur when drums fall or are crushed. Such leaks are often unable to be contained by patching the container. Deterioration leaks are caused by rust in metal drums and degradation in plastic drums. Such leaks are often small and numerous.

14.5.2.4.2 Using Drum Features to Control a Leak The easiest technique to control a leak in a drum is to use its inherent shape to stop the leak. This simply involves positioning the drum so that the hole is above the fluid level. For example, if a leak occurs in the bottom of a drum, turning the drum over usually solves the problem. If the leak is on one side, rolling the drum on its opposite side will stop the leak. However, before moving a drum, check its integrity. Severely rusted drums may collapse when repositioned.

Bung leaks can often be controlled by tightening the bung or rim clamp, or by replacing the washer.

14.5.2.4.3 Plugging and Patching Drums to Control Leaks Basic equipment for plugging and patching consists of a variety of low-tech items including washers, metal screws, wooden wedges, golf tees, duct tape, non-sparking hammers and wrenches, glues, rubber patches, rubber plugs, etc.

Techniques largely depend on the size and shape of the hole. For small round holes, sheet metal screws and washers are often effective. Inserting a neoprene gasket between the washer and the drum will help to form a tight seal. Rubber plugs, small wooden plugs, and even golf tees are effective. Golf tees form a better seal if they are dipped in wax prior to use. Gently push or hammer these plugs into the hole.

For larger holes, such as those made by a fork lift, position the hole above the product level, then clean the surface, place a wedge in the hole, tape over the wedge, and epoxy over the tape. Prepare the surface by cleaning off paint and dirt with a rag, wire brush, or other compatible material. Drive a wedge into the hole and cut off the excess portion of the protruding wedge. Place duct tape over the plug and extend it at least two inches in all directions. Finally, coat the area with epoxy.

An irregular hole or gash can be plugged by gluing neoprene to sheet metal, placing this homemade gasket over the hole, and clamping the gasket to the drum using a clamp strap.

A relatively new patching technique uses a vinyl belt with a built-in inflatable bladder. Positioned so that the bladder is over the hole, the belt is then secured to the drum and the bladder inflated with a carbon dioxide cartridge.

After the drum has been patched, it is usually placed in a recovery drum (Figure 14-2). A recovery drum is specifically designed to receive a drum of a slightly smaller size (for example, an 85-gallon recovery drum is designed to receive a 55-gallon drum). This process, called overpacking, can be accomplished by rolling the patched drum onto wood planks or PVC pipes, then pushing the patched drum into the recovery drum. The planks or pipe act as a rolling surface for the drum. Overpacking should proceed cautiously because the patched drum cannot tolerate much stress.

14.6 Communications

In an emergency, crucial messages must be conveyed quickly and accurately. Even through noise and confusion, site staff must be able to communicate information, such as the location of injured personnel, orders to evacuate the site, and notice of blocked evacuation routes. Outside support sources must be contacted, help must be obtained, and measures for public notification and safety must be ensured. To do this, a separate set of internal emergency signals should be developed and rehearsed daily. External communication systems and procedures should be clear and accessible to all workers.

14.6.1 INTERNAL COMMUNICATIONS

Internal emergency communication systems are used to alert workers to danger, convey safety information, and maintain site control. Any effective system or combination of systems may be employed. Radios or field telephones are often used when work teams are far from the command post. Alarms or short, clear messages can be conveyed by audible signals, such as bullhorns, megaphones, sirens, bells, whistles, or visual signals, such as colored flags, lights, and hand or whole-body movements. The primary system must have a backup. For example, hand signals may be used as a backup if radio communications fail. All internal systems should be clearly understood by all personnel, checked and practiced daily, and intrinsically safe (spark-free).

When designing and practicing communication systems, remember that:

- background noise on-site will interfere with talking and listening;

- wearing PPE will impede hearing and speech and limit vision (such as the ability to recognize hand and body signals); and
- inexperienced radio users may need practice in speaking clearly.

14.6.2 EXTERNAL COMMUNICATIONS

Off-site sources must be contacted to obtain assistance or to inform officials about hazardous conditions that may affect public or environmental safety. The telephone is the most common mode of off-site communication. Phone hookups are considered a necessity at all but the most remote sites. When planning the external communication aspect of the contingency plan, the following items must be considered:

- The NRC (1-800-424-8802) should be contacted in the event of a significant chemical release.
- Designated persons should be authorized by site management for off-site communications and contacting public emergency aid teams such as fire departments, ambulance units, and hospitals.
- If there is no site telephone system, all personnel must know the location of the nearest cellular and public telephones.

14.7 Safe Distances and Places of Refuge

No single recommendation can be given for evacuation or safe distances because of the wide variety of hazardous substances and releases found at sites. For example, a "small" chlorine leak may call for an isolation distance of only 500 feet, while a "large" leak may require an evacuation distance of 1,500 feet or more.

Safe distances can only be determined at the time of an emergency based on a combination of site and incident-specific factors. However, planning and outlining potential emergency scenarios will help familiarize personnel with points to consider. Factors that influence safe distances include the:

- toxicological properties of the substance;
- physical state of the substance;
- quantity released;
- rate of release;
- vapor pressure of the substance;
- vapor density relative to air;
- wind speed and direction;
- atmospheric stability;
- height of release;
- air temperature and temperature change with altitude; and
- local topography (for example, barriers may enhance or retard a cloud or plume, but reduce the effects of a blast).

Evacuation distance guidelines are contained in the North American Emergency Response Guidebook, available through the DOT's Research and Special Program Administration.

On-site refuges (safety stations) can be set up for localized emergencies that do not require site evacuation. The refuge should be located in a relatively safe area within the EZ. These refuge areas should never be used for activities such as eating, drinking, or air changes. Typical items located in refuge areas might include shade, decon liquid, a wind indicator, a communication system with the command post, first aid supplies, special monitoring devices, tools, and fire extinguishers.

If an incident threatens the safety or health of the surrounding community, the public will need to be informed and possibly be evacuated from the area or protected in place. Site management should plan for this scenario and coordinate with the appropriate local, state, and federal groups. Note that evacuation of the public is the responsibility of local authorities. Site personnel should not attempt large-scale public evacuation.

14.8 Evacuation Routes and Procedures

A severe emergency, such as a fire or explosion, may cut workers off from the normal exit near the command post. Therefore, alternate routes for evacuating victims and endangered personnel should be established in advance, marked, and kept clear. Routes should be directed from the EZ through an upwind CRZ to the Support Zone, and from the Support Zone to an off-site location in case conditions necessitate a general site evacuation. The following guidelines will help establish safe evacuation routes:

- place the evacuation routes in a predominantly upwind direction of the EZ;
- run the evacuation routes through the CRZ. Even if there is not enough time to process the evacuees through decontamination procedures, there should be a mechanism for accounting for all personnel;
- consider the accessibility of potential routes. Take into account obstructions such as locked gates, trenches, pits, tanks, drums, or other barriers, and the extra time or equipment needed to maneuver around or through them;
- develop two or more secondary routes that lead to safe areas and that are separate or remote from each other. Multiple routes are necessary in case one is blocked by a fire, spill, or vapor cloud;
- mark routes "safe" or "not safe" on a daily basis according to wind direction and other factors;
- mark evacuation routes with materials such as barricade tape, flagging, or traffic cones. Equally important, mark areas that do not offer safe escape or that should not be used in an emergency;
- consider the mobility constraints of personnel wearing protective clothing and equipment. They will have difficulty crossing even small streams and going up and down banks; and
- make escape routes known to all who go on-site.

14.9 Decontamination

Decontamination during site emergencies is similar to the decontamination procedures discussed in a previous unit. However, there is one important exception—victim decontamination. Some authorities suggest that the medical condition of the victim be handled before decontamination, and that the victim be transported to the hospital before decontamination. Even when hospitals have a decontamination system, it will place a severe strain on the system to receive a contaminated patient and especially for multiple patients. A program of initial field decontamination followed by a secondary hospital decontamination is the best way to ensure that the victim is clean, and that hospital and emergency transport personnel are not harmed by contaminated victims.

The following basic points should be considered when assessing victim decontamination:

- solid or particle contaminants should be brushed off as completely as possible prior to washing. Heavy liquid contaminants should also be blotted from the body prior to washing;
- remove contaminated clothing and rinse victim with water;
- wash victim, if possible, with tincture of green soap or other mild soap. Pay special attention to hair, nail beds, and skin folds. Soft brushes and sponges may be used. Be careful not to abrade the skin and use extra caution over bruised or broken skin areas. The victim should be rinsed with large quantities of water;
- if possible, use warm water for more extensive washing. If cold water must be used, there is a risk of hypothermia; and
- contain all runoff.

Basic care of the victim (airway, breathing, and circulation) may need to be conducted prior to decontamination in life-threatening situations. This practice will mandate that a person trained in basic first aid be protected by proper PPE.

14.10 Medical Treatment and First Aid

In emergencies, toxic exposures and hazardous situations that cause injuries and illnesses will vary from minor to serious. Medical treatment may range from bandaging of minor cuts and abrasions to life-saving techniques. In many cases, essential medical help may not be immediately available. For this reason, it is vital to train on-site emergency personnel in first aid treatment techniques, to establish and maintain telephone contact with medical experts, and to establish liaisons with local hospitals and ambulance services. When designing this program, the following essential points should be included:

- train a cadre of personnel in emergency treatment such as first aid and CPR. Training should be thorough, repeated frequently, and geared to site-specific hazards;

- establish liaison with local medical personnel. Inform and educate these personnel about site-specific hazards so that they can be helpful if an emergency occurs. Develop procedures for contacting them and familiarize all on-site emergency personnel with these procedures; and
- set up on-site emergency first-aid stations. See that they are kept well supplied and restocked immediately after each emergency.

14.11 Documentation

The Project Team Leader should initiate a follow-up investigation and document all aspects of the emergency. This is important in all cases, but especially so when damage occurs to the surrounding environment. Documentation may be used to help avert recurrences, as evidence in future legal action, for assessment of liability by insurance companies, and for review by government agencies. Methods of documenting can include a written transcript taken from tape recordings made during or after the emergency or from field books. Whatever document form is used, all must be:

- Accurate—all information recorded objectively;
- Authentic—a chain-of-custody procedure should be used. Each person making an entry must date and sign the document. Keep the number of documentors to a minimum (to avoid confusion and because they may have to give testimony at hearings or in court). Nothing should be erased. If details change or revisions are needed, the person making the notation should mark a horizontal line through the old material and initial the change.
- Complete—At a minimum, the following should be included:
  —chronological history of the incident;
  —facts about the incident and when they became available;
  —title and names of personnel and composition of teams;
  —decisions made and by whom; orders given and to whom, by whom, and when they were made; actions taken, including who did what, when, where, and how;
  —types of samples and test results, including air monitoring results;
  —possible exposures of site personnel; and
  —history of all injuries or illnesses during or as a result of the emergency.

14.12 Sources of Information

Bronstein, Alvin C. and P.L. Currance: *Emergency Care for Hazardous Materials Exposure.* 2nd ed. St. Louis, MO: Mosby Year Book, Inc., 1994.

Currance, P.L.: Staging Decon Operations. *Rescue* (July/August 1989).

Dennison, Mark S.: *Hazardous Waste Regulation Handbook: A Practical Guide to RCRA and Superfund.* New York: John Wiley & Sons, 1994.

Einolf, D.M.: Complying with HAZWOPER & Understanding Your Training and Planning Requirements. *ECON* (January 1993).

Harper, S.F.: LEPCs and Title III. *Hazmat World* (September 1989).

"Hazardous Waste Operations and Emergency Response: Final Rule," *Federal Register 29*: 1910 (6 March 1989) (updated through 1998).

Hopkins, R. and E. Blair: Industrial Emergency Management Requires More than Lucky Breaks. *Occupational Health and Safety* (March 1993).

Koren, Herman: *Handbook of Environmental Health and Safety Principles and Practices.* Vol. II. Chelsea, MI: Lewis Publishers, 1991.

Sullivan, Thomas S. (ed.): *Environmental Law Handbook.* 14th ed. Rockville, MD: Government Institutes, Inc., 1997.

Wheeler, John R.: *Toxic Substance Control Act Compliance Guide and Service,* 3 Vols. Environmental Books, 1997.

Worobec, Mary Devine and C. Hogue: *Toxic Substances Controls Guide.* 2nd ed. Washington, DC: Bureau of National Affairs, 1992.

FIFTEEN

Example of a Site Health and Safety Plan (HASP)

15 | CONTENTS

A site-specific HASP is required by 29 CFR 1910.120. This plan helps ensure that all safety issues of site operations are thoroughly examined prior to commencing field work. The site HASP should be modified as needed to include every stage of site activity.

Because planning requires information, planning and site characterization should be coordinated. An initial HASP should be developed so that site characterization can proceed in a safe manner. The information from this characterization can then be used to refine the HASP so that future site activities can proceed safely. The plan should be revised whenever new information about site hazards is obtained.

15.1 Development and Content

Development of a site HASP should involve both off-site and on-site management and be reviewed by industrial hygiene and occupational safety experts, physicians, chemists, or other appropriate personnel. At a minimum, the plan should:

- name key personnel and alternates responsible for site safety;
- describe the risks and control measures associated with each operation conducted;
- list training that is required for personnel to perform their job responsibilities and to handle the specific hazardous situations they may encounter;
- describe the protective clothing and equipment to be worn by personnel during various site operations;
- describe any site-specific medical surveillance requirements;
- describe the requirements for periodic air monitoring, personnel monitoring, and environmental sampling, if needed;
- describe the actions to be taken to mitigate existing hazards (such as containment of contaminated materials) to make the work environment less hazardous;
- define site control measures and include a site map that also shows the routes to medical facilities;
- establish decontamination procedures for personnel and equipment;
- set forth a Contingency Plan for safe and effective response to emergencies; and
- set forth the site's SOPs. SOPs are those activities that can be standardized (such as decontamination, sampling, and respirator fit-testing). These procedures should be:
  - —prepared in advance;
  - —based on the best available information, operational principles, and technical guidance;
  - —field-tested by qualified health and safety professionals and revised as appropriate;
  - —appropriate to the types of risk at the site;
  - —easy to understand and use;

—provided in writing to all site personnel, followed by a briefing on their use; and
—included in training programs for site personnel.

15.2 Contents of a Site HASP

The HASP should contain specific information as described in this section. A sample HASP is also presented at the end of this unit. Keep in mind, all information should be adapted to a specific site and based on the preliminary characterization findings of that site.

15.2.1 INTRODUCTION

The introduction sets forth the scope and applicability of the site HASP to the intended location and tasks. It also defines who is allowed on-site, such as workers, supervisors, visitors, and the media, and tells where they may be while on-site.

15.2.2 KEY PERSONNEL AND HAZARD COMMUNICATIONS PLAN (29 CFR 1910.120(b)(2))

This section should include the names of key personnel such as the Project Manager, the Field Operations Leader, the Site Supervisor, the SSHO, and their alternates. The plan should also identify communications procedures and provide for briefings to be held prior to a site activity being initiated. Such meetings should be held at any time they may appear necessary to ensure that employees are adequately apprised of the health and safety procedures being followed on the site. This is especially important when new hazards are located and PPE and tasks are changed accordingly.

15.2.3 HEALTH AND SAFETY RISK ANALYSIS (29 CFR 1910.120 (b)(4))

Health and safety risk analyses should be established for each task and operation identified in the site-specific work plan. Discussions should include the identification of chemical contaminants, affected media (land, surface water, air, groundwater), concentrations of hazardous chemicals, and potential routes of exposure for use in risk analysis. The risk analysis section should also address anticipated on-site operations and tasks and associated safety problems.

15.2.4 SITE CONTROL MEASURES (29 CFR 1910.120 (d))

The site control program in the HASP specifies the procedures that will be used to minimize employee exposure to hazardous substances before the cleanup operations begin and during the operations. The program must be developed during the planning stages of a hazardous waste cleanup operation, and must be modified as any new information becomes available. The site control program should include a site map, designation of work zones, site communication, safe work practices, identification of the nearest medical assistance, and a description of the "buddy system" for site operations.

15.2.5 EMPLOYEE TRAINING ASSIGNMENTS (29 CFR 1910.120 (e))

Training assignments should address the employee's initial health and safety training, annual health and safety refresher training, on-the-job training, supervisory training, and first-aid and CPR training. Employees should not be allowed to supervise field activities until they have received training commensurate with their responsibilities. Training and briefings on new material or legislation should be held when necessary.

15.2.6 MEDICAL SURVEILLANCE (29 CFR 1920.120 (f))

The medical surveillance program is required for monitoring the health status of personnel who are potentially exposed to hazardous substances in the field and who wear respirators 30 days or more per year. It must include initial and periodic medical examinations, examination upon termination of employment, and medical recordkeeping.

15.2.7 PERSONAL PROTECTIVE EQUIPMENT (29 CFR 1910.120 (g))

The HASP must describe the different PPE ensembles that will be used to address potential hazards during site activities. The HASP should also include or refer to a comprehensive PPE program that addresses site hazards; duration of site activities; limitations of PPE during temperature extremes; PPE selections, maintenance, storage, and decontamination; and training for PPE use, inspection, and monitoring. PPE should be used only when engineering controls and work practices are insufficient to adequately protect workers against exposure.

15.2.8 AIR AND PERSONNEL MONITORING (29 CFR 1910.120 (h))

The HASP must describe the employee and air monitoring equipment and environmental sampling techniques and instrumentation that will be used on-site for evaluating potential exposure to contaminants that result from site activities. The monitoring program must include procedures for initial entry monitoring, periodic monitoring, and monitoring of high risk employees.

15.2.9 SPILL CONTAINMENT PROGRAM (29 CFR 1919.120 (j))

The HASP should include any elements of the spill containment program that may be relevant to the site. It should also provide procedures on containing and isolating hazardous substances spilled in the course of a transfer, major spill, or an on-site release.

15.2.10 CONFINED SPACE ENTRY PROCEDURES (29 CFR 1910.120 (b)(9))

If confined space entry is anticipated on-site, the HASP should describe procedures for entry. Such procedures ensure the safety of site personnel who must enter areas where natural ventilation is insufficient to reduce contaminant concentrations.

15.2.11 DECONTAMINATION PROCEDURES (29 CFR 1910.120 (k))

The HASP should include decontamination procedures, for both individuals and equipment on-site, as well as in places where there is a potential for exposure to a hazardous substance. These procedures should explain how to minimize contact with hazardous substances and how to conduct personnel and equipment decontamination when leaving a contaminated area.

15.2.12 EMERGENCY RESPONSE/CONTINGENCY PLAN (29 CFR 1910.120 (l))

The emergency response plan in the HASP must include a description of how anticipated emergencies would be handled at the site and how the risks associated with a response would be minimized. Such information should include personnel roles and lines of authority, emergency recognition and prevention, evacuation routes and procedures, the emergency contact and notification system, emergency medical treatment procedures, and emergency equipment and facilities. This plan must be developed prior to beginning site operations.

15.2.13 EMERGENCY COMMUNICATION

In the event that a spill takes place, a worker is exposed, or other emergencies arise, lines of communications between the work site and local emergency agencies should be in place. Hospitals should be aware of potential needs, as should fire departments, HAZMAT teams, LEPCs and other local assistance.

In any given plan, more or less detail will be required depending upon the hazards involved, the number and types of tasks involved, the discovery of previously undetected hazards, changes in remediation plans, and other circumstances. In order to demonstrate how the abovementioned elements can be put together into a viable HASP, an example plan has been created and is presented at the end of this unit. While based on an actual remediated site, all names, telephone numbers, and locations are imaginary. Specific product names are included only because they were the actual products used at the site in the example. No endorsement of these products is being made by the publication of this sample HASP.

15.3 Sources of Information

Environmental Protection Agency: *Standard Operating Safety Guides* (Publication 9285.1–03, PB92–963414). Washington, DC: Office of Emergency and Remedial Response, 1992.

Site-Specific Health and Safety Plan (Example)

HB&G RR TIE TREATING PLANT: LUDLOW, IDAHO
(This HASP is based on an actual site, but names and locations have been changed.)

This health and safety plan (HASP) has been developed by HB&G Railroad to provide minimum guidelines for employees and contractors working at the Ludlow Tie Plant site. HB&G requires all contractors to develop and administer their own job-specific HASPs. Contractors shall be responsible for the health and safety of their own employees working on the site.

This is a general plan and is not intended to address health and safety issues associated with specific site activities. Since different activities on-site will have different health and safety requirements, each major activity requires its own activity-specific site HASP and standard operating procedures (SOPs). Companies responsible for implementing site activities such as on-site construction, drilling, remedial action, pilot studies, etc., are responsible for developing and administering a site HASP for their own employees. Activity-specific site safety plans shall be prepared in accordance with the requirements set forth in 29 CFR 1910.120. These HASP(s) must address each work task. A job safety analysis must be prepared and appended to the HASP for specific tasks. Contractors are required to submit site HASPs to HB&G for file purposes only prior to initiation of any site activities. HB&G has a Site Operator employed full time at the site whose duties include overseeing of contractors and attending to visitors.

All contractors working on the site must comply with OSHA training and medical monitoring requirements stated in 29 CFR 1910.120. HB&G requires that all contractors submit a "Contractor Medical and Training Certification Letter" following the format presented in Figure X-1.

A 1.0 FACILITY DESCRIPTION

The Ludlow Tie Plant treated railroad ties from 1896 to 1983. Initially, a zinc chloride treatment process was used. In the 1920s, the process was replaced with a creosote and oil treatment, and in more recent years pentachlorophenol was used. The site was placed on the National Priorities List (NPL) by the EPA in 1992. Approximately 140 acres of surface area is contaminated. Hazardous contaminants at the site include chemicals used in the wood-treating operations performed at the plant. Some of the organic wood-treating chemicals used include pentachlorophenol (PCP), creosote, and the aliphatic hydrocarbon oils. Inorganic wood-preserving chemicals include zinc chloride and copper arsenate.

In April 1997, buildings were demolished and a groundwater reverse gradient cutoff wall contaminant isolation system was completed as a first operable unit under CERCLA. This system, a subsurface soil bentonite cutoff wall which surrounds the site, consists of a groundwater management system maintaining an inward hydraulic gradient across the wall, and an activated carbon treatment system which treats the contaminated groundwater that backs up against the wall. The objective of the reverse gradient cutoff wall contaminant isolation system is to prevent off-site migration of contaminated groundwater. HB&G is currently engaged in a process development program to achieve in-situ treatment of on-site contaminants.

The Ludlow Tie Plant site is located just south of Ludlow, Idaho. Services of the town of Ludlow are readily accessible to the site. A map of the site is shown in Figure X-2. The water treatment building serves as the Health and Safety Officer's office. City water, electricity, and telephone service are available in the water treatment building. The primary access roads to the site are shown on the map. A road built on top of the cutoff wall surrounds the site.

A 2.0 HAZARD CHARACTERIZATION

Various hazards may be present during activities at the Ludlow site including physical injuries, hypothermia, frostbite, heat stroke and exhaustion, drowning, oxygen deficiency, cave-in of trenches, electric shock, and chemical exposure. Activity-specific site safety plans must address the hazards and evaluate the risks associated with implementation of specific tasks.

Exposure to the hazardous contaminants at the Ludlow site is a potential risk in many of the site activities. Heavy contamination is visually identified by its

TO: (Site Operator)

FROM: (Contractor)

DATE:

SUBJECT: Authorized Personnel List
(Name of Site)

This letter certifies that the personnel listed below are (Contractor)____________________ employees who are scheduled to perform work on the (Site Name) ______________________________ during (Month/Year) ____________________. Each of the employees named below participates in the medical surveillance program according to 29 CFR 1910.134, has beeen examined by a physician within the past 12 months, and has no medical restrictions for respirator use for personal protective equipment use, or for work on a hazardous waste site. In addition, each employee named below has beem trained according to the provisions of 29 CFR 1910.120 (e) established for the protection of hazardous waste site workers.

Name of Employee | Job Function

The physician to be notified in case of emergency is:

____ (___)____________

Contractor's Consulting Physician | Phone Number

NOTE: In case of emergency, if the above information is not provided, Contractor employees will be taken to the emergency room, as identified in the site Health and Safety Plan.

__________________________________ ____________________

Signature of Contractor | Date

Title:____

Figure X-1 Example of a Contractor Medical and Training Certification Form

black oily appearance and characteristic odor. Exposure to contamination by any pathway, including inhalation, ingestion, and skin contact, should be avoided.

The wood treating chemicals used in the tie-treating operations included many organic compounds. Creosote is a complex hydrocarbon substance produced from the distillation of coal tar and consists of over 200 major and several thousand minor organic compounds.

The hazardous compounds of concern associated with wood treating contamination include polynuclear aromatic hydrocarbons (PNAs); benzene; alkyl benzenes, consisting of toluene, ethyl benzene, xylenes, and isopropyl benzene; pentachlorophenol; phenol;

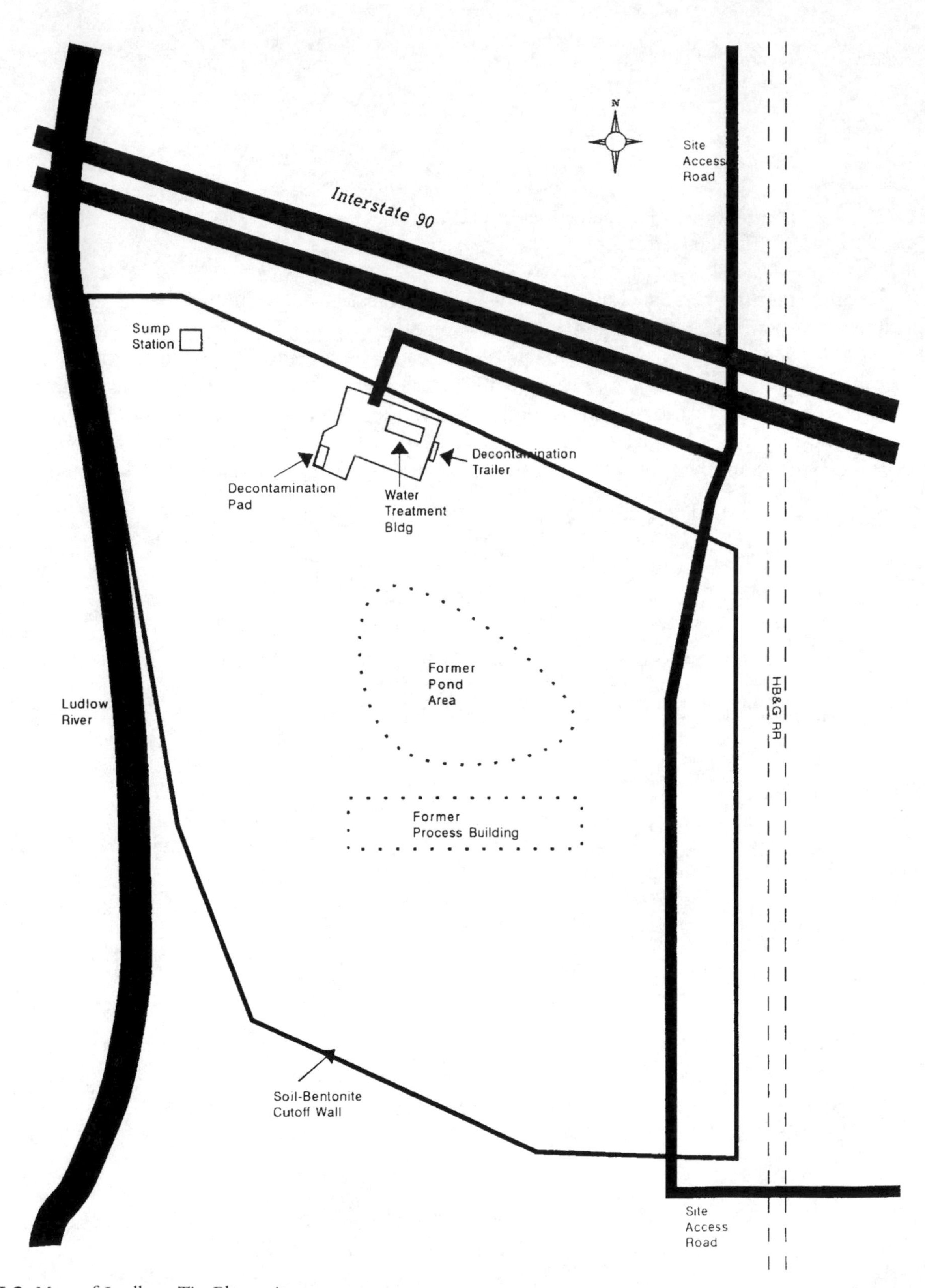

Figure X-2 Map of Ludlow Tie Plant site

naphthalene; and the polychlorinated dibenzo-dioxins (PCDDs) and dibenzo-furans (PCDFs). A brief summary of the toxicant profiles relevant to human exposure concerns for these compounds is presented below.

A 2.1 POLYNUCLEAR AROMATIC HYDROCARBONS (PNAS)

PNAs comprise a large and diverse class of multiple-ring, aromatic, organic compounds, some of which have proven to be carcinogenic. Carcinogenicity is the most serious manifestation of chronic exposures. Primary routes of exposure include dermal absorption, ingestion, and inhalation of dusts contaminated with PNAs.

A 2.2 AROMATIC HYDROCARBONS

Aromatic hydrocarbons include benzene and the alkyl benzenes listed above, such as toluene. They comprise a class of single ring aromatic hydrocarbons with attached carbon chains. Benzene can be absorbed through intact skin. Acute exposure may produce central nervous system depression and irritation to skin, eyes, and the respiratory tract. Benzene can cause abnormal changes in blood, and is a human carcinogen. Primary routes of exposure are inhalation and dermal absorption.

A 2.3 PHENOL

Phenol is a single ring aromatic hydrocarbon that is a type of alcohol. It is highly toxic. It is readily absorbed through intact skin. Symptoms of exposure may include vomiting, difficulty in swallowing, nervous system disorders, fainting, and skin rashes. However, such symptoms are generally not associated with low-level exposure.

A 2.4 PENTACHLOROPHENOL

This is a compound that includes chlorine atoms attached to phenol. A wide variety of acute toxic effects can be caused by ingestion or dermal exposure, including fever, vomiting, dizziness, and irritation to eyes, nose, throat, and skin. It can cause a serious dermatitis.

A 2.5 NAPHTHALENE

This is a common two-ring aromatic hydrocarbon abundant in coal tar pitch and is used in other industries to make moth balls. Naphthalene is detectable at very low concentrations by smell and this characteristic makes it a good indicator of contamination. Chronic exposure may produce cataracts of the eyes.

A 2.6 POLYCHLORINATED DIBENZO-P DIOXINS AND POLYCHLORINATED DIBENZOFURANS

These are two similar chemical families consisting of two chlorinated benzene rings bridged together. These compounds are undesirable by-products of the manufacturing process involving chlorinated phenols. Of the 210 possible types (isomers) of these two groups, most of the toxicological studies have been performed on compounds 2, 3, 7, 8-tetrachloro dibenzo-dioxin (TCDD) and 2, 3, 7, 8-tetrachloro dibenzo-furan (TCDF), which are believed to be some of the most toxic chemicals known based on studies being carried out by the EPA. These have shown mixed results concerning long-term health effects. Surprisingly, no evidence of adverse effects have been seen in rats and mice. Dioxins and furans have been detected in contaminated materials from the Ludlow site, although suspected 2, 3, 7, 8-TCDD and 2, 3, 7, 8-TCDF have not been detected on the site.

A 2.7 METALS

In addition to hazardous organic compounds, some inorganics may be present in the contaminated soils. Arsenic and cadmium are considered to be carcinogens by the NIOSH and ACGIH. Arsenic causes lung, skin, and lymphatic cancers while cadmium is associated with lung, prostatic, and kidney cancers. Zinc, copper, cadmium, chromium, and arsenic are potential contaminants of concern, especially in the pond areas. During pond closure activities, pond sediments were excavated and disposed. Therefore, most of the inorganic contaminants may have been removed from the site.

A 2.8 EXPOSURE LIMITS

Exposure limits for various contaminants detected on-site are presented in Table X-1. Limits shown represent concentrations in air which are believed to be acceptable for repeated worker exposure based on an 8-hour work day and 40-hour work week. OSHA is tasked with promulgation and enforcement of occupational

Table X-1 **Exposure Limits for Contaminants at HB&G Ludlow Tie Treating Plant Pond**

Contaminant	OSHA[a] PEL	NIOSH[b] REL	ACGIH[c] TLV
Single-Ring Aromatics			
Benzene	1.0 ppm	0.1 ppm (Ca)	0.5 ppm (Ca)
1, 2-Dichlorobenzene	50 ppm	50ppm	25 ppm
1, 4-Dichlorobenzene	75 ppm	(Ca)	10 ppm
Ethyl Benzene	100 ppm C	100ppm C	100 ppm
Toluene	200 ppm	100 ppm	50 ppm
Phenolics			
Pentachlorophenol+	0.5 mg/m^3	0.5 mg/m^3	0.5 mg/m^3
Phenol	5 ppm	5 ppm	5 ppm
Polynuclear Aromatic Hydrocarbons			
Coal Tar Pitch Volatiles+	0.2 mg/m^3	0.1 mg/m^3 (Ca)	0.2mg/m^3 (Ca)
Naphthalene	10 ppm	10ppm	10 ppm
Polychlorinated Dibenzo-p-Dioxin (PCDD)	—	lowest concentration	—
Polychlorinated Dibenzofuran (PCDF)	—	feasible	—
Metals			
Arsenic+	0.01mg/m^3 (C)	0.002mg/m^3(Ca)	0.01 mg/m^3 (Ca)
Cadmium+	0.005 mg/m^3	(Ca)	0.01 mg/m^3
Chromium+	1 mg/m^3	0.5 mg/m^3	0.5 mg/m^3
Copper (dust)+	1 mg/m^3	1 mg/m^3	1 mg/m^3
Zinc oxide (dust)+	15 mg/m	15 mg/m^3 (C)	10 mg/m^3

Ca = causes cancer in humans

C = Ceiling value

+Not detectible with FID analyzer.

[a]Occupational Safety and Health Administration permissible exposure limits (eight-hour time-weighted averages.)

[b]National Institute for Occupational Safety and Health recommended exposure limits (eight- or ten-hour, time weighted averages.)

[c]American Conference of Governmental Industrial Hygienists threshold limit values, time weighted average for 8-hour workday and 40-hour workweek.

References:

1. U.S. Department of Health and Human Services, *NIOSH Pocket Guide to Chemical Hazards*, US Government Printing Office, Washington, DC, 1997.
2. American Conference of Governmental Industrial Hygienists, *2000 TLVs and BEIs, Threshold Limit Values for Chemical Substances and Physical Agents,* ACGIH, Cincinnati, OH, 2000.

health standards for workplace exposures. These legally enforceable PELs are found in 29 CFR 1910, Subpart Z, "General Industry Standards for Toxic and Hazardous Substances." Unless otherwise noted, the exposure limits are 8-hour, TWA concentrations. NIOSH has recommended additional exposure limits for compounds beyond those regulated by OSHA. ACGIH has also established exposure guidelines in the form of TLVs for worker exposure. ACGIH recommendations are updated yearly based on the most recent scientific data available. OSHA and NIOSH limits are not updated yearly. OSHA may adopt, by reference, the NIOSH and ACGIH exposure recommendations (29 CFR 1910.1200).

Inhalation of airborne contaminants is not expected to be an important exposure pathway in open, outside areas at the site. The dominant organic contaminants have low vapor pressures and do not volatilize to a great extent. Air monitoring was performed during pond closure activities in July through October of 1997.

These samples were taken under worst-case conditions for volatilization of organics because highly contaminated sludges were exposed to the air and were pumped and heated. Six compounds or classes of compounds were monitored, including arsenic, coal tar pitch volatiles, naphthalene, benzene, pentachlorophenol, and toluene. For most of the samples collected, concentrations were below the detection limit. Benzene, pentachlorophenol, and toluene were not detected in any samples. The maximum concentrations detected for arsenic, coal tar pitch volatiles, and naphthalene were below the TLV, as shown:

AIR CONCENTRATIONS DURING THE PARTIAL POND CLOSURE

	ACGIH TLV	Maximum Reported Concentration
Arsenic	0.1 mg/m^3	0.00243 mg/m^3
Coal tar pitch volatiles	0.2 mg/m^3	0.09 mg/m^3
Naphthalene	10 ppm	1.4 ppm

Naphthalene, which was detected as high as 1.4 ppm, is a good indicator of airborne contamination because of its low threshold odor concentration. Naphthalene can be smelled in concentrations as low as 1 to 10 parts per billion (ppb). Naphthalene has the characteristic odor of mothballs, which are made from naphthalene.

Dusty conditions could create the potential for inhalation or ingestion of contaminated soil particles. Full facepiece APRs with MSA brand cartridge with at least an R-95 rating should be worn if the potential for ingestion and inhalation of airborne contaminated soil exists.

The primary exposure pathway of concern is through skin contact. All on-site personnel must minimize exposure to wood-treating contaminants by wearing the proper personal protective clothing. If contaminants come into contact with the skin, the area must be thoroughly washed with soap and water. Workers should shower at the end of the day before leaving the job site. If eye contact occurs, the eyes should be thoroughly flushed with large amounts of water or with an eye wash kit.

Table X-2 **Key Personnel**

Site Safety and Health Officer (Site Operator)	Keri Masden (XXX) XXX-XXX
Alternate Site Operator	Ruth Gray (by contract)
Director of Health, Safety, and Environment	Doug O'Donald (XXX) XXX-XXXX

A 3.0 KEY PERSONNEL

The HB&G Site Operator is also the designated SSHO, who is responsible for site safety and security (see Table X-2). When contractors or visitors enter the site, he informs them of hazards and established safety precautions. He verifies that contractors performing work with potential for exposure to hazardous wastes or hazardous substances have implemented a health and safety program that meets the requirements of 20 CFR 1910.120. The Site Operator initiates required emergency response operations. All injuries or emergencies must be reported to the Site Operator.

The HB&G Director of Health, Safety, and Environment (HS&E) in Denver, Colorado, has overall responsibility for the Ludlow Tie Plant site. Any safety concerns noted by the Site Operator are forwarded to the HB&G Director of HS&E. The Site Operator will report all injuries or emergencies to the Director of HS&E.

Contractors to HB&G engaged in site activities will designate an individual as the safety and health officer in their site HASP. This individual is responsible for implementing the work rules and provisions of their HASP.

A 4.0 SITE CONTROL

The purpose of site control is to minimize potential contamination of workers, protect the public from the site's hazards, prevent the spread of contamination from the site, and prevent vandalism and theft. The OSHA requirements for site control, presented in 29 CRF 1910.120(d), require as a minimum:

- a site map;
- site work zones;
- use of a "buddy system";
- site communications;
- SOPs or safe work practices; and
- identification of the nearest medical assistance.

The site control map presented in Figure X-3 identifies major features of the site control plan.

The concept of site work zones is intended to prevent the spread of hazardous substances by workers from contaminated areas to clean areas. Contamination exists throughout most of the area contained in the cutoff wall. Clean areas (support zone) of the site are shown in Figure X-3. A fence surrounds the site to prevent public access and vandalism. All visitors to the site must check in with the Site Operator at the water treatment building. Workers on-site must use the appropriate decontamination facilities and procedures to prevent taking contamination off-site or into the water treatment building and the clean zone of the decontamination trailer.

Some activities may be performed in highly contaminated areas or may bring contaminants to the surface where risk of exposure is increased. In such cases, the work zones are either flagged off or fenced, and access is limited to authorized personnel wearing the appropriate personal protective clothing. Work zones will be addressed in the activity-specific site HASPs.

The "buddy system" requires that work activities be performed with a partner who can provide assistance in case of an emergency. Personnel should never perform field work without one or more additional people present. For example, work in highly contaminated areas, excavating contaminated soil, sloping or shoring any excavation that personnel must enter, and changing out carbon vessels should not be performed alone. Any work activities performed in extreme weather conditions should not be performed alone.

Site communications systems include two-way radios and telephones. The primary means of communication between the on-site and off-site personnel is the telephone. The site operator carries a two-way radio when he is on-site away from his office.

SOPs and safe work practices are addressed in activity-specific site HASPs. When developing the site HASP, a hazard assessment of each activity is made, and the appropriate precautions must be identified and incorporated into the plan. Work limitations include the following, as a minimum:

- there will be no eating, drinking, chewing, or smoking in contaminated portions of the site;
- no worker will enter trenches or confined spaces without taking appropriate precautions in accordance with regulations and published practices for such entry;
- all workers will observe personnel protective clothing guidelines;
- all workers will observe decontamination guidelines;
- workers will perform air monitoring to characterize the site, especially when working in highly contaminated areas or when strong odors are apparent;
- when temperatures or wind chill are below freezing, a warm-up break in a heated area will be scheduled for 10 minutes every 2 hours or as necessary. Drink noncaffeinated, nonalcoholic beverages during each break. Do not become overheated during breaks. Remove or open outer clothing to prevent sweating. Monitor buddies for signs of frostbite. If severe shivering occurs, this is a sign of danger and requires immediate warming;
- personnel protective clothing can put workers at considerable risk for heat stress. Monitor workers for signs of heat stress (excessive sweating, elevated heart rate, pale or red skin, cramps, dizziness, nausea). Maintain body fluids at normal levels by drinking about 8 ounces of water or Gatorade every ½ hour. Schedule rest periods in safe areas where workers can open impermeable clothing and cool off in the shade as necessary to prevent heat stress; and
- work in dusty conditions could create a potential for ingestion and inhalation of contaminated soil particles. If the SSHO feels there is a potential for ingestion and inhalation of airborne contaminated soils, all team members are required to wear full face piece APRs with MSP cartridges.

All site HASPs must include provisions for emergency response.

A 5.0 PERSONAL PROTECTIVE EQUIPMENT

The use of personal protective equipment (PPE) is the most important factor in preventing chemical exposure while working in contaminated areas. Site personnel with potential for exposure must be trained in the appropriate selection and use of PPE. This training must be documented in writing. The levels of protec-

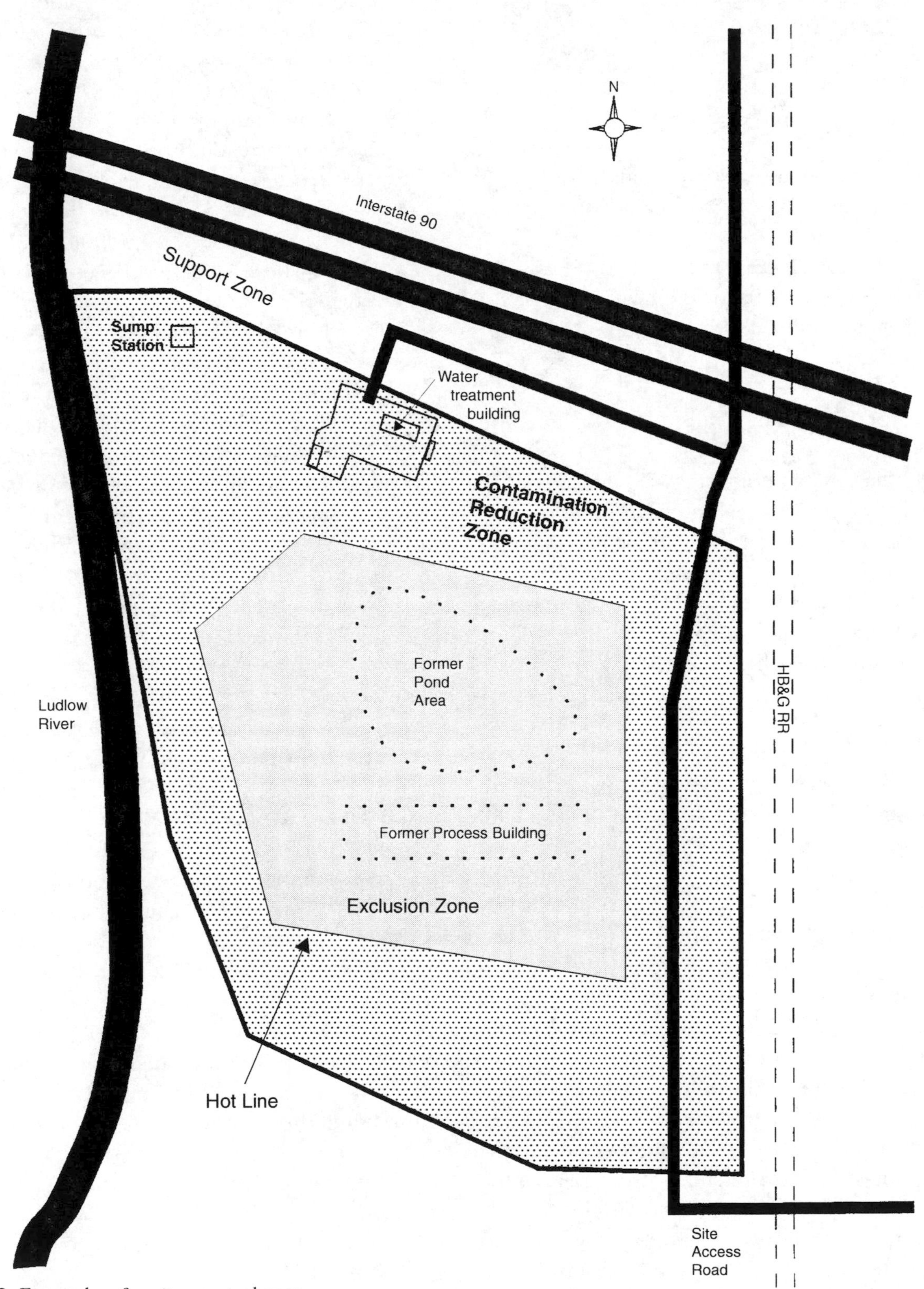

Figure X-3 Example of a site control map

tion required at the Ludlow site include Level D, Modified Level D, and Level C.

A 5.1 LEVEL D

Level D is only to be used when working in uncontaminated areas in the CRZ and EZ where no potential exists for contact with contaminated materials. Required PPE includes cotton coveralls, hard-hat, and safety glasses.

A 5.2 MODIFIED LEVEL D

This level of PPE is to be used when there is a potential for skin contamination, but respiratory protection is not required. Required PPE includes:

- Tyvek or Saranex coveralls;
- hard-hat;
- safety glasses or chemical splash goggles;
- rubber boots with steel toe/steel shank;
- disposable boot covers if working in contaminated soils;
- inner surgical gloves and outer neoprene gloves; and
- a face shield, if the potential for splashing exists.

A 5.3 LEVEL C

Level C is to be used when respiratory protection and skin protection are both required. This level of PPE includes Modified Level D clothing with the addition of an APR, with a full face piece and canister-equipped (MSA Cartridges).

Level D is appropriate for many of the operations at the Ludlow site. However, an upgrade to Modified Level D may be necessary when activities may involve contact with contaminated media, such as excavating contaminated soils, sampling contaminated streams, or working near operations where contaminants are brought to the surface. Personnel should also upgrade to Modified Level D protection whenever a potential for direct contact of contaminants with skin or clothing exists. Personnel should upgrade to Level C (respiratory protection) whenever atmospheric hydrocarbon concentrations are over background as determined by monitoring instruments or as instructed by the SSHO. If there is a potential for splashing of liquid contaminants, personnel should wear a face shield or full face APR, depending on the atmospheric hydrocarbon level.

The need for respiratory protection (upgrade to Level C) is determined by monitoring the atmosphere with a photoionization detector (PID) analyzer. The PID is a direct-reading, nonspecific instrument that detects organic compounds in the air. The instrument provides a direct readout proportional to the concentration of detected contaminants in terms of an equivalent concentration of benzene or other standards such as cyclohexane. The detection limit is approximately 1 to 2 ppm. The PID should be used to monitor the air in situations where hydrocarbon vapors are anticipated, such as during drilling or excavation, in confined spaces, when vapors are detected by smell, or when workers indicate irritation or other concerns. Typically, the smell of naphthalene will be present in atmospheres that are well below the detection capability of the PID. Background concentration is determined by taking a reading upwind of the site. If the monitored atmosphere exceeds background on the PID, personnel should wear an APR. The PID is not reliable in atmospheres exceeding 95 percent relative humidity and, therefore, may not be reliable when determining action levels in humid atmospheres such as vessels and sumps. Other nonionizing devices, such as absorbent charcoal tubes, may be needed to identify some low-level contaminants such as benzene.

A 6.0 DECONTAMINATION

Decontamination is the process of removing or neutralizing contaminants that have accumulated on personnel and equipment. The purpose of decontamination is to protect workers from hazardous substances and to minimize the transfer of hazardous substances to clean areas or off-site.

The first step in decontamination is to observe work practices and procedures that will minimize contact with contaminants. The general guidelines listed below should be followed:

- do not walk through areas of obvious contamination;
- avoid contact of clothing with hazardous substances;
- do not sit or kneel in contaminated areas;
- keep contaminated areas covered;
- wear disposable outer garments and use disposable equipment where appropriate;

- keep contaminated items in disposable bags until they are decontaminated;
- cover the interior of vehicles with disposable plastic to prevent contamination of seats, steering wheel, etc.;
- when wearing impermeable clothing, make sure there are no pathways for entry by zipping up, taping, etc.;
- make sure impermeable clothing has no cuts or punctures; and
- make sure any open wounds, such as cuts or scratches, are covered with an impermeable material.

Once clothing, PPE, tools, equipment, vehicles, etc., have been contaminated, they must be disposed of or decontaminated before leaving the site or entering the clean zone of the decontamination trailer. Two decontamination facilities are present at the site. One is a large decontamination pad and the second is a personnel decontamination facility. The large decontamination pad is a 20-foot by 50-foot concrete slab for large equipment and vehicles. The personnel decontamination facility consists of a small concrete slab and a trailer, which includes a dirty zone locker room, a shower, sink, and toilet facilities, and a clean zone locker room.

Other miscellaneous decontamination equipment includes the following:

- large steam cleaner for large decontamination pad;
- long-handled brushes for cleaning large equipment and vehicles;
- galvanized buckets and tubs;
- brushes;
- scrapers;
- detergent;
- trash cans and plastic liners for disposable items;
- disposable wipes; and
- soap, washcloths, and towels for personnel.

The SSHO is responsible for monitoring decontamination procedures and evaluating their effectiveness. If the procedures are found ineffective, appropriate steps must be taken to correct any deficiencies.

Contaminated large equipment and vehicles must be steam cleaned with high pressure sprayers before leaving the site. Personnel operating the steam cleaner must wear Modified Level D protective equipment with a face shield to prevent exposure or possible burns from splashing. A long brush can be used, if necessary, to help remove mud or contamination.

Small equipment items can either be steam cleaned on the large decontamination pad or can be cleaned with a brush, detergent, and water in a galvanized tub on the personnel decontamination pad. All equipment and tools must be cleaned before removing any personal protective clothing.

The process described below is required for personnel decontamination for Level C or Modified Level D protection:

- dispose of outer boots. Scrub rubber boots with soapy water and a brush in a galvanized tub;
- scrub outer gloves with detergent and water in a galvanized tub;
- rinse with water;
- remove boots and outer gloves first, then the Tyvek suit. Discard the Tyvek suit. Place gloves and boots in separate plastic bags;
- remove and clean the APR with disinfectant/sanitizer and water. Rinse well with clean water;
- remove and dispose of inner gloves;
- enter the dirty zone locker room and remove clothing. Store all equipment in a locker;
- wash thoroughly in the shower; and
- enter the clean zone locker room to put on street clothes.

To prevent the possibility of contaminating personal clothing, work clothes must be worn when the potential for contamination exists. At a minimum, this is Level D PPE. Work clothes should be laundered on-site or at a special, dedicated industrial laundry facility. They are not to be taken home for cleaning! Street clothes should only be worn to and from the site, or when performing activities where no potential for contamination exposure exists, such as in office areas or in the Support Zone.

If any permeable clothing becomes wet with contamination, the clothing should be removed and a shower should be taken immediately. Any contaminated clothing should be disposed of or decontaminated and washed before removing from the site. Nondisposable coveralls worn under Tyvek suits should be washed daily if potential exposure has occurred.

All discarded contaminated materials should be placed in appropriately labeled drums or other suitable, designated containers and stored for proper off-site disposal.

A 7.0 EMERGENCY RESPONSE

The SSHO or Site Operator is responsible for notifying the proper authorities in case of an emergency. The SSHO may also formally designate another site person to contact authorities in case of an emergency. Possible emergency situations at this site include process upsets and mechanical failure, fire and explosion, personal injury, chemical release, and break-in, theft, or vandalism.

A list of authorities to be contacted is presented in Table X-3. This list must be kept near the phone in a visible location. Any emergencies must be reported to the SSHO and the Director of HS&E.

Examples of process upsets include a burst pipe, a leaking valve or pump seal, overpressure of a vessel, etc. The Site Operator should shut down the system or stop the source of pressure immediately, and then contact the Director of HS&E.

Small fires may be put out with a fire extinguisher or water hoses. All fires should be reported to the Ludlow Fire Department. If a fire is in a location that is designated as a confined space and the space is filled with smoke, personnel should not enter the building. Immediately call the fire department from the nearest phone.

In case of personal injury, appropriately trained personnel should apply first aid and get medical attention immediately. The Ludlow Hospital should be contacted if ambulance service is required, or the injured person should be driven to the emergency room.

Any incidents of chemical exposure, including inhalation of contaminants in concentrations above the TLV or significant skin contact, should be reported to the occupational physician. MSDSs should be consulted for toxicity information and first aid procedures for chemical exposure. MSDSs are kept on file in the office of the Site Operator.

The Site Operator will immediately notify the Northumberland County Sheriff's Department in case of break-in, theft, or vandalism.

Table X-3 **Emergency Response List**

Report all emergencies to the HB&G Railroad Site Operator immediately. If the Site Operator is not available, notify the Director of HS&E.

HB&G Railroad Health and Safety Officer (Site Operator):
Keri Madsen
Phone: (XXX) XXX-XXXX

HB&G Railroad Director of Health, Safety and Environment:
Doug O'Donald
Phone: (XXX) XXX-XXXX

In case of fire or explosion, contact:
Ambulance
Phone: 911
Ludlow Fire Department
Phone: 911

In case of personal injury, contact:
Ludlow Hospital
Phone: (XXX) XXX-XXXX

In case of break-in, theft, or vandalism, contact:
Northumberland County Sheriff's Department
Phone: 911

SIXTEEN

Containers

16 | CONTENTS

Accidents can occur during the handling of drums and other hazardous waste containers. Hazards include detonations, fires, explosions, vapor generation, and physical injury resulting from moving heavy containers or from working around stacked and deteriorated drums. While these hazards are always present, following the proper procedures can lessen the risks to site personnel.

This unit describes practices and procedures for the safe handling of drums and other hazardous waste containers. It is intended to aid the Project Team Leader in developing a waste container handling program. In addition to the regulations in this unit, the Project Team Leader should also be aware of other pertinent regulations, including:

- OSHA regulations 29 CFR Parts 1910.120(j), (p)(6) and 1926.252, .407, and .903, which describe general requirements and standards for storing, containing, and handling chemicals and containers and for maintaining equipment used for handling materials;
- EPA regulation 40 CFR Part 265, Appendix IV and Appendix V, which stipulate requirements for types of containers, maintenance of containers and containment structures, and design and maintenance of storage areas;
- DOT regulation 49 CFR Parts 171 through 178, which stipulate requirements for containers and procedures for shipment of hazardous wastes; and
- OSHA regulation 29 CFR Part 1910.1030, which deals with precautions, training, handling, and housekeeping of materials considered to be potentially infectious or bloodborne pathogens.

16.1 Inspection

The appropriate procedures for drum handling depend on the drum contents. Therefore, prior to any handling, drums should be visually inspected to gain as much information as possible about their contents. The inspection crew should always look for the following:

- symbols, manufacturing labels, words, or other marks on the drum indicating its contents;
- signs of deterioration, such as corrosion, rust, and leaks;
- indications that the drum is under pressure, such as swelling, bulging, or hissing;
- drum type and possible contents based on drum type (refer to Table 16-1); and

Figure 16-1 Bulging drums must be approached with caution

- configuration of the drumhead
  - —a removable lid is designed to contain solid material or another entire drum;
  - —a lid with a bung is designed to contain a liquid;
  - —a lined drum may contain a highly corrosive or otherwise hazardous material.

Conditions in the immediate vicinity of the drums may provide information about drum contents and their associated hazards. For example, if the air near the drums is high in organic vapors, the drums may contain fuels or organic solvents. Monitoring should be conducted around the drums using instruments such as a radiation survey instrument, organic vapor monitors, and combustible gas meters.

As a precautionary measure and until their contents are characterized, personnel should assume that unlabelled drums contain hazardous materials. They should also remember that drums are frequently mislabeled, particularly drums that are reused. A drum's label may not describe its contents accurately.

If there is the possibility of buried drums, ground-penetrating systems such as radar, electrical resistively, magnetometry, and metal detection can be used to estimate their location and depth.

16.2 Planning

Since drum handling is dangerous, every step of the operation should be carefully planned based on all available information, including drum contents, integrity of the sides and ends, as well as drum weight. A full drum can weigh 500 pounds. Moving drums improperly can quickly cause broken fingers, broken feet, and injured backs. The results of the preliminary inspection can be used to determine hazards and the appropriate response, as well as which drums need to be moved, opened, or sampled. A preliminary plan should be developed that specifies the extent of handling necessary, the personnel selected for the job, and the most appropriate procedures for handling the drum(s) based on the hazards associated with the probable contents as determined by visual inspection. This plan should be revised as new information is obtained during drum handling.

16.3 Handling

Handling may or may not be necessary depending on how the drums are positioned at a site. The purpose of handling is to:

- respond to any obvious problems that might impair worker safety, such as radioactivity, leakage, or the presence of explosive substances;
- unstack and orient drums for sampling;
- if necessary, organize drums into different areas on-site to facilitate characterization;
- orient drums for leak control or for emptying; and
- conduct removal of drums or other remedial action.

Table 16-1 **Special Drum Types and Their Associated Hazards**

Type	Contents
Polyethylene or PVC-Lined Drums	These drums often contain strong acids or bases. If the lining is punctured, the substance usually quickly corrodes the steel, resulting in a significant leak or spill. Polyethylene and PVC can also be degraded by ultraviolet light and attacked by some chemicals, such as carbon tetrachloride or nitrobenzene, over time.
Non-ferrous Drums	These are very expensive drums that usually contain an extremely dangerous material or a material of very high value. They can be made of aluminum, nickel, stainless steel, or other unusual metals.
Steel Drums	These drums are generally used for non-corrosive liquids, sludges, and solids—a general all-purpose type of container.
Overpack Drums	Overpack drums are used to pack whole, leaking, or deteriorated drums. Overpacks should be approached with care because materials inside may have leaked and begun breaking down into other hazardous products.
Fiber Drums	These drums are used to dispose of or to store dry solids, such as soap, asbestos, pesticide granules, and sodium hydroxide (caustic) crystals. Some can be lined to contain liquids, so they are not limited to dry materials. Many are used for disposal of asbestos or lead paint wastes removed from structures.

Figure 16-2 Using a drum cart to move a drum

Since accidents occur frequently during handling, particularly initial handling, drums should only be handled if absolutely necessary. Prior to handling, all personnel should be warned about the hazards and instructed to minimize handling as much as possible. In all phases of handling, personnel should be alert for new information about potential hazards. These hazards should be dealt with before continuing with more routine handling operations. Overpack drums (larger drums in which leaking or damaged drums are placed for storage or shipment (49 CFR Part 173.3(c))) and an adequate volume of absorbent should be kept near areas where minor spills may occur. Where major spills may occur, a containment berm that is adequate enough to contain the entire volume of liquid in the drums, should be constructed before any handling takes place. If the drum contents spill, personnel trained in spill response should be enlisted to isolate and contain the spill.

Several types of equipment can be used to move drums, including:

- a drum grappler attached to a hydraulic excavator;
- a small front-end loader that can be either loaded manually or equipped with a bucket sling;
- a rough terrain forklift;
- a roller conveyor equipped with solid rollers; or
- drum carts designed specifically for drum handling.

Figure 16-3 Moving a drum by hand is dangerous

Figure 16-4 Examples of overpack (or salvage) drums

The drum grappler is the preferred piece of equipment for drum handling. It keeps the operator removed from the drums, thus lessening the likelihood of injury if they detonate or rupture. If a drum is leaking, the operator can stop the leak by rotating the drum and immediately placing it into an overpack. In case of an explosion, grappler claws help protect the operator by partially deflecting the force of the explosion.

The least desirable method of moving drums is by hand. Even tipping and rolling an intact drum presents a physical hazard because of the drum's weight, which can be over 500 pounds. This practice of tipping and rolling drums should also be avoided since it can cause injury to the back and extremities. However, sometimes drums must be moved manually.

The following procedures can be used to maximize worker safety during any drum handling and movement operations:

- train personnel in proper lifting and moving techniques to prevent back injuries;
- ensure that the vehicle selected has sufficient rated load capacity to handle the anticipated loads, and that the vehicle can operate smoothly on the available road surface;
- supply operators with appropriate respiratory protective equipment when needed. Normally, either a combination SCBA/SAR with an air tank fastened to the vehicle, or an airline respirator with an escape SCBA are used due to the high potential hazards of drum opening;
- have overpacks ready before any attempt is made to move drums;
- before moving a drum, determine if the drum is sound and capable of being handled by mechanical means;
- determine the most appropriate sequence in which the various drums and other containers should be moved. For example, small containers may have to be removed first to permit heavy equipment to enter and move the larger drums;
- exercise extreme caution in handling drums that are not intact and tightly sealed; and
- ensure that the operator has a clear view of the roadway when carrying drums. Where necessary, have "load spotting" workers available to guide the operator's motions.

16.3.1 DRUMS CONTAINING RADIOACTIVE WASTE

If the drum exhibits radiation levels above background, immediately contact the health and safety office to seek specialized assistance such as a health physicist. Do not handle any drums determined to be radioactive until persons with expertise in this area have been consulted.

16.3.2 DRUMS THAT MAY CONTAIN EXPLOSIVE OR SHOCK-SENSITIVE WASTE

If a drum is suspected of containing explosive or shock-sensitive waste as determined by visual inspection or site characterization data, seek specialized assistance before performing any handling. The following procedures are recommended for any containers that may contain explosive or shock-sensitive waste.

- Handle drums with extreme caution and by a pre-determined procedure.
- Prior to handling these drums, make sure all nonessential personnel have moved a safe distance away.
- Use remote opening techniques or a grappler unit constructed for explosive containment when initially handling these drums.
- Palletize the drums prior to transport and secure the drums to the pallets.
- Use an audible siren signal system, similar to those employed in conventional blasting operations, to signal the commencement and completion of explosive waste handling activities.
- Have an all-purpose A-B-C fire extinguisher on hand to control small fires.
- Maintain continuous communication with the SSHO and/or the Command Post until handling operations are complete.

16.3.3 BULGING DRUMS

Pressurized drums are extremely hazardous. Wherever possible, do not move drums that may be under internal pressure, as evidenced by bulging or swelling. If a pressurized drum has to be moved, handle the drum with remote equipment or a grappler unit constructed for explosive containment, whenever possible. Either move the bulging drum only as far as necessary to allow seating on firm ground, or carefully overpack the drum. Exercise extreme caution when working with, or adjacent to, potentially pressurized drums. If possible, handle pressurized drums in the morning after they have cooled during the night. Also, ice can be packed around the drum to reduce pressure.

16.3.4 DRUMS CONTAINING PACKAGED LABORATORY WASTES (LAB PACKS)

Laboratory packs (such as drums containing individual containers of laboratory substances normally surrounded by cushioning, absorbent material) can be ignition sources for fires at hazardous waste sites. Until otherwise characterized, these containers should be considered to hold explosive/flammable or shock-sensitive materials. If handling is required, the following precautions are among those that should be taken.

- Prior to handling or transporting lab packs, make certain all nonessential personnel have moved a safe distance away.
- Whenever possible during initial handling, use a grappler unit constructed for explosive containment.
- Maintain continuous communication with the SSHO and/or the Command Post until handling operations are complete.
- Once a lab pack has been opened, have a chemist inspect, classify, and segregate the bottles inside without opening them. Bottles should be classified according to the hazards of the wastes. An example of a system for classifying lab pack wastes is provided in Table 16-2. The objective of a classification system is to ensure safe segregation of the contents. Once this has been done, the bottles and other containers should be packed with sufficient cushioning and absorption materials to prevent excessive movement of the bottles and to absorb all free liquids, and then shipped to an approved disposal facility.
- If crystalline material is noted at the neck of any bottle, consider it to be a shock-sensitive waste due to the potential presence of picric acid or other similar material. Get expert advice before attempting to handle it.
- Palletize the repacked drums prior to transport, securing the drums to pallets.

16.3.5 MEDICAL WASTES

Wastes from hospital and research facilities may contain disease-causing organisms. These etiological agents may be dispersed into the environment through water, air, or direct contact. Therefore, the proper containment and handling of such materials is necessary to prevent the contamination of workers and the general population. Medical wastes include blood and body fluid-contaminated hospital waste, surgically-generated tissues and parts, laboratory animal carcasses and their wastes, experimental and commercial drugs that have not met laboratory and/or government specifications or are outdated, microbiological wastes, blood-

Table 16-2 **Example of Lab Pack Content Classification System for Disposal**

Classification	Examples
Inorganic acids	Hydrochloric Sulfuric
Inorganic bases	Sodium hydroxide Potassium hydroxide
Strong oxidizing agents	Ammonium nitrate Barium nitrate Sodium chlorate Sodium peroxide
Strong reducing agents	Sodium thiosulfate Oxalic acid Sodium sulphite
Anhydrous organics and Organometallics	Tetraethyl lead Phenylmercuric chloride
Anhydrous inorganics and Metal hydrides	Potassium hydride Sodium hydride Sodium metal Potassium
Toxic organics	PCBs Insecticides
Flammable organics	Hexane Toluene Acetone
Inorganics	Sodium carbonate Potassium chloride
Inorganic cyanides	Potassium cyanide Sodium cyanide Copper cyanide
Organic cyanides	Cyanoacetamide
Toxic metals	Arsenic Cadmium Lead Mercury

contaminated bandages, cultures of infectious agents, liquid blood and blood products, human pathological products, used and unused sharps, isolation wastes, and other materials of this type.

In response to public concern about medical waste washing up on national beaches in the 1980s, Congress enacted the Medical Waste Tracking Act of 1988 as Subtitle J to RCRA. It set up a program to monitor all medical waste and advise on methods of containment and disposal. Among the recommendations were the use of distinctive, well-labeled packaging; the segregation of infectious wastes from noninfectious materials; the use of a universal biological hazard symbol; specific packing materials (impervious, tear-resistant, colored, tightly-capped flasks for liquids and puncture-resistant containers for sharps); noncompaction of filled packages; minimal storage time; limited access to storage areas; and treatment by incineration, irradiation, or thermal inactivation.

Other biological wastes include biological warfare agents, which are produced, stored, and sometimes buried at various military or laboratory centers. Background work during site characterization should include a check for these activities, and extra care should be taken when such a site is entered.

Since standards have only recently been set for the packaging and disposal of medical wastes, it is possible for workers to encounter improperly labeled, packaged, and stored biological/infectious materials. Proper protective precautions and training should be undertaken if these wastes are suspected at waste cleanup sites.

16.3.6 DETERIORATED DRUMS

If a drum containing a liquid cannot be moved without likelihood of rupture, immediately transfer its contents to a sound drum using a pump designed for transferring that liquid. The drum must be grounded prior to pumping. Pumping ungrounded drums may produce a spark from static electricity. When transfer is complete, the pump should be decontaminated or disposed of properly.

16.3.7 BURIED DRUMS

When buried drums are encountered, in-situ control measures may be preferable to drum removal. Methods used to contain wastes on-site are often less costly than drum excavation and removal. These methods also minimize the exposure of field personnel to toxic or hazardous wastes. On-site control measures include capping, surface water controls, groundwater contaminant pumping, subsurface drains, and slurry walls. These measures require long-term security and site control since the waste products are still in place.

If necessity dictates that buried drums are to be removed, several safety factors must be considered. First, the area should be assessed for underground utilities, such as buried electric, telephone, gas, water, and sewer lines. Second, prior to initiating subsurface excavation, a detection system (such as ground-penetrating radar) should be used to estimate both the location and depth of the drums. Next, the soil should be removed with caution in order to minimize the potential for drum rupture. Finally, no actions used in dislodging or disturbing buried materials should be done without appropriate safety backup for fire, explosion, or potential exposures.

16.4 Opening Containers

Drums are usually opened and sampled during site investigations. To enhance the efficiency and safety of drum-opening personnel, the following procedures should be instituted.

- If a SAR protection system is used, place a bank of air cylinders outside the work area, and supply air to the operators via airlines and escape SCBAs. This reduces physical stress and heat stress on workers when they do not have to carry the full weight of SCBAs, and can extend work time.
- Protect personnel by keeping them at a safe distance from drums being opened.
- When opening a drum, do not lean over the bung. Instead, lean away from the bung to avoid any escaping vapor.
- If a drum is leaning, work on the uphill side to help minimize spills and to lessen the chance of a drum falling on personnel or equipment.
- If possible, continuously monitor during opening. Place sensors of monitoring equipment, such as colorimetric tubes, dosimeters, radiation survey instruments, explosion meters, organic vapor analyzers, and oxygen meters, as close as safely possible to the source of contaminants.

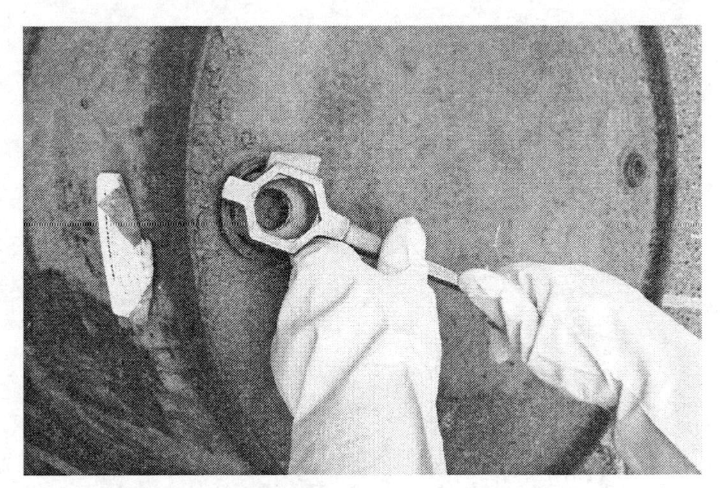

Figure 16-5 A brass bung wrench is an appropriate tool to open drums

- Whenever possible, use drum opening equipment to minimize worker exertion. These include remote-controlled devices such as:
  —a pneumatically-operated impact wrench to remove drum bungs;
  —hydraulically or pneumatically-operated drum piercers; and
  —backhoes equipped with bronze spikes for penetrating drum tops in large scale operations.
- Do not use picks, chisels, and firearms to open drums.
- If the drum shows signs of swelling or bulging, perform all steps slowly. Relieve excess pressure prior to opening and, if possible, from a remote location using devices such as a pneumatic impact wrench or hydraulic penetration device. If pressure must be relieved manually, place a barrier, such as explosion-resistant plastic sheeting, between the worker and the bung to deflect any gas, liquid, or solids that may be expelled as the bung is loosened.
- Drums that show no sign of swelling or bulging may still be under pressure, especially on a warm day. If a drum starts venting (usually accompanied by a hissing sound), stop the opening procedure, step back (upwind if possible), monitor the air, and let the drum release its pressure. This usually takes only a few seconds. Then proceed with the opening process.

- Open non-ferrous metal drums and polyethylene or polyvinyl chloride (PVC)-lined drums by removing or drilling the bung. Exercise extreme caution when manipulating these containers.
- Do not open or sample individual containers within laboratory packs.
- As soon as possible, reseal open bungs and drill openings with new bungs or plugs to avoid explosions and/or vapor generation. If an open drum cannot be resealed, place it into an overpack. Plug any openings in pressurized drums with pressure-venting caps set to a 5-pounds per square inch (psi) release to allow for the venting of vapor pressure. Special filters, such as a HEPA filter, are sometimes installed on waste drum bungs to contain radioactive particle releases through a pressure-venting cap.
- Measures should be taken to contain and mitigate any spills.
- Decontaminate equipment after each use to avoid mixing incompatible wastes.

16.5 Sampling Contents of Containers

Drum sampling can be one of the most hazardous activities for workers because it often involves direct contact with unidentified wastes. Prior to collecting any sample, the following steps should be completed.

- Research background information about the waste.
- Determine which drums should be sampled.
- Select the appropriate sampling device(s) and container(s).
- Develop a sampling plan that includes the number, volume, and locations of samples to be taken.
- Develop SOPs for opening drums, sampling, sample packaging, and transporting.
- Develop an SOP detailing the appropriate personal equipment for each activity.
- Never stand on drums. Use mobile steps or other stable platforms to achieve the height necessary to safely sample from the drums.
- Obtain samples using appropriate equipment. The most common sampling device is a glass rod (drum thief). Do not use contaminated items, such as discarded rags, for sampling. These items may contaminate the sample, producing erroneous test results. They may also react with the waste in the drum, producing a more hazardous situation.

16.6 Characterization of Contents

The goal of characterization, through methods such as sampling, is to obtain the data necessary to determine how to safely and efficiently package and transport the wastes for treatment and disposal. If wastes are bulked, they must be sufficiently characterized to determine which of them can be safely combined (see "Bulking" later in this unit). The first step in obtaining data should be to use standard tests to classify the wastes into general categories including reactives, water reactives, inorganic acids, organic acids, heavy metals, pesticides, cyanides, inorganic oxidizers, and organic oxidizers (Table 16-3). In some cases, further analysis may be needed to precisely identify the waste materials.

When possible, samples should be characterized using an on-site laboratory. This provides data as rapidly as possible and minimizes the time lag before appropriate action can be taken to handle any hazardous materials. Also, it precludes any potential problems associated with transporting samples to an off-site laboratory (for example, sample packaging, waste incompatibility, fume generation). If samples must be analyzed off-site, they should be packaged in accordance with appropriate regulations, such as those promulgated by DOT, EPA, commercial shipping companies, and the U.S. Postal Service.

16.7 Staging Containers

Although every attempt should be made to minimize drum handling, drums must sometimes be staged (for example, moved in an organized manner to predesignated areas) to facilitate characterization and remedial action, and to protect them from potentially hazardous site conditions (for example, movement of heavy equipment and high temperatures that might cause explosion, ignition, leaks, or pressure buildup). Stag-

Table 16-3 **Classification, Sampling, and Screening Form For Drum Content Characterization**

SITE: ____

DRUM SIZE:
0 unknown ____
1 55 gal. ____
2 30 gal. ____
3 other ____
specify ____

DRUM NO. ____

DRUM OPENING:
0 unknown ____
1 ring top ____
2 closed top ____
3 open top ____
4 other ____
specify ____

SAMPLE NO. ____

DRUM TYPE:
0 unknown ____
1 metal ____
2 plastic ____
3 fiber ____
4 glass ____
5 other ____
specify ____

SCREENING RESULTS (AREA):
0 unknown ____
1 radioactive ____
2 acid/oxidizer ____
3 caustic/reducer/cyanide ____
4 flammable organic ____
5 nonflammable organic ____
6 peroxide ____
7 air or water reactive ____
8 inert ____

DRUM COLOR:	PRI	SEC
0 unknown	____	____
1 cream	____	____
2 clear	____	____
4 white	____	____
5 red	____	____
6 green	____	____
7 blue	____	____
8 brown	____	____
9 pink	____	____
10 orange	____	____
11 yellow	____	____
12 gray	____	____
13 purple	____	____
14 amber	____	____
15 green-blue	____	____

DRUM CONTENTS COLOR:		
0 unknown	____	____
1 cream	____	____
2 clear	____	____
3 black	____	____
4 white	____	____
5 red	____	____
6 green	____	____
7 blue	____	____
8 brown	____	____
9 pink	____	____
10 orange	____	____
11 yellow	____	____
12 gray	____	____
13 purple	____	____
14 amber	____	____
15 greeen-blue	____	____

DRUM CONDITION:
0 unknown ____
1 good ____
2 fair ____

DRUM MARKING KEYWORD 1
____

DRUM MARKING KEYWORD 2
____

DRUM MARKING KEYWORD 3
____

DRUM CONTENTS STATE:	PRI	SEC
0 unknown	____	____
1 solid	____	____
2 liquid	____	____
3 sludge	____	____
4 gas	____	____
5 trash	____	____
6 dirt	____	____
7 gel	____	____

DRUM CONTENT AMOUNT:	
0 unknown	____
1 full	____
2 part	____
3 empty	____

CHEMICAL ANALYSIS:	YES	NO
radiation	____	____
ignitable	____	____
water reactive	____	____
cyanide	____	____
oxidizer	____	____
organic vapor	____	ppm
pH	____	

SCREENING DATA:	YES	NO	
Radioactive	____	____	≥ 1 mR over background
Acidic	____	____	pH ≥ 12
Air Reactive	____	____	Reaction of ≥ 10°F temp. change
Water Reactive	____	____	Reaction of ≥ 10°F temp. change
Water Soluble	____	____	Dissolves in water
Water Bath OVA	____	____	Reading = ____ ≥ 10 ppm = Yes
Combustible	____	____	Catches fire when torched in water bath
Halide	____	____	Green flame when heated with copper
Inorganic	____	____	Water Bath OVA and COMBUSTIBLE = No
Organic	____	____	INORGANIC = No
Alcohol/Aldehyde	____	____	Water Bath OVA, Water Soluble and COMBUSTIBLE = Yes
Cyanide	____	____	Drager tube over water bath ≥ 2 ppm
Flammable	____	____	COMBUSTIBLE = Yes, and SETA flashpoint ≤ 140° F
Oxidizer	____	____	Starch iodine paper shows positive reaction
Inert or Other	____	____	Everything "no" except INORGANIC or ORGANIC

Extracted from NIOSH/OSHA/USCG/EPA *Occupational Health and Safety Guidance Manual For Hazardous Waste Site Activities*, p. 11-8.

ing involves a trade-off between the increased hazards associated with drum movement and the decreased hazards associated with the enhanced organization and accessibility of the waste materials.

The number of staging areas necessary depends on site-specific circumstances such as the scope of the operation, the accessibility of drums in their original positions, and the perceived hazards. Investigation usually involves little (if any) staging, while remedial and emergency operations can involve extensive drum staging. The extent of staging must be determined specifically for each site and should be kept to a minimum. Up to five separate types of areas may be used, including the following.

- Initial staging area, where drums can be organized according to type, size, and suspected contents, and stored prior to sampling;
- Opening area, where drums are opened, sampled, and resealed. Locate this area a safe distance from the original waste disposal or storage site and from all other staging areas to prevent a chain reaction in case of fire or explosion. An added precaution would be to berm two sides for greater safety from explosions;
- Secondary sampling area, where, during large-scale remedial or emergency tasks, a separate sampling area may be set up some distance from the opening area in order to reduce the number of people present and to limit potential casualties in case of an explosion;
- Holding area (also known as staging area), where drums are temporarily stored after sampling pending characterization of their contents. Do not place unsealed drums with unknown contents in the holding area in case they contain incompatible materials. Either remove the contents or overpack the drum.
- Bulking area (also known as the final staging area), where substances that have been characterized are bulked for transport to treatment or disposal facilities. Locate the bulking area as close as possible to the site's exit, and:
  —grade the area and cover it with plastic sheeting;
  —construct approximately one-foot-high dikes around the entire area, or as needed to contain a spill; and
  —segregate drums according to their basic chemical categories (acids, heavy metals, pesticides, etc.) as determined by characterization. Construct separate areas for each type of waste present to preclude the possibility of intermingling incompatible chemicals when bulking.

16.8 Bulking Containers

Wastes that have been thoroughly characterized are often mixed together and placed in bulk containers such as tanks or vacuum trucks for shipment to treatment or disposal facilities. This increases the efficiency of transportation, reduces hazards to workers by reducing the number of times they must handle drums, and also reduces the potential of physical injury. Bulking should be performed only after thorough waste characterization by trained and experienced chemists and other personnel is complete. The preliminary tests described earlier under "Characterization" provide only a general indication of the nature of the individual wastes. In most cases, additional sampling and analysis to further characterize the wastes is necessary. This should include compatibility tests in which small quantities of different wastes are mixed together under controlled conditions and observed for signs of incompatibility. Vapor generation and heat of reaction tests should be conducted. Bulking is performed at the final staging area by using the following procedures.

- Inspect each tank trailer and remove any residual materials from the trailer prior to loading any bulked materials. This will prevent reactions between incompatible chemicals. Remember to dispose of any residual materials properly.
- To move hazardous liquids, use pumps that are properly rated by the National Fire Protection Association and that have a safety relief valve with a splash shield. Make certain the pump hoses, casing, fittings, and gaskets are compatible with the material being pumped. Include grounding provisions for hoses and couplings, as needed, to avoid dangerous static discharge.
- Because the pumping process may provide an ignition source (such as static electricity generated by moving liquid), the drum may require bonding and grounding.
- Inspect hose lines before beginning work to ensure that all lines, fittings, and valves are intact and contain no weak spots.
- Take special precautions when handling hoses because they often contain residual material that can splash or spill on the personnel operating the hoses.
- Protect personnel against accidental splashing.
- Protect lines from vehicular and pedestrian traffic.

- Store flammable liquids in approved containers.

16.9 Shipping Containers

Shipment of materials to off-site treatment, storage, or disposal facilities involves the entry of waste-hauling vehicles on the site. Hazardous waste shipments must comply with DOT regulations (49 CFR Parts 171-178) and EPA regulations (40 CFR Part 263). The following guidelines can enhance the worker's ability to safely comply with these regulations:

- Locate the final staging (bulking) area as close to the site exit as possible.
- Prepare a traffic control plan that minimizes conflict between decontamination teams and waste haulers. Install traffic signs, lights, and other control devices as necessary.
- Provide adequate areas for vehicles to turn around. When necessary, build or improve on-site roads.
- Stage hauling vehicles in a safe area until ready for loading. Drivers should remain in the cabs. The time that drivers spend in hazardous areas should be minimized.
- Outfit the driver with appropriate protective equipment and provide appropriate training.
- Drums to be shipped should be tightly sealed prior to loading, with all leaking or deteriorated drums overpacked. Nonreactive, absorbent material (such as vermiculite) sufficient to absorb 55 gallons of liquid should be placed in and around the salvage drum for further protection against spills. Make certain that truck beds and walls are clean and smooth to prevent damage to drums. Drums should not be stacked and should be secured to prevent shifting during transport.
- Keep bulk solids several inches below the top of the truck container. Cover loads with a layer of clean soil, foam, and/or tarp. Secure the load to prevent shifting or release during transport.
- Periodically weigh vehicles to ensure that vehicle and road weight limits are not exceeded.
- Decontaminate the vehicle prior to leaving the site to ensure that contamination is not carried onto the Support Zone public roads.
- Develop procedures for responding quickly to off-site vehicle breakdown and accidents to ensure minimal impact to the general public.

16.10 Sources of Information

"Bloodborne Pathogens," *Federal Register (Amended) 56:235* (30 July 1999).

Fox, David (managing ed.): *Occupational Safety and Health Standards for General Industry (29 CFR Part 1910), with amendments.* Chicago, IL: CCH Inc., 1998.

Koren, Herman: *Industrial Dictionary of Environmental Health and Occupational Safety.* Chelsea, MI: Lewis Publishers, Inc., 1995.

Koren, Herman and M.S. Bisesi: *Handbook of Environmental Health and Safety: Principles and Practices.* 3rd ed. Chelsea, MI: Lewis Publishers, 1995.

National Institute for Occupational Safety and Health, Occupational Safety and Health Administration, U.S. Coast Guard, Environmental Protection Agency: *Occupational Safety and Health Guidance Manual for Hazardous Waste Site Activities.* Washington, DC: U.S. Government Printing Office, 1985.

Sullivan, Thomas F. (ed.): *Environmental Law Handbook.* 14th ed. Rockville, MD: Government Institutes, Inc., 1997.

Wagner, K., et al: "Drum Handling Practices at Hazardous Waste Sites." EPA Project Summary from the Hazardous Waste Engineering Research Laboratory, Cincinnati, OH, August 1986.

Wheeler, John R.: *Toxic Substance Control Act Compliance Guide and Service.* Plano, TX: Environmental Books, 1997.

Worobec, Mary Devine and C. Hogue: *Toxic Substances Controls Guide.* 2nd ed. Washington, DC: Bureau of National Affairs Books, 1992.

SEVENTEEN

Transportation of Hazardous Materials and Wastes

17 | CONTENTS

As the transporting of hazardous materials increases in the United States, accidents involving these shipments also increase. During transportation of hazardous materials between 1991 and 2000, the spillage of more than one million pounds of hazardous materials was recorded. Of this, the greatest amount occurred during trucking, with fewer spills connected with movement by railroad and all other forms of transportation. One out of every two trucks on the road carries some type of hazardous material. From 1991 through 2000, a total of 140,304 incidents, 224 deaths, and 4,749 injuries occurred involving hazardous materials moved by all types of transportation. More than 3 billion tons of regulated hazardous materials are transported each year and over 800,000 shipments of this material occur each day. About 2 percent of these materials move by air. Illegal hazardous materials shipments pose unique risks to aircraft because of their ability to trigger in-flight emergencies and crashes, such as the 1996 crash of a ValueJet DC-9 that killed 110 persons. A major part of the problem, which takes place during all forms of transportation, is the illegal shipment or inaccurate marking of materials placed in vehicles for movement from one place to another. In 1998, Federal Aviation Administration (FAA) data showed 1,596,941 hazardous materials enforcement cases, of which 59 percent involved undeclared hazardous materials.

Public concern over the risk of chemical exposure to hazardous materials during transport have prompted both local and federal governments to promulgate regulations, increase enforcement, and train emergency response teams. This unit gives an overview of the regulations governing the transport of hazardous materials as set forth in 49 CFR Part 171 (General Information, Regulations, and Definitions), Part 172 (Hazardous Materials Tables and Hazardous Materials Communication Regulations), Part 173 (General Requirements for Shipments and Packaging), and 40 CFR, Part 262 (standards applicable to generators of hazardous waste) and Part 263 (standards applicable to transporters of hazardous waste). Since these regulations cover approximately 700 pages, this unit discusses only the general requirements for transporting materials by road, and excludes a multitude of details that must be followed. In addition, the laws change rapidly, making it necessary to closely check current regulations and changes as noted in the Federal Register and CFRs. States also have the option of imposing additional regulations on the transport of hazardous materials within their boundaries.

This unit focuses mainly on DOT regulations since they cover nearly all types of hazardous materials and their handling, packaging, labeling, and movement from site to site. In 1991, roadside inspections in Pennsylvania found that nearly 20 percent of the trucking violations for hazardous materials transportation were caused by improper placarding or manifests, drivers forgetting to change or apply placards, vandals changing or stealing them, and shippers displaying less dangerous markings to enable them to use restricted tunnels and bridges.

17.1 Federal Regulations

The first federal law regulating the transport of hazardous materials was passed in 1866 and covered shipments of explosives and flammables. Other laws followed, but it was not until 1966, when the DOT was formed, that safety standards for a wide range of hazardous materials were promulgated. The DOT was organized into five basic administrative units (Coast Guard, FAA (Federal Aviation Administration), Federal Highway Administration (FHA), Federal Railroad Administration (FRA), and Research and Special Programs Administration (RSPA)).

The HMTA of 1975, as amended, is the major transportation-related statute affecting the DOT, giving the agency regulatory authority over the transportation of hazardous materials, clarifying regulatory and enforcement responsibilities, and covering any person who transports or packages hazardous materials. The HMTA is responsible for the designating and classifying hazardous materials, prescribing safety standards for containers, establishing requirements for markings, labels, and placards, as well as issuing packaging exemptions, inspection, and enforcement. Enforcement of the HMTA is shared by the following administrations under delegations from the Secretary of the DOT:

- RSPA—responsible for container manufacturers, reconditioners, and retesters; shares authority over shippers of hazardous materials;
- FHA—enforces all regulations pertaining to motor carriers;
- FRA—enforces all regulations pertaining to rail carriers;
- FAA—enforces all regulations pertaining to air carriers; and
- Coast Guard—enforces all regulations pertaining to shipments by water.

Hazardous materials regulations under these auspices are subdivided by function into four basic areas:

- procedures and/or policies—49 CFR, Parts 101, 106, and 107;
- material designations (shipping papers, package markings, package labeling, vehicle placarding)—49 CFR, Part 172;
- packaging requirements—49 CFR, Parts 173, 178, 179, and 180; and
- operational rules—49 CFR, Parts 171, 173, 174, 175, 176, and 177.

In order to help clarify the conflicting state, local and federal regulations, Congress enacted the Hazardous Materials Transportation Uniform Safety Act (HMTUSA) in 1990. This act requires the Secretary of Transportation to promulgate regulations concerning the safe transport of hazardous materials in intrastate, interstate, and foreign commerce. The Secretary also retains the authority to designate materials as hazardous when they pose reasonable risks to health, safety, or property. The statute includes provisions to encourage uniformity among different state and local highway routing regulations, to develop criteria for the issuance of federal permits to motor carriers of hazardous materials, and to regulate the transport of radioactive materials.

Several agencies have overlapping authorities for regulating shipments of radioactive materials. The DOE and the EPA share responsibility for the transportation of hazardous waste or radioactive and hazardous waste mixtures generated at facilities operated by the DOE. The DOE has also agreed to comply with EPA requirements for hazardous waste transporters, requiring such transporters to obtain EPA identification numbers for the waste, comply with the manifest system, and deal with hazardous waste discharges. Such regulations incorporate and require compliance with DOT provisions for labeling, marking, placarding, proper container use, and discharge reporting.

A number of amendments and new rulings have been applied to the transportation of hazardous materials in the past several years. The following summaries present some of the more notable ones. In order to be current with such changes, the authors recommend that the reader consult the Federal Register for updates.

On October 1, 1991, a new ruling was issued by

the DOT's RSPA. Known as HM-181, it represented a major change in hazardous materials transportation rules. In brief, it comprehensively revised the hazardous materials regulations with respect to hazard communication; it adopted a performance-oriented approach to packaging; and it changed the classification system of hazardous materials, basing the new system on the United Nations' structure. Since going into effect on October 1, 1994, packages not meeting performance tests cannot be marketed or certified for hazardous materials service.

An additional ruling, HM-126-F, specifies that no hazardous materials employee can work in hazardous materials packaging, marking and labeling, documentation, shipping, or transportation unless that employee has been trained and certified by their employer. Covered employees are required to receive some general training in the various classes of hazardous material and information included on labels and truck placards. The regulation also mandates function-specific training, radiation safety training, safety training for anyone likely to be exposed to hazardous materials in an accident, and specifics on driver training for covered classes of employees. Under DOT HM-126-F, each hazardous material employee must be trained and recertified every three years. New employees must be trained within 90 days of their starting dates. In addition, DOT guidelines clearly place the responsibility for training compliance on employers, who must also keep records and proof of employee certification.

Another ruling, HM-206, issued in January 1997 (modified 1997 and 1999), better identifies hazardous materials in transportation. Changes included:

- adding a new "POISON INHALATION HAZARD" label placard to enhance the identification of materials that are poisonous if inhaled;
- lowering the quantity for a specific class of hazardous materials to be loaded on transport vehicles; and
- expanding requirements for transport vehicles and freight containers that have been fumigated.

Such improvements in the identification of hazardous materials in transportation assists emergency response personnel in responding to and mitigating the effects of incidents involving hazardous materials. It also improves the safety of transportation workers and the general public.

A final ruling in March 1999 consolidated HM-215C and HM-217 (labeling requirements for poisonous materials). In this instance, the regulations were designed to better align with international standards by incorporating changes to proper shipping names, hazard classes, packing groups, air transport quantity limitations, and vessel stowage requirements, among others. It eliminated the "KEEP AWAY FROM FOOD" label for poisonous materials in Division 6.1. This ruling was also necessary to facilitate the transport of hazardous materials in international commerce.

A notice in the Federal Register on May 24, 1999, advised that the RSPA of the HMTA was revising regulations applicable to the transportation and unloading of liquified compressed gases. The revisions included new inspection, maintenance, and testing requirements for cargo tank discharge systems such as delivery hose assemblies. It also included revised attendant requirements that apply to the loading and unloading of liquified petroleum gas and anhydrous ammonia. Training of attendants may be necessary in order for them to quickly identify and stop an unintentional release. In addition, a performance standard for passive emergency discharge control equipment that could shut down unloading operations without human intervention was addressed.

Concern for the safe handling and transportation of hazardous materials over the road, and the need for a uniform system of regulations, prompted the HMTUSA of 1990. Since then, the DOT has effectively extended federal interstate hazardous regulations to all intrastate transportation. HM-200 went into effect in October 1997 and required intrastate transporters to follow DOT regulations on packaging, placards, training, shipping names, and all other details. The intended effect of the rule was to raise the level of safety in the transportation of hazardous materials by applying a uniform system of safety regulations to all hazardous materials transported in commerce throughout the United States.

In conjunction with the DOT, the EPA regulates the shipment of hazardous wastes. Through Subtitle C of RCRA, the EPA established standards for transporters

Figure 17-1 Example of DOT label (left) and an EPA marking (right)

of hazardous wastes per 40 CFR 263 and coordinated these regulations with the DOT. With respect to transportation, RCRA identifies hazardous waste, specifies markings on containers (Figure 17-1), and tracks the movement of hazardous wastes through the use of a manifest system. DOT requirements for labeling, placarding, and container use, among other things, have been adopted by the EPA, making the hazardous waste transportation requirements consistent between the two agencies.

17.2 Definition and Classification of Hazardous Materials

The DOT defines a hazardous material as a substance or material that has been determined by the Secretary of Transportation to be capable of posing an unreasonable risk to health, safety, and property when transported in commerce. The DOT has categorized hazardous materials into hazard classes as defined in Table 17-1.

If a hazardous waste meets any of the hazard class definitions presented in Table 17-1, it is also classified as a hazardous material by DOT and transport of this material must follow DOT regulations. However, all shipments of hazardous waste must be manifested and marked in accordance with RCRA regulations. Therefore, the shipment of hazardous waste always mandates that RCRA regulations be followed, but the shipment is under DOT regulations only if the waste characteristics meet DOT specifications.

17.3 Definition of Hazardous Waste Generators and Transporters

Under RCRA, a generator is defined as ". . . any person by site, whose act or process produces hazardous waste . . . or whose act first causes a hazardous waste to become subject to regulation." In other words, the generator is the entity who creates waste that fits the definition of hazardous. When transport is involved the DOT considers the generator as the shipper.

The transporter moves the hazardous waste from one location to another. DOT usually refers to the transporter as the "carrier." Some actions of the transporter can change their status to generator, thus subjecting them to a host of additional regulations under RCRA. This can happen if a transporter mixes hazardous waste, retains the waste for longer than the specified time, imports hazardous waste, or discharges the waste rather than transporting it to an approved facility.

17.4 Responsibilities of the Generator in Transporting Hazardous Waste

Prior to transport of a hazardous waste, the generator must fill out a form known as a Uniform Hazardous Waste Manifest (EPA Form 8700-22). This form substitutes for a shipping paper. Some states have taken this federal form and have added requirements for additional information. In such cases, the state Uniform Hazardous Waste Manifest must be used when shipments originate or terminate in that state. In addition, the transporter must make certain that the manifest being used is current. The expiration date of the manifest is in the upper right-hand corner.

The generator is responsible for correctly filling out the hazardous waste manifest. In addition, the generator must receive and retain a copy of the manifest, signed by the designated receiving facility. The manifest must be retained for 3 years and is used as the basis for biennial reports. If the generator does not receive a fully signed copy of the manifest from the designated facility within a specified time, a follow-up investigation must be initiated by the generator according to EPA guidelines.

Figure 17-2 illustrates the EPA Hazardous Waste Manifest. Specific instructions for filling out this manifest are found in the Appendix to 40 CFR, Part 262.

Table 17-1 **DOT Hazard Classes, Definitions, and Examples Per 49 CFR 172.101 Hazardous Materials Table**

Hazard Class	Definitions	Examples
Class 1		
Explosives	Any substance or article including a device that is designed to function by explosion (the extremely rapid release of gas and heat) or which, by chemical reaction within itself, is able to function in a similar manner even if not designated to function by explosion	
1.1	Mass explosion hazard—affects almost the entire load instantaneously	Nitroglycerin; black powder; blasting explosives
1.2	Projection hazard, but not a mass explosion hazard	Flares; torpedoes with charge; rocket warheads; cartridges (blank)
1.3	Fire hazard and either a minor blast hazard and/or minor projection hazard; not a mass explosion hazard	Tracers for ammunition; smoke signals; rocket motors; solid propellants; some fuses
1.4	Minor explosion hazard; largely confined to the package; little or no projection	Explosive substances (Not Otherwise Specified); explosive rivets; primer caps; practice grenades; detonators
1.5	Very insensitive explosives; substances which have a mass explosion hazards but are so insensitive that there is very little probability of initiation or of burning or detonation under normal conditions of transport	Blasting caps
1.6	Extremely insensitive explosives; no mass explosion hazard; negligible probability of accidental initiation or propagation	No specific examples given; product must meet definition under manufacturer's product characteristics; see MSDS sheet
Forbidden (No #)	Shall not be offered for transportation; contains chlorate; also containing ammonium salt; acidic substances; leaking damaged package; unstable propellants; condemned or deteriorated propellants; nitroglycerin and other non-authorized liquid explosives; loaded firearms; fireworks containing an explosive and a detonator; fireworks containing yellow or white phosphorous; certain forbidden explosives that contain an explosive and a detonator	Ammonium nitrate; hot coke; liquid nitroglycerin; vinyl nitrate polymer; acetylacetone peroxide; naphthalene diozomide
Class 2		
Gases		
2.1	Flammable gas; material that is a gas at 68°F (20°C) or less and meets certain flammability requirements	Liquefied petroleum gases; methyl fluoride; ethylmethyl ether; lighters or lighter refills
2.2	Nonflammable, nonpoisonous. Compressed gas including liquefied gas, pressurized cryogenic gas, compressed gas in solution, asphyxiant gas, and oxidizing gas	Rare gases and oxygen mixtures; nitrous oxide; trifluoromethane, refrigerated liquid
2.3	Gas, poisonous to humans by inhalation	Liquefied sulfur dioxide; stibine; phosphorous; parathion
No #	Nonliquefied compressed gas; charged pressure package; entirely gaseous at 68°F (20°C)	Covers a wide variety of gases shipped under the defining conditions

Continued

Table 17-1 *Continued*

Hazard Class	Definitions	Examples
Class 2 *Con't.*		
No #	Cryogenic liquid; refrigerated liquefied gas having a boiling point colder than –130°F (–54°C)	Refrigerated liquid ethylene; refrigerated neon; refrigerated liquid xenon
No #	Refrigerant gas or dispersant gas; must be nonpoisonous to humans	Fluorocarbons and mixtures of fluorocarbons
No #	Liquefied compressed gas; a gas which in a packaging under the charged pressure is partially liquid at 68°F (20°C)	Covers a variety of gases shipped under the defining conditions
No #	Compressed gas in solution: a non-liquefied compressed gas which is dissolved in a solvent	Covers a variety of gases shipped under the defining conditions
Class 3		
Flammable and combustible liquid	Flammable liquid; flammable—having a flash point less than 141°F (60.5°C), or any material in a liquid phase with a flash point at or above 100°F (37.8°C) that is intentionally heated and transported at or above its flashpoint in bulk packaging Combustible liquid; any liquid that does not meet the definition of any other hazard class specified in the Class 3 subchapter and has a flash point above 141°F (60.5°C) and below 200°F (93°C)	Distilled spirit of 140 proof or lower; toluene; petroleum spirit; paint; alcohols; epoxy; acetone oils; 1-Hexane; xylene
Class 4		
4.1	Flammable solid; desensitized explosives; reactive metal powders; solids that may cause a fire through friction; readily combustible solids; self-reactive materials that may decompose at high transport temperatures; wetted explosives	Matches; urea nitrate; molten sulfur; picric acid; amorphous phosphorous
4.2	Spontaneously Combustible Material; a pyrophoric material—a liquid or solid that, even in small quantities and without an external ignition source can ignite within 5 minutes after coming in contact with air; self-heating materials—when in contact with air and without an energy supply, is liable to self-heat and spontaneously ignites if the temperature of the material exceeds 392°F (200°C)	Dry titanium powder; metal catalysts; zirconium powder; anhydrous potassium sulfide
4.3	Dangerous-when-wet material; by contact with water, is liable to become spontaneously flammable or to give off flammable or toxic gas	Sodium; rubidium; potassium metal alloys
Class 5		
5.1	Oxidizer; material that may, by giving up oxygen, cause or enhance the combustion of other materials	Sodium nitrate; potassium nitrate; inorganic perchlorates; hydrogen peroxide
5.2	Organic peroxide; any organic compound containing oxygen in the bivalent –O–O– structure and which may be considered a derivative of hydrogen peroxide where one or more of the hydrogen atoms have been replaced by organic radicals	Polyester resin kits; organic peroxide types 1 through 7

Continued

Table 17-1 *Continued*

Hazard Class	Definitions	Examples
Class 6		
6.1	Poisonous materials; a material, other than a gas, which is known to be so toxic to humans as to afford a hazard to health during transportation; is presumed to be toxic to humans through laboratory animal tests by oral, dermal, or inhalation exposures of specific quantities; is irritating with properties similar to tear gas, causing extreme irritation	Tear gas; certain nitrophenol pesticides; strychnine; arsenic; phenyl mercaptan; mercury compounds
6.2	Infectious substances; any viable microorganism or its toxin that causes or may cause disease in humans or animals; includes agents listed by the Department of Health and Human Services; any other agent that causes or may cause severe, disabling, or fatal disease; the terms infectious substance and etiologic agent are synonymous; a diagnostic specimen—any animal or human material, including, but not limited to, excreta, secreta, blood, blood components, tissue and tissue fluids being shipped for the purpose of diagnosis; biological product—material that is prepared and manufactured under specific regulations; regulated medical waste—waste or reusable material, other than a culture or stock of infectious substance, that contains an infectious substance and is generated in the diagnosis, treatment or immunization of humans or animals, its research or production testing	Regulated medical waste; polio virus; salmonella; cholera toxins; clostridium botulinum
Class 7	Radioactive material; any material having a specific activity greater than 0.002 microcuries per gram, including ores, fissile material, natural, enriched, or depleted materials, mill tailings; water with certain tritium concentrations	Uranium products; thorium products; plutonium; uranyl nitrate as a solid
Class 8	Corrosive material; liquid or solid that cause visible destruction or irreversible alterations in human skin tissue at the site of contact; a liquid that has a severe corrosion rate on steel or aluminum under certain criteria	Zinc chloride; sulfuric acid; some liquid dyes; chromic fluoride; ammonium sulfide
Class 9	Miscellaneous hazardous material; a material which presents a hazard during transportation but which does not meet the definition of any other hazard class; any material which has an anesthetic, noxious, or other similar property which could cause extreme annoyance or discomfort to a flight crew member so as to prevent the correct performance of assigned duties; any material that meets the definition in §171.8 of this subchapter for an elevated temperature material, a hazardous substance, a hazardous waste, or a marine pollutant	Asbestos; electric wheel chairs (batteries); molten sulfur; PCBs; lithium batteries; dry ice; formaldehyde solutions
No #	Other regulated materials—ORM—D; a material such as a consumer commodity, which presents a limited hazard during transportation due to its form, quantity, and packaging	Cartridges; small arms; consumer commodities (nail polish, etc.)

REPORTABLE QUANTITY VALUE RQ's - 5000/1000/100/10/1		REPORT ANY "RQ" DISCHARGE TO NATIONAL RESPONSE CENTER (800) 424-8802, AND 911 EMERGENCY NUMBER OR LOCAL OPERATOR EMERGENCY CONTACT: CHEMTREC (800) 424-9300	PLACARDS PROVIDED	EMERGENCY RESPONSE GUIDE NUMBER	a.	b.
a. RQ =	c. RQ =				c.	d.
b. RQ =	d. RQ =					

Please print or type (*Form designed for use on elite (12-pitch) typewriter.*) *Form Approved OMB No. 2050-0039 Expires 9-30-96*

UNIFORM HAZARDOUS WASTE MANIFEST

1. Generator's US EPA ID No. — Manifest Document No.
2. Page 1 of

Information in the shaded areas is not required by Federal law.

GENERATOR

3. Generator's Name and Mailing Address — A. State Manifest Document Number; B. State Generator's ID
4. Generator's Phone ( )
5. Transporter 1 Company Name — 6. US EPA ID Number — C. State Transporter's ID; D. Transporter's Phone
7. Transporter 2 Company Name — 8. US EPA ID Number — E. State Transporter's ID; F. Transporter's Phone
9. Designated Facility Name and Site Address — 10. US EPA ID Number — G. State Facility's ID; H. Facility's Phone

11. US DOT Description (*Including Proper Shipping Name, Hazard Class, and ID Number*)	12. Containers No.	Type	13. Total Quantity	14. Unit Wt/Vol	I. Waste No.
a.					
b.					
c.					
d.					

J. Additional Descriptions for Materials Listed Above

K. Handling Codes for Wastes Listed Above

15. Special Handling Instructions and Additional Information

AGENCY DISPLAY OF ESTIMATED BURDEN

"Public reporting burden for this collection of information is estimated to average: 37 minutes for generators, 15 minutes for transporters, and 10 minutes for treatment, storage and disposal facilities. This includes time for reviewing instructions, gathering data, and completing and reviewing the form. Send comments regarding the burden estimate, including suggestions for reducing this burden, to: Chief, Information Policy Branch, PM-223, U.S. Environmental Protection Agency, 401 M Street, SW., Washington, DC 20460; and to the Office of Information and Regulatory Affairs, Office of Management and Budget, Washington, DC 20503."

16. **GENERATOR'S CERTIFICATION:** I hereby declare that the contents of this consignment are fully and accurately described above by proper shipping name and are classified, packed, marked, and labeled, and are in all respects in proper condition for transport by highway according to applicable international and national government regulations.

If I am a large quantity generator, I certify that I have a program in place to reduce the volume and toxicity of waste generated to the degree I have determined to be economically practicable and that I have selected the practicable method of treatment, storage, or disposal currently available to me which minimizes the present and future threat to human health and the environment, **OR**, if I am a small quantity generator, I have made a good faith effort to minimize my waste generation and select the best waste management method that is available to me and that I can afford.

Printed/Typed Name	Signature	Month Day Year

TRANSPORTER

17. Transporter 1 Acknowledgement of Receipt of Materials

Printed/Typed Name	Signature	Month Day Year

18. Transporter 2 Acknowledgement of Receipt of Materials

Printed/Typed Name	Signature	Month Day Year

FACILITY

19. Discrepancy Indication Space

20. Facility Owner or Operator: Certification of receipt of hazardous materials covered by this manifest except as noted in Item 19.

Printed/Typed Name	Signature	Month Day Year

Figure 17-2 Example of an EPA Hazardous Waste Manifest

17.5 Responsibilities of the Transporter

17.5.1 GENERAL REQUIREMENTS

Prior to accepting hazardous waste for shipment, the transporter must obtain an identification number from the EPA. The transporter must use only RCRA-permitted facilities and cannot store waste for more than 10 days without becoming subject to regulations concerning storage facilities. Generator regulations may apply if the transporter mixes hazardous wastes.

DOT regulations specify that transporters cannot accept hazardous materials that have not been properly identified, packaged, marked, labeled, and manifested. Before leaving the generator's facility with a load of hazardous waste, the transporter must sign and date the manifest to acknowledge acceptance of the hazardous waste from the generator. The transporter must return a signed copy to the generator before leaving the generator's property. The transporter must ensure that the remaining pages of the manifest accompany the hazardous waste.

A transporter who delivers a hazardous waste to another transporter or to the designated disposal facility must:

- obtain the date of delivery and the handwritten signature of that transporter or of the owner or operator of the designated facility on the manifest;
- retain one copy of the manifest in accordance with recordkeeping requirements described in section 17.5.3; and
- give the remaining copies of the manifest to the accepting transporter or designated facility.

Note that a transporter who ships hazardous waste from a source that generates between 100 and 1,000 kilograms (220 and 2,200 pounds) of hazardous waste in a calendar month (a Small Quantity Generator) need not comply with these requirements, provided that the transporter records on a manifest the following information for each shipment:

- the name, address, and EPA identification number of the generator of the waste;
- the quantity of waste accepted;
- all DOT-required shipping information; and
- the date the waste is accepted.

Conditionally Exempt Small Quantity Generators (CESQG), or those that generate no more than 100 kilograms of hazardous waste per month, must comply with three waste management requirements to remain exempt from full hazardous waste regulations that apply to generators of larger quantities. These requirements include:

- identifying the hazardous waste;
- not storing more than 2,200 pounds of hazardous waste on the site at any time; and
- ensuring delivery of hazardous waste to a government-approved offsite treatment or disposal facility.

Some states have additional requirements for CESQGs. For example, they may be required to follow some of the Small Quantity Generator requirements, such as obtaining an EPA identification number or complying with storage standards.

The transporter carries the manifest when transporting waste to the disposal facility. In addition, the transporter retains these records for a period of at least three years after termination or expiration of the agreement.

17.5.2 COMPLIANCE WITH THE MANIFEST

The transporter must deliver the entire quantity of hazardous waste that has been accepted from a generator (or another transporter) to the facility designated on the manifest. If the hazardous waste cannot be delivered to the designated facility for any reason, the transporter must deliver the entire quantity of hazardous waste to the alternate facility designated on the manifest, if such a facility is designated. If the hazardous waste cannot be delivered in accordance with the manifest, the transporter must contact the generator for further directions and must revise the manifest according to the generator's instructions.

17.5.3 RECORDKEEPING

A transporter of hazardous waste must keep a copy of the manifest that was signed by the generator, the driver, and the next designated transporter (if applicable), or the owner or operator of the designated facility.

The manifest must be retained for a period of three years from the date the hazardous waste was accepted by the initial transporter. In addition to the EPA manifest, the transporter may carry shipping papers that duplicate many items in the manifest. However, retention of shipping papers is not required by the EPA.

17.5.4 DISCHARGES DURING TRANSPORT

Federal regulations define a discharge as "the accidental or intentional spilling, leaking, pumping, pouring, emitting, or dumping of hazardous waste into or on any land or water." The EPA requires that transporters take appropriate emergency response action to contain and clean up all discharges that occur during transport. In addition, appropriate EPA-mandated mitigation efforts may be needed to reduce impacts on human health and the environment.

The transporter is required to immediately notify the NRC if the spill exceeds reportable quantities as determined by the table in 40 CFR 302.4 and Appendices A and B of the same regulation. Releases of reportable quantities of extremely hazardous substances (40 CFR 355) also require immediate notification of the local and state emergency planning committees. These lists are the same as the "RQ" list in SARA.

Following the emergency, the EPA requires that all normal regulations for the final disposition of the hazardous waste be followed. The DOT requires a written report of the incident.

On August 17, 1990, an emergency response communication rule was published with a mandatory implementation date of December 31, 1990. The requirements of this rule are summarized below.

- The shipping paper or uniform hazardous waste manifest must include a 24-hour emergency response telephone number. This is the generator's (or shipper's) responsibility. The generator can make arrangements with an agency like CHEMTREC to act as their agent to provide this information and be in compliance.
- Information necessary for emergency response must be written on or attached to the uniform hazardous waste manifest or shipping paper. This requirement could be satisfied with an MSDS, a copy of the DOT Emergency Response Guidebook, or other comparable information. In addition, the motor carrier must have emergency response information available, including a basic description of the hazardous materials, immediate hazards to health, risks of fire or explosion, immediate precautions to be taken in the event of an accident, immediate methods for handling fires, initial methods for handling spills or leaks in the absence of fire, and preliminary first aid measures.

17.6 Responsibilities of the Receiving Facility

The hazardous waste receiving facility must continue the cradle-to-grave tracking system that was started with the generator and continued with the transporter. Key components are discussed below.

When a facility receives hazardous waste accompanied by a manifest, the owner or operator must:

- check the manifested hazardous wastes to ensure that the shipped waste can be accepted at the receiving facility;
- analyze wastes to make certain that substance identities are as specified on the manifest;
- sign and date each copy of the manifest to certify that the hazardous waste covered by the manifest was received;
- note any significant discrepancies in each copy of the manifest;
- immediately give the transporter one copy of the signed manifest;
- send a copy of the manifest to the generator within 30 days after delivery; and
- retain a copy of each manifest at the facility for at least three years from the date of delivery.

17.7 DOT Labeling and EPA Marking

Prior to shipping hazardous materials, containers must be appropriately labeled and marked according to DOT and EPA specifications. However, the EPA now specifies that all materials be labeled, marked, and placarded in accordance with the requirements of the DOT (49 CFR part 172.500). The purpose of the DOT label is to identify the general hazards associated with virgin materials. The EPA marking is specific to hazard-

HAZARDOUS
WASTE
FEDERAL LAW PROHIBITS IMPROPER DISPOSAL
IF FOUND, CONTACT THE NEAREST POLICE, OR PUBLIC SAFETY AUTHORITY, OR THE U.S. ENVIRONMENTAL PROTECTION AGENCY
PROPER D.O.T. SHIPPING NAME ____ UN OR NA# ____
GENERATOR INFORMATION:
NAME ____
ADDRESS ____
CITY ____ STATE ____ ZIP ____
EPA ID NO. ____ EPA WASTE NO. ____
ACCUMULATION START DATE ____ MANIFEST DOCUMENT NO. ____
HANDLE WITH CARE!
CONTAINS HAZARDOUS OR TOXIC WASTES

Figure 17-3 EPA marking for hazardous waste containers

ous wastes and identifies the origin and nature of the waste.

The DOT defines a label as a small warning sign placed on individual containers that will be transported in a vehicle. The label may provide guidance during transport for placarding, special care, and response if discharged. The design of the label is specified by the DOT. In addition, a "marking" is applied to the outside of containers of 110 gallons or less prior to transport. The "marking" contains cautions or descriptive information useful for the proper handling, storage, and disposal of hazardous waste. The EPA has a specified "marking" for containers of hazardous waste (Figure 17-3).

17.7.1 DOT LABELS

17.7.1.1 Special Provisions

No person may transport any container bearing a DOT label unless the labeled container actually contains a hazardous material or hazardous residue and the affixed label accurately reflects the hazard class associated with the materials being shipped. In addition, no person may offer for transportation, and no carrier may transport a package bearing any marking or label that, by its color, design, or shape, could be confused with or conflict with a DOT or EPA hazardous material label or hazardous waste marking.

A label or marking must be used for each separate classification of hazard within the container. All labels/markings should be applied to the same place on all containers to provide consistency in locating information.

When hazardous materials in different hazard classes are grouped within the same packaging or within the same outside container or overpack, the packaging, outside container, or overpack must be labeled as required for each class of hazardous material contained therein.

Each metal barrel or drum containing a flammable liquid with a vapor pressure between 16 and 40 p.s.i.a. at 100°F must have a BUNG label that signifies the potential danger involved when loosening the bung nut.

17.7.1.2 Placement of Labels

Each label must be printed on or affixed to the surface of the package near the marked, proper shipping name. When two or more labels are required, they must be displayed or affixed next to each other. Each label must be affixed to a background of contrasting color, or have a dotted or solid outer border. Finally, a label must not be obscured by markings or attachments.

17.7.1.3 Label Specifications

Each label that is affixed to or printed on a package must be durable and weather-resistant. Each label must be at least four inches on each side and each side must have a black, solid-line border, one-quarter (1/4) inch from the edge. A label may contain the appropriate International Classification System hazard class number, United Nations' classification code, and, when appropriate, the division number. The printing, inner border, and symbol on each label must be specified by the DOT. The DOT has distinctive labels corresponding to the hazard classes.

The United States and other member countries of the United Nations have taken steps to standardize the transport of hazardous materials, including the hazardous communications requirements for international business. These nations are attempting to develop a Globally Harmonized System (GHS) for the classification and labeling of chemicals. Such a system is intended to consolidate classification criteria for all chemicals and would standardize hazard warning symbols, hazard labels, MSDSs, and other communication information. Some of the U.S. agencies participating in this program include the Consumer Product Safety Commission, the Department of Commerce, the DOT, the FDA, the EPA, and OSHA. Representatives from the private sector, industry and trade associations, worker representatives, health and safety professionals, and public interest groups are also participating in the GHS. To date, participants have agreed on a set of classification criteria for hazardous materials transportation. These criteria are now in the United Nations' Committee of Experts for possible adoption into the United Nations Recommendations for the Transport of Dangerous Goods.

NAFTA, with Canada, the United States, and Mexico as partners, has standardized most labels, placards and transportation regulations for inter-country shipments.

17.7.2 EPA MARKINGS

Markings are provided to assist motor carrier personnel in recognizing packages and to properly respond to emergency situations. Before transporting hazardous waste or offering hazardous waste for transportation off-site, the EPA requires that the generator mark each container of 110 gallons or less with a label such as that shown in Figure 17-3.

Information that may be requested on the marking includes the proper DOT shipping name (United Nations shipping number), the generator's EPA identification number, the EPA waste number, accumulation start date (the date hazardous waste was first put in the container), and the Uniform Hazardous Waste Manifest number.

If the material is classified as "ORM-D," no labels are required because ORM-D waste does not meet any of the DOT hazard classes. The designation "ORM" means Other Regulated Materials, such as consumer commodities including hair spray in aerosol containers or cases of nail polish.

Finally, the receiving facility may require specific information, such as the waste profile number. This information is commonly stenciled on the container.

17.8 DOT Placarding

The DOT has 23 basic placards (Figure 17-4), including special designations for railroad cars, that generally correspond to the hazard classes and labels previously discussed. Prior to 1991, generalized charts could be produced showing shipping quantities, hazard classes and their corresponding labels and placards. Adoption of the United Nations' classification system and a redefinition of hazards has produced more detailed and item-specific listings. Nearly 170 pages in length, the Hazardous Materials Table of 49 CFR, Part 172.101 is set up to address numbering, classes, compatible classes, packaging, labeling, item names/descriptions, quantity limitations, and special provisions among other characteristics. Reference should be made to this table and others when deciding on proper labeling compliance. The placard shows the United Nations or NA number of the hazardous material across the center of the placard when hazardous materials are shipped in tanks.

The DOT requires certain vehicles transporting hazardous materials to have signs (or placards) on the outside of the vehicles specifying the type of hazardous materials being shipped. No person may affix a placard on a transport vehicle unless:

- The material being transported is a hazardous material; and
- The placard accurately represents the hazard class of the material being transported.

In addition, no person may affix or display any sign or other device on a transport vehicle that, by its color, design, shape, or content, could be confused with any placard required by the DOT.

17.8.1 GENERAL REQUIREMENTS

Any vehicle containing hazardous material must have the appropriate placard displayed on each end and each side. A transport vehicle or freight container with two or more classes of materials that require different placards may be placarded "DANGEROUS" in place of the separate placarding. However, when 5,000 pounds or more of one class of material are loaded at one loading facility, the placard specified for that class must be applied. When the gross weight of all hazardous materials is less than 1,001 pounds, a placard is usually not required.

For transport vehicles, freight containers, and rail cars requiring a placard, the placard must contain the

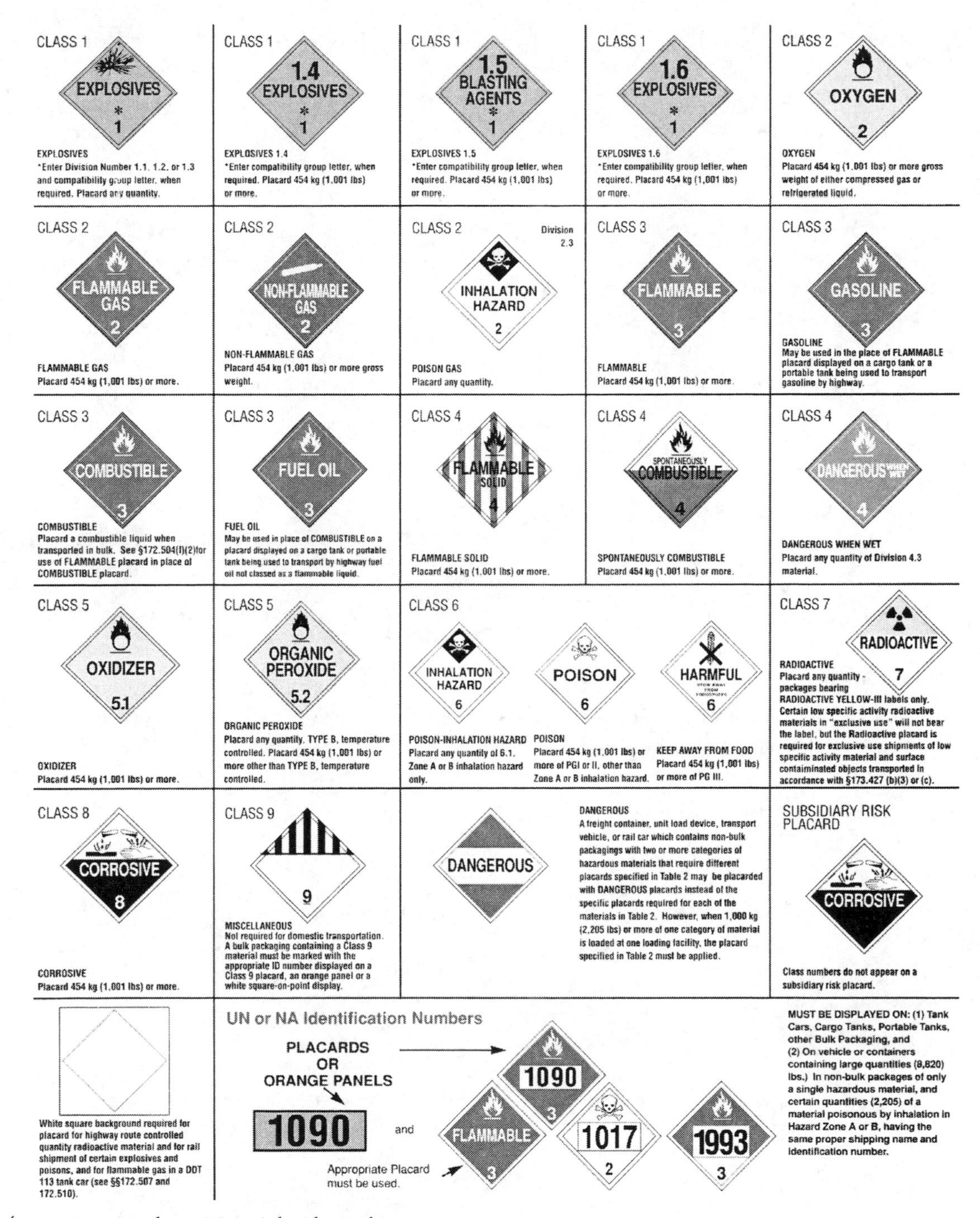

Figure 17-4 DOT Hazardous Materials Placards

hazard class name and symbol and the International Classification Number. Placards for tank cars, cargo tanks, portable tanks, and bulk packaging consists of the hazard class symbol, the four-digit United Nations number, and the International Classification Number. The NA number—most familiar on trucks—indicates the type of emergency response required by that class of material and is usually the main information first-responders (in an emergency) need to see.

17.8.2 AFFIXING PLACARDS

Any person offering a motor carrier a hazardous material for transportation by highway must provide the motor carrier with the required placards for the material being offered. This must be done prior to, or at the same time, the material is offered for transportation, unless the carrier's motor vehicle is already placarded for the material being transported. No motor carrier may transport a hazardous material in a motor vehicle unless the placards for the hazardous material are affixed as required by the DOT.

17.8.3 VISIBILITY AND DISPLAY OF PLACARDS

Each placard used on a motor vehicle must be readily visible from the direction it faces. While the placard is required on the front of a motor vehicle, this may include the front of a truck-tractor instead of, or in addition to, the front of the cargo body to which the truck-tractor is attached.

Each placard on a transport vehicle, portable tank, or freight container must:

- be securely attached or affixed in a holder;
- be located away from other devices such as ladders, pipes, doors, and tarpaulins;
- be located, if possible, so that dirt and water are not sprayed on it from the wheels of the transport vehicle;
- be located away from any marking (such as advertising) that could substantially reduce its effectiveness;
- have the words or identification number (when authorized) printed on it horizontally, reading from left to right; and
- be well maintained by the carrier so that the format, legibility, color, and visibility of the placard will not be reduced due to damage, deterioration, or by being obscured with dirt or other matter.

Shipments of hazardous waste that are classified as ORM-D do not require placarding, but do require EPA markings and any appropriate DOT labeling. Shipments of hazardous waste that meet one of the hazard classes (other than ORM-D) previously listed require the appropriate hazard class placard and labels.

Specific regulations are in place regarding the placarding and marking of railroad or over-the-road tank cars. Information on the required identification numbers on placards, orange panels, or white square-on-point configurations, as appropriate, must be provided to the carrier by each person offering a portable tank containing a hazardous material. Such placards must be located on each side and at each end of the transport vehicle or freight container in which the hazardous material is located. Cargo tanks carrying gases must be marked at each end and both sides of the tank in letters at least two inches high. They must also include the proper shipping name for the gas (49 CFR, Part 172.101 (Table)), an appropriate common name for the material, and markings to indicate if the cargo tank is constructed of quenched and tempered steel. This type of identification assists emergency response personnel responding to incidents and improves the safety of transportation workers and the public.

17.9 Sources of Information

American Trucking Association: *Driver's Guide to Hazardous Materials.* Alexandria, VA: American Trucking Association, 1992.

American Trucking Association: *Handling Hazardous Materials, Department of Safety* by the Safety Department Staff. Alexandria, VA: American Trucking Association, 1995.

Environmental Protection Agency: *Understanding the Hazardous Waste Rules, A Handbook for Small Businesses* by the Office of Solid Waste. Washington, DC: U.S. Government Printing Office, 1996.

Fabey, Michael: Trucking Disaster: Bound to Happen? *Philadelphia Business Journal,* 11:1 (1992).

Gose, Joe: Railroads Say They Take Undeserved Rap in Toxic Spills. *The Kansas City Business Journal,* 11:19 (1992).

"Hazardous Material Transportation Safety Reauthorization Act of 1999—Section-By-Section Analysis," *U.S. Code,* Title 49, Section 5101 *et seq.*

Hoffman, Stanley: Hazmat—The New Facts of Life. *Chilton's Distribution,* 92:40 (1993).

Kaberline, Brian: New Hazardous Materials Reg Requires Training For All Workers (Hazardous Materials Transportation Uniform Safety Act of 1990). *The Kansas City Business Journal 11:8* (1993).

Kaye, Ken and A. Friedberg: The Crash of ValueJet Flight 592: Path to Disaster. *Sun-Sentinel,* Ft. Lauderdale, FL, pp. 1A+ (August 27, 1993).

McConville, D.J.: Tracking the Hazmat Express. *Chilton's Distribution,* 91:51 (1992).

McKenna, J.T.: Team Critiques Hazmat Oversight. *Aviation Week and Space Technology,* pp. 54–55 (1999).

Meter, Eugene: *Chemistry of Hazardous Materials.* 3rd ed. Englewood Cliffs, NJ: Prentice-Hall, 1997.

Parrish, Michael: Dangerous Cargo: Highways, Railways and Hazardous Materials, America's Poisons on the Move. *Los Angeles Times,* Los Angeles, CA, pp. 1A+ (September 29, 1992).

Rondy, John: Lesson Learned: Go By the Book When Shipping Hazardous Materials. *The Business Journal-Milwaukee,* 11:20B (1993).

Solinger, Tim and J.J. Keller and Associates, Inc. Staff (eds.): *Hazardous Waste Regulatory Guide,* Neenah, WI: J.J. Keller Publishing Society, (1998).

U.S. Department of Transportation: *A Guide for the Inspection of Hazardous Waste Shipments by Motor Vehicle or at Freight Facilities.* Washington, DC: U.S. Government Printing Department, 1988.

U.S. Department of Transportation: *A Guide for the Inspection of Hazardous Waste Shipments* by Science Applications International Corporation. Washington, DC: U.S. Department of Transportation, 1988.

U.S. Department of Transportation: *An Overview of the Federal Hazardous Materials Transportation Law* by the Research and Special Programs Administration Office. Washington, DC: U.S. Department of Transportation, Office of Hazardous Materials Initiatives and Training, 1999.

U.S. Department of Transportation, Hazardous Material Safety, Research and Special Programs Administration, "Hazardous Materials Information System," August 2001. www.hazmat.dot.gov/

U.S. Department of Transportation, Research and Special Programs Administration: *Year 2000 North American Emergency Response Guidebook (NAERG 2000).* Neenah, WI: J.J. Keller & Associates, Inc., 2000.

U.S. Seeks Permits for Transportation of Some Metals. *The Wall Street Journal,* p. A16(W) and p. A9(E). (23 June 1993).

Wentz, C.A.: *Hazardous Waste Management.* 2nd ed. McGraw-Hill Publishing Co., 1995.

Wheeler, John R.: *Toxic Substance Control Act Compliance Guide and Service.* 3 Vols. Plano, TX: Environmental Books (1997).

Worobec, Mary D. and C. Hogue. *Toxic Substances Controls Guide.* 2nd ed. Washington, DC: Bureau of National Affairs, 1992.

EIGHTEEN

Confined Space

18 | CONTENTS

Confined spaces can be deadly. OSHA estimates that nearly 4.8 million times a year, workers will enter permit-required confined spaces. On April 15, 1993, OSHA made a rule that addresses the safety of workers who must, by necessity of their jobs, enter confined spaces. This standard is called the "Permit-Required Confined Spaces for General Industry." OSHA expects this standard to prevent 54 fatalities and more than 5,000 serious injuries per year. Nearly 240,000 workplaces are covered by the standard, most of which must make at least some changes to comply with the ruling. In the implemented standard, the employer must document and control all confined space entries including training entrants. The number of nonpermit space entries each year is unspecified. Unfortunately, many workers do not recognize the presence of confined spaces nor appreciate the dangers when such areas are entered. This unit is designed to provide assistance in confined space recognition, pre-entry procedures, safety techniques, and the law of worker safety standards set out in 29 CFR 1910.146. This unit does not qualify an individual to enter confined spaces. Specialized training is required by law before confined spaces are entered. This law sets forth safety requirements, including a permit system, for entry into specifically-designated confined spaces that pose special dangers for individuals entering them. This comprehensive program for confined space entry came about following events such as those described below.

- Case History #1
  A worker at a steel mill, assigned to clean a blockage in a degreaser vessel dust collector, entered the vessel through an access manhole and proceeded to clear the obstruction. A coworker assigned to assist, left the area to locate an electrical receptacle. About 10 or 15 minutes later, the coworker returned and found the first worker unconscious. The second worker was able to remove the man and call for assistance. The worker died. An oxygen test of the vessel showed 10 percent oxygen.
- Case History #2
  An employee of a zinc refinery was working in a zinc dust condenser when he collapsed. Another employee donned a SCBA and attempted to enter the condenser for rescue. He was not able to fit through the portal wearing the SCBA so he removed it, handed it to another employee, and then entered the condenser. Before he could be handed the SCBA to continue with the rescue, he collapsed and fell into the condenser. The first employee was declared dead at the scene. The would-be rescuer died two days later. The toxic gas was determined to be carbon monoxide.
- Case History #3
  An employee of a trailer service company entered an 8,500-gallon cargo tanker to weld a leak on the interior wall. Despite the pres-

ence of strong lacquer thinner fumes (the material previously carried in the tanker), and even though the written company safety policy required the use of an explosion meter, the welder decided to proceed with repairs. When he began welding, an explosion occurred. The employee was removed from the tank and declared dead.

- Case History #4
  Two foundry employees entered a sand bin to clear a jam. Sand that had adhered to the sides of the bin began to break loose and fall on them. One employee quickly became buried up to his chest, just below his armpits. The other employee left the bin to obtain a rope. Returning to the bin, he tied the rope around the partially buried employee and tried to pull him free. During the attempted rescue, additional sand fell, completely covering and suffocating the employee who had been entrapped.
- Case History #5
  A workman entered a bag house in the dust collection system of a basic-oxygen, steel-making furnace on routine inspection. He stepped onto the dust conveyer, which was not supposed to be operating at the time, and was caught in the machinery. The employee died before rescuers could remove him from the auger pipe conveyer.
- Case History #6
  Three men entered a 20-foot deep underground vault next to a major highway, apparently to conduct an inspection prior to construction. They carried no air monitoring equipment, were not wearing protective equipment, and placed no barriers or other indicators next to the roadway to indicate their presence. All three men were found dead at the bottom of the vault the next day. Investigators speculated that the men likely died in "domino" fashion, one after the other. The first man in the confined space may have fallen into the water after being overcome by fumes. The second man may have been overcome trying to rescue the first one. The third victim, probably above ground directing traffic around the manhole, likely noticed something was wrong and entered the confined space in an attempt to rescue the others. None had left word with their employer where they were going or what they had planned to do.

These are examples of the significant risks and hazards that may be encountered in confined spaces. An evaluation by OSHA of some 20,000 reports covering industrial accidents produced a list of the principal confined space hazards, which are discussed below. Emphasis is placed on the atmosphere present within the work space and the ability of this atmosphere to asphyxiate, become toxic, or be flammable. Other hazards presented in section 18.2, are no less fatal, but also cause occupational fatalities.

18.1 Atmospheric Hazards

18.1.1 ASPHYXIATING ATMOSPHERES

The most common cause of confined space deaths is from asphyxiating atmospheres. The definition of asphyxiating atmosphere does not consider the presence of toxic elements, but is based on the absence of oxygen. Any space having less than 19.5 percent oxygen is considered to have an asphyxiating atmosphere since any concentration less than this is judged to be inadequate for a worker's respiratory needs when performing physical labor.

The oxygen level in a space may decrease from absorption by substances such as activated charcoal or be consumed by chemical reactions such as rusting. Bacterial action (fermentation) causes oxygen depletion as do welding, cutting, and brazing. Oxygen can also be displaced by another gas, such as carbon dioxide or helium. Total displacement of oxygen will quickly result in unconsciousness, followed by death. Asphyxiating atmospheres are the most common cause of confined space fatalities.

18.1.2 TOXIC ATMOSPHERES

Toxic atmospheres are the second leading cause of death to workers in confined spaces. A toxic atmosphere refers to an atmosphere containing gases, vapors, or fumes known to have poisonous physiological effects. These are taken into account independently of oxygen concentration. The most commonly encountered toxic gases are carbon monoxide and hydrogen sulfide.

Some toxic atmosphere materials have chronic health effects. Some kill quickly, while others can produce both immediate and delayed effects. For example, carbon disulfide at low concentrations may produce no signs of exposure but can cause permanent and cumulative brain damage as a result of repeated low level exposures. At higher concentrations, it can kill quickly.

Toxic substances can come from a variety of sources including the product previously or currently stored in the space, the work being performed in the confined space, and the conditions of areas adjacent to confined spaces. The product can be absorbed into the walls of a container and give off toxic gases when removed. For example, when sludge is removed from a tank, decomposed material can give off hydrogen sulfide gas.

Toxic atmospheres can be generated in-situ by various processes, including welding, cutting, brazing, painting, scraping, sanding, and degreasing. The vapors from cleaning solvents used in many industries are very toxic in confined spaces.

Toxicants produced by work in an area adjacent to a confined space can enter and accumulate. For example, a slug of illegal chemical dumped into a sewer can carry vapors with it and can change the working environment very quickly.

18.1.3 FLAMMABLE OR EXPLOSIVE ATMOSPHERES

OSHA considers an atmosphere to pose a serious fire or explosion hazard in a confined space if a flammable gas or vapor is present in a concentration greater than 10 percent of its lower flammable limit or if a combustible dust is present at a concentration greater than or equal to its lower flammable limit. This category includes such gases as methane and acetylene, vapors of solvents or fuels such as mineral spirits, gasoline, or toluene, or combustible dusts such as coal dust or grain dust. If a source of ignition, such as a sparking electrical tool, is introduced into a space containing a flammable atmosphere, an explosion may result. An oxygen-enriched atmosphere (above 23.5 percent for a confined space) will cause flammable materials, such as clothing and hair, to burn violently when ignited.

18.2 Other Confined Space Hazards

"Engulfment" refers to situations where an individual is entrapped or enveloped, usually by dry, bulk materials and is in danger of suffocation or crushing. In some cases, the engulfing material may be so hot or corrosive that the victims sustain fatal chemical or thermal burns but are never totally buried.

Extremely hot or cold temperatures can present problems. For example, if the space has been steamed or refrigerated, the temperature should be allowed to normalize before workers enter.

Accidents occur in confined spaces when employers fail to isolate equipment within the space from the power sources that run them. This includes augers, paddles, and stirring devices. Deaths result from mechanical force injury, such as the crushing, amputation, or decapitation of the victim's body.

Noise within a confined space can be amplified because of the design and acoustic properties of the space. Excessive noise can not only damage hearing, but can also affect communication, causing a shouted warning to go unheard.

Slips and falls can occur on a wet surface causing injury or death to workers. A wet surface will also increase the likelihood of electrical shock in areas where electric circuits, equipment, and tools are used.

Workers in confined spaces should take precautions to prevent the possibility of falling objects, particularly in spaces that have topside openings for entry, and where work is being done concurrently above the worker.

Work in some confined spaces, such as underground vaults or petrochemical processing vessels may involve fall hazards since these areas can be many feet high. Adequate fall protection devices are needed for workers.

18.3 Definition of Confined Space

According to OSHA, "confined space" means a space that:

- has adequate size and configuration for employees' entry;
- has limited means of access or egress; and
- is not designed for continuous employee occupancy.

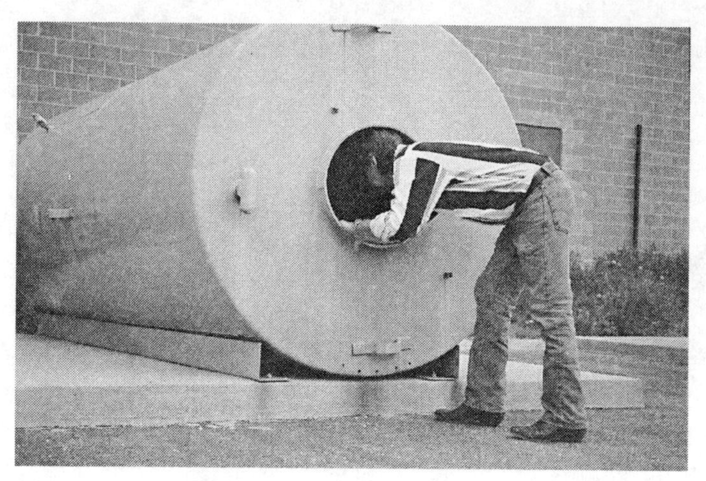

Figure 18-1 Simply placing your head in a confined space constitutes entry

"Adequate size and configuration" describes a space that can be entered by a person. A worker is determined to have entered a confined space when any part of the entrant's body breaks the plane crossing the access opening. The entire body does not have to be inside the confined space to constitute entry.

"Limited access" refers to openings into such spaces that are limited primarily by size or location. Openings may be small in size and difficult to move through easily, especially when wearing protective clothing and respiratory gear. Small openings may also make it difficult to get lifesaving equipment in or out of the spaces when rescue is needed. Some openings can be very large and still be difficult to access, as in open-topped spaces such as pits, degreasers, mixers, and excavations. Access to open-topped spaces may require the use of ladders, hoists, or other devices, and escape from such areas may be very difficult, especially in emergency situations.

The term "not designed for continuous occupancy" relates to those spaces not designed for entry on a routine basis. They have probably been made to store a product, enclose materials or processes, or transport products or substances. Occasional entry for inspection, maintenance, repair, cleanup, or similar tasks is often difficult and dangerous due to chemical or physical hazards within the space. Air may not move in or out of confined spaces readily, thereby trapping deadly gases, particularly if the space has been used to store or process chemicals or organic substances that may decompose. There may not be enough oxygen inside the confined space to support life, or the air could be oxygen-rich increasing the chance of fire or explosion if a source of ignition is present.

The OSHA standard applies to all of general industry, including agricultural services (such as grain elevators), manufacturing, transportation, utilities, wholesale trade, food stores, hotels and other lodgings health services, museums, botanical gardens, and zoos, just to name a few. It also applies to boilers storage vessels, furnaces, railroad tank cars, manholes sewers, wells, pits, and cooking and processing vessels, among others. If a space meets any of the criteria described above, it can then be further divided into either a space requiring a full entry permit or one that does not.

18.4 OSHA Confined Space Standards

18.4.1 SCOPE AND APPLICATION

While the OSHA ruling is designed to protect employees in general industry from the hazards of entry into permit-required confined spaces, this ruling does not apply to agriculture, construction, or shipyard employment. However, OSHA may enforce confined space sites in these work classifications under the regulations provided in 29 CFR 1910.269, revised 1915, and 1910.5. If there are hazards present that are not addressed in these regulations, then the general standard (29 CFR 1910.146) applies. This confined space standard does not apply to marine terminals (29 CFR 1917) or to longshoring (29 CFR 1918), which fall under other confined space safety regulations. OSHA also has requirements for confined spaces in the workplace, such as monitoring, testing, and communications (training, warning signs, etc.).

The rule applies except where it is superseded, in whole or in part, by other industry-specific regulations.

18.4.2 GENERAL REQUIREMENTS OF THE STANDARD

General requirements are aimed at the employer and set out the specific procedures that must be taken in order to determine workplace applicability to permit-required confined space definitions. It also determines what the employer's duties are to protect workers who enter confined spaces. The procedure consists of several steps, beginning with an evaluation of the work-

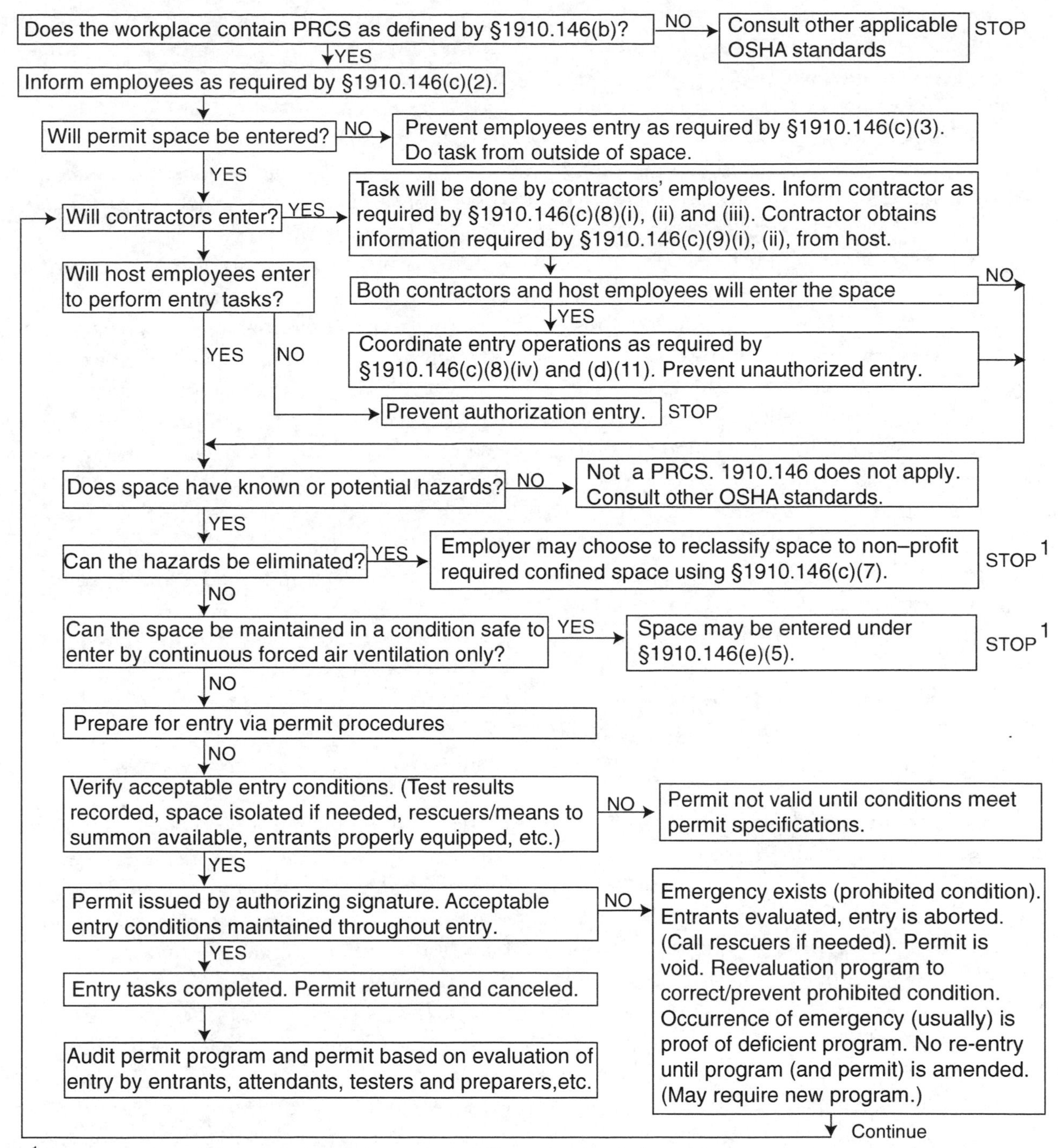

[1] Spaces may have to be evacuated and re-evaluated if hazards arise during entry.

Figure 18-2 Permit-required confined space decision flow chart

place to determine permit-required spaces. The distinction between permit-required confined spaces and nonpermit confined spaces is discussed below.

18.4.2.1 Permit-Required Confined Spaces

Under the standard, a permit-required confined space has one or more of the following characteristics:

- contains, or has the potential to contain, a hazardous atmosphere, which may include oxygen deficiency, oxygen enrichment, toxic gases such as hydrogen sulfide and/or flammable, or combustible gases such as methane;
- contains a material such as grain that has the potential for engulfment;

- its internal configuration is shaped so that an entrant could be trapped or asphyxiated by inwardly converging walls or by a floor that slopes downward and tapers to a smaller cross-section;
- contains any other recognized serious safety or health hazard, such as, but not limited to, electrical hazards, temperature, noise, slick/wet surfaces, or falling objects.

If hazards are present, danger signs must be posted, or the space must be completely blocked off. If entry is necessary, the employer must develop and implement a written permit space entry program and make it available to the employees. Reevaluation of the space must be done if its use changes, or new or increased hazards develop. Altering the use or configuration of the space, making it safe for entry may change its designation from permit-required to nonpermit, providing all of the listed procedures are followed. Finally, if outside contract personnel are hired to work in permit-required spaces, the employer still has the final responsibility for identifying each space and for making certain that the contractor is fully qualified to do this type of work.

18.4.2.2 Nonpermit Spaces

Nonpermit spaces are confined spaces that do not contain or, with respect to atmospheric hazards, do not have the potential to contain any hazard capable of causing death or serious physical harm to the worker. Spaces that require neither an entry permit nor an attendant to be stationed outside the space, are commonly found in the telecommunications and electrical utilities industries. These include manholes and unvented vaults. However, telecommunications employers do have obligations under the confined space standard. The standard does require them to evaluate the workplace to determine if there are any permit-required spaces. Manholes are addressed under 29 CFR 1910.268 (o), and manholes suspected of posing other hazards must be evaluated under 29 CFR 1910.146 (d)(2). This does not mean that manholes and work situations are not covered under the general standard. OSHA can cover situations where manholes become contaminated with toxins or other hazardous chemicals. If the work area cannot be made safe before entry, through compliance with 29 CFR 1910.268 (o)(2)(i)(B), any entries would be performed under the provisions of the general standard.

OSHA requires individual employers to classify and document the types of confined spaces their businesses contain. This determines the level of personal protection and procedures necessary for entry. It is possible for the same types of spaces to be classified differently by different employers, depending on what is being done inside those spaces. A space classified as permit-required during times when operations, such as painting, degreasing, or welding, are in progress could be classified as nonpermit if steps are taken to remove the hazards. The standard allows the employer to turn permit-required spaces into nonpermit spaces by installing or using proper ventilation. This is permissible if the only potential hazard is one relating to atmospheric conditions, and if this hazard was brought under control with improved ventilation.

18.4.3 THE PERMIT-REQUIRED CONFINED SPACE PROGRAM

Certain requirements must be met by the employer to control and regulate employee entry into permit-required spaces. In addition to identifying, evaluating, and controlling hazards prior to entry, the employer must also implement any measures necessary to prevent unauthorized entry. This includes all proper isolation procedures, such as barriers, and the protection of entrants from external hazards, such as objects that may blow or fall into the permit space. Employer's must provide safe and fully operational equipment at no cost to the employees. This includes equipment used for testing/monitoring, ventilating, communications, personal protection, and lighting and shielding. It also includes all equipment necessary for entering or exiting the spaces, such as ladders and rescue and emergency materials.

Following the completion of preliminary functions, employers must continuously monitor the space to ensure that the proper number of outside space attendants are available, and that only qualified and properly trained individuals are assigned duties within the space. Procedures must also be in place for emergency and rescue services, whether they are in-house or are contracted from outside sources. The employer must be able to immediately suspend permits, as well as control contract workers so that the operations of one company do not harm workers from another company. Finally, all permits and work procedures must be reviewed in order to ensure that all employees participating in permit space entry are properly protected.

The permit is signed by the entry supervisor before entry begins, and must be available to all authorized entrants. The permit must be posted so that any entrant can be certain that pre-entry preparations have been completed.

18.4.4 THE ENTRY PERMIT

The entry permit is the document that records compliance with regulations and authorizes entry into a permit-required space. It is often a two-page form as shown in Figures 18-3 and 18-4. It also details the employer's efforts to identify and control conditions. The permit must identify the space to be entered, the purpose of entry, and the date and duration of the entry. It must list authorized entrants specifically so that it can quickly be determined who is inside the space at any given time. The names of all attendants and supervisors present or those who authorized the entry are recorded. All hazards are listed, as are the measures used to isolate the space. Also listed in a concise summary are the tests used for monitoring and their results, all rescue and emergency services, all communication procedures, and necessary safety equipment. In addition, the entry supervisor has the authority to terminate entry and cancel the permit when operations have been completed, or when a situation not designated on the permit occurs in or near the entry space. If the entry is canceled for any reason, reentry cannot be made without securing and properly filing another permit. Finally, a list must be made of any additional permits, such as those needed for hot work. The employer must keep each canceled entry permit on file for at least one year.

18.4.5 TRAINING

The employer is required to provide training to all applicable employees so that these personnel acquire the awareness, understanding, knowledge, and skills necessary for the safe performance of their assigned duties. The employees are to be trained so that they can, with proficiency, carry out any of the duties required in the OSHA standard. Individual elements of training are further specified in sections dealing with authorized entrants, attendants, entry supervisors, and rescue personnel. Personnel are to be briefed and retrained as new or revised procedures are announced.

The employer must also provide proof certification that employees have received the required training. Such certification must contain the employee's name, the signatures of the trainers, training topics, and the dates of training.

18.4.6 DUTIES OF AUTHORIZED ENTRANTS

An authorized entrant is an employee sanctioned by the employer to enter a permit-required space. This person faces the greatest risk of death or injury from exposure to the hazards that may be contained within the space. The permit program is intended to provide protection for entrants, but the entrants themselves must also perform duties to help ensure their own safety. The employer is responsible for ensuring that these duties are performed by training the worker and by effectively communicating the work rules. Before workers can be authorized to enter permit-required spaces, they should be aware of the hazards they may face, including the signs, symptoms, and consequences of exposure. They should be proficient in the use of proper equipment, have adequate communication procedures worked out with all attendants in case evacuation is necessary, and know how to exit the space quickly. Entrants should alert the attendant when they notice a change in conditions within the space. Their most critical contact is the attendant, who must be stationed outside the space and be alert to any problems.

18.4.7 DUTIES OF ATTENDANTS

Providing an attendant outside a permit-required space is a method of monitoring the status of the entrant, monitoring conditions within the space, and summoning rescue services if they are required. In order for an employee to be a qualified attendant, the employer must ensure a specific amount and type of training. The attendant must know of the hazards that could be present and must also be aware of possible behavioral effects that hazards may have on the entrant. Attendants must keep an accurate account of who is in the space. It is part of their duty to remain outside the space until relieved by another qualified attendant. They must maintain communications with the entrants

Confined Space Entry Permit

Date and Time Issued: ______________ Date and Time Expires: ________
Job site/Space I.D.: ______________ Job Supervisor: ______________
Equipment to be worked on: ________ Work to be performed: ________
Stand-by personnel: ______________ ______________ ______________

1. Atmospheric Checks: Time ________
   Oxygen ________%
   Explosive ________% L.F.L.
   Toxic ________PPM

2. Tester's signature: ______________________________

3. Source isolation (No Entry):

	N/A	Yes	No
Pumps or lines blinded,	( )	( )	( )
disconnected, or blocked	( )	( )	( )

4. Ventilation Modification:

	N/A	Yes	No
Mechanical	( )	( )	( )
Natural Ventilation only	( )	( )	( )

5. Atmospheric check after isolation and ventilation:
   Oxygen ________% > 19.5 %
   Explosive ________% L.F.L < 10 %
   Toxic ________PPM < 10 PPM H(2)S
   Time ________
   Testers signature: ______________________________

6. Communication procedures: ______________________________
____

7. Rescue procedures: ______________________________
____
____
____

8. Entry, standby, and back up persons:

	Yes	No
Successfully completed required training?		
Is it current?	( )	( )

9. Equipment:

	N/A	Yes	No
Direct reading gas monitor - tested	( )	( )	( )
Safety harnesses and lifelines for entry and standby persons	( )	( )	( )
Hoisting equipment	( )	( )	( )
Powered communications	( )	( )	( )
SCBA's for entry and standby persons	( )	( )	( )
Protective Clothing	( )	( )	( )
All electric equipment listed Class I, Division I, Group D and Non-sparking tools	( )	( )	( )

10. Periodic atmospheric tests:

Oxygen	____%	Time	____	Oxygen	____%	Time	____
Oxygen	____%	Time	____	Oxygen	____%	Time	____
Explosive	____%	Time	____	Explosive	____%	Time	____
Explosive	____%	Time	____	Explosive	____%	Time	____
Toxic	____%	Time	____	Toxic	____%	Time	____
Toxic	____%	Time	____	Toxic	____%	Time	____

We have reviewed the work authorized by this permit and the information contained here-in. Written instructions and safety procedures have been received and are understood. Entry cannot be approved if any squares are marked in the "No" column. This permit is not valid unless all appropriate items are completed.

Permit Prepared By: (Supervisor)______________________________
Approved By: (Unit Supervisor)______________________________
Reviewed By (Cs Operations Personnel) :

______________________________ ______________________________
(printed name) (signature)

This permit to be kept at job site. Return job site copy to Safety Office following job completion.

Copies: White Original (Safety Office)
Yellow (Unit Supervisor)
Hard(Job site)

Figure 18-3 Confined Space Evaluation Prior To Entry Form

ENTRY PERMIT

PERMIT VALID FOR 8 HOURS ONLY. ALL COPIES OF PERMIT WILL REMAIN AT JOB SITE UNTIL JOB IS COMPLETED

DATE: - - SITE LOCATION and DESCRIPTION ______________________
PURPOSE OF ENTRY ______________________
SUPERVISOR(S) in charge of crews Type of Crew Phone #
____
____
COMMUNICATION PROCEDURES ______________________
RESCUE PROCEDURES (PHONE NUMBERS AT BOTTOM) ______________________
____

* BOLD DENOTES MINIMUM REQUIREMENTS TO BE COMPLETED AND REVIEWED PRIOR TO ENTRY*

REQUIREMENTS COMPLETED	DATE	TIME
Lock Out/De-energize/Try-out	____	____
Line(s) Broken-Capped-Blanked	____	____
Purge-Flush and Vent	____	____
Ventilation	____	____
Secure Area (Post and Flag)	____	____
Breathing Apparatus	____	____
Resuscitator - Inhalator	____	____
Standby Safety Personnel	____	____
Full Body Harness w/"D" ring	____	____
Emergency Escape Retrieval Equip	____	____
Lifelines	____	____
Fire Extinguishers	____	____
Lighting (Explosive Proof)	____	____
Protective Clothing	____	____
Respirator(s) (Air Purifying)	____	____
Burning and Welding Permit	____	____

Note: Items that do not apply enter N/A in the blank.

**RECORD CONTINUOUS MONITORING RESULTS EVERY 2 HOURS

CONTINUOUS MONITORING** TEST(S) TO BE TAKEN	Permissible Entry Level								
PERCENT OF OXYGEN	19.5% to 23.5%	___	___	___	___	___	___	___	___
LOWER FLAMMABLE LIMIT	Under 10%	___	___	___	___	___	___	___	___
CARBON MONOXIDE	+35 PPM	___	___	___	___	___	___	___	___
Aromatic Hydrocarbon	+ 1 PPM * 5PPM	___	___	___	___	___	___	___	___
Hydrogen Cyanide	(Skin) * 4PPM	___	___	___	___	___	___	___	___
Hydrogen Sulfide	+10 PPM *15PPM	___	___	___	___	___	___	___	___
Sulfur Dioxide	+ 2 PPM * 5PPM	___	___	___	___	___	___	___	___
Ammonia	*35PPM	___	___	___	___	___	___	___	___

* Short-term exposure limit: Employee can work in the area up to 15 minutes.
+ 8 hr. Time Weighted Avg.: Employee can work in area 8 hrs (longer with appropriate respiratory protection).
REMARKS: ______________________

GAS TESTER NAME & CHECK #	INSTRUMENT(S) USED	MODEL &/OR TYPE	SERIAL &/OR UNIT #
__________	__________	__________	__________
__________	__________	__________	__________

SAFETY STANDBY PERSON IS REQUIRED FOR ALL CONFINED SPACE WORK

SAFETY STANDBY PERSON(S)	CHECK #	CONFINED SPACE ENTRANT(S)	CHECK #	CONFINED SPACE ENTRANT(S)	CHECK #
__________	______	__________	______	__________	______
				__________	______

SUPERVISOR AUTHORIZING - ALL CONDITIONS SATISFIED ______________________
DEPARTMENT/PHONE ______________________
AMBULANCE 2800 FIRE 2900 Safety 4901 Gas Coordinator 4529/5387

(*Source: 29 CFR 1910, Permit-Required Confined Spaces for General Industry, Final Rule. *Fed. Reg.* 58: 9 [1 Dec. 1998].)

Figure 18-4 Confined Space Entry Permit; used Subsequent to Confined Space Evaluation Form

to monitor their status and alert the entrants in the event of evacuation. Attendants summon rescue and other emergency services if assistance is warranted. They keep unauthorized persons out of the space and perform non-entry rescues, if specified. Otherwise, attendants should not perform other tasks that would interfere with their primary duty of monitoring and protecting the authorized entrants.

18.4.8 DUTIES OF ENTRY SUPERVISORS (PERMIT ISSUER)

Under the confined space regulations, the burden of employee safety is placed on employers. This is primarily accomplished by naming an entry supervisor who has overall accountability for safe entry operations. The entry supervisor should know what hazards employees may face in any permit-required space, as well as the signs, symptoms, and consequences of exposure to these hazards. The supervisor will verify that the permit has been fully and correctly completed, that all necessary tests have been conducted, and that all procedures, equipment, and emergency services are in place before finally signing the permit and allowing entry to begin. They have the right to terminate the entry, remove unauthorized individuals from the premises, and can also determine whether the procedures being performed are consistent with the acceptable conditions described in the terms of the permit.

18.4.9 RESCUE AND EMERGENCY SERVICES

Over 50 percent of the workers who die in confined spaces are attempting to rescue other workers. In spite of all the precautions taken to ensure that employees can safely enter and work inside confined spaces, hazards may arise so quickly and conditions change so rapidly that entrants are unable to escape from the confined space without some type of assistance. Revisions to the OSHA ruling, issued December 1, 1998, and effective February 1, 1999, clarified employer and employee responsibilities for the development and implementation of confined space rescue plans. Under these regulations, employees must be allowed to observe any testing of a confined space done prior to entry, must be allowed to review confined space entry plans, and must be allowed to participate in the development and implementation of the plan.

Employers are required to ensure that each authorized confined space entrant uses a harness with retrieval line that would facilitate the successful removal of the entrant in an emergency situation. The employer must also train employees to perform rescue duties, must schedule practice drills at least once a year, and must evaluate and ensure a rescuer's ability to retrieve victims from identified confined spaces.

One of the more problematic parts of rescue in confined spaces is reliance upon off-site rescue services. The employer is required to evaluate a prospective rescuer's ability to respond to a rescue in a "timely manner." The employer must also identify and evaluate on-site hazards and thoroughly inform the rescue services, including outside services, of all possible hazards. Such services should also have access to all permit-required spaces for training and review.

In addition, employers must provide PPE to employees who enter confined spaces, and train them so they are proficient in the equipment's use. The permit-required space must be reevaluated if requested by the employees (results of the evaluation being provided). Employees working in confined spaces must be trained in basic first-aid and CPR with at least one member of the rescue team having current certification in first aid and CPR at all times.

The employer must have, as part of its plan, the use of non-entry rescue retrieval systems and must determine whether such equipment may increase the overall risk to the entrant. Rescuers must know the specifications for such retrieval systems and cross-check this information with the equipment that is available. Finally, if an injured entrant is exposed to a substance for which there is an MSDS or other written information, such information shall be kept at the site and given to the medical facility treating the exposed worker.

In spite of all the warnings, regulations, and required training procedures, confined space fatalities still occur. The confined space standard was put on the record in its final form on January 14, 1993, and went into effect April 15, 1993. A newspaper, printed November 12, 1993, recorded that a firm in Hudson, Colorado, was fined $245,000 for safety violations discovered while investigating a fatality from a tank car accident. OSHA reported seven alleged violations, including the use of unapproved respirators, failure to inspect safety equipment, and failure to train and supervise employees in how to enter a confined space safely. The violations were largely the result of improperly trained workers entering a railroad tank car to rescue a fellow worker who had fallen into the car. The company relied on outside rescue provisions that were not readily available for this type of accident.

This incident points out what may be the standard's weakest link—rescue provisions. Some employers, especially small companies, may consider it unrealistic and overly burdensome to have their own rescue services. OSHA gives employers the option to have an onsite rescue team or to use an offsite service such as an outside contractor or local rescue squad to perform a rescue. Workers in a severely oxygen-deficient atmosphere must be rescued within 5 minutes if they are to survive. A slow or inadequate response by rescuers, especially outside services, may cause loss of valuable minutes to effect a live rescue when things go wrong in a confined space.

18.5 Sources of Information

"29 CFR 1910, Permit-Required Confined Spaces for General Industry: Final Rule," *Federal Register 58*:9 (1 December 1998).

Finnegan, Lisa (ed.): OSHA Clarifies Confined Space Rule. *Occupational Hazards*. p. 15 (January 1999).

LaBar, Gregg: Confined Space Safety: Getting Into Compliance. *Occupational Hazards*. pp. 31–35 (March 1993).

LaBar, Gregg: Taking the Danger Out of Confined Spaces. *Occupational Hazards*. pp. 21–25 (February 1991).

Lewis, David: Railcar Firm Fined $245,000 in Fatality. *Rocky Mountain News* (19 November 1993).

Occupational Safety and Health Administration: *Questions and Answers for PRCS Standard Classification*. In www.osha-slc.gov/html/faq-confined spaces.html. Washington, DC: U.S. Department of Labor, August 1999.

NINETEEN

Excavating, Trenching, and Shoring

19 | CONTENTS

19.1 Introduction

Two employees were installing storm drain pipes in a trench approximately 20 to 30 feet long, 12 to13 feet deep, and 5 to 6 feet wide. The side walls consisted of unstable soil undermined by sand and water. There was 3 to 5 feet of water in the north end of the trench and 5 to 6 inches of water in the south end. At the time of the accident, a backhoe was being used to clear the trench. The west wall of the trench collapsed and one of the employees was crushed and killed. A summary of this fatality and the general work situation showed that the weather was clear and that the crew of three had approximately four months experience at this type of work. They had been on the job for five minutes that day, had not had training or education for work hazards, no regular worksite inspection was provided, a safety and health program was not in place, and a competent safety monitor was not on site. Although this incident had many earmarks of a tragedy ready to happen, it is not an isolated instance of improper work conditions or ignorance of the law. In 2000, 123 workers were killed by being caught in or crushed by collapsing material. In that same year, 40 individuals were killed in excavation or trenching cave-ins while 45 died when they were caught in or crushed by a collapsing structure. This type of danger is greatest in the construction industry, with 78 percent of the total excavation fatalities in 1998 occurring there. Other industries with the same type of fatalities, though in lesser numbers, are agriculture and mining. Additionally, OSHA statistics show that 79 percent of the workers killed in trenching accidents were working in trenches less than 15 feet deep.

Excavating is recognized as one of the most hazardous construction operations. OSHA issued its first excavating and trenching (E&T) standard in 1971 to protect workers from excavation hazards. Since then, OSHA has amended the standard several times to increase worker protection and reduce the frequency and severity of excavation accidents and injuries. Despite these efforts, excavation-related accidents resulting in injuries and fatalities continue to occur. In 1985, OSHA issued a Directive (CPL 2.69) with a special emphasis on trenching and excavation. However, by the beginning of the 1990s, deaths in this field were still nearly 100 per year and yearly work-related injuries numbered over 1,000. These numbers led to a reevaluation of existing laws and requirements for training and enforcement. OSHA completely updated the standard in 29 CFR 1926 to simplify many of the existing provisions, added and clarified definitions, and eliminated duplicate provisions and ambiguous language. The following is a summary of the current regulations found under Construction Industry Regulations, 29 CFR, Subpart P—Excavations, sections 1926.650 through 652, with accompanying Appendices A through F.

19.2 Excavating and Trenching Regulations

E&T procedures are performed thousands of times daily across the United States. Such work can be observed in diverse situations. Earth moving and open excavations are common in the installation and repair of utility lines, replacement of water and sewer lines, the installation of footings for bridges or piers, in demolition projects, in domestic and commercial swimming pool construction, and even in the preparation of graves. E&T work also occurs at hazardous waste sites, such as during the removal of buried drums, during the removal of contaminated soils, and during the replacement of LUSTs. Due to their location (below ground) and the wide variety of external factors that can affect them, such as traffic, weather, and soils, these activities have the potential for disaster. Constant change in the nature of the work, unstable working conditions, poor equipment maintenance, hurried deadlines, transient and frequently unskilled laborers, as well as exposure to other hazards encountered in digging make for a varied list of dangers. Injuries experienced in E&T operations far surpass fatalities. Falls, twisted ankles, sprains, cuts, being struck by equipment or falling tools, chemical exposure, and heat and cold stress are a few of the causes of injuries. Training workers to be on the lookout for physical hazards is a large part of E&T regulations. Observation and comprehension of hazards with avoidance or abatement of such hazards is the core of E&T safety.

19.2.1 DEFINITIONS

As with many other regulations, 29 CFR 1926.650-652 has definitions specific to its nature. Terms most encountered in this unit are defined below. These and other definitions apply to all open excavations made in the earth's surface.

- Benching—a method of protecting employees from cave-ins by excavating the sides of an excavation to form one or a series of horizontal levels or steps, usually with vertical or near-vertical surfaces between levels (Figure 19-1).
- Cave-in—the separation of a mass of rock or soil from the sides of an excavation, or the loss of soil from under a trench shield or support system, and its sudden movement into the excavation, either by falling or sliding, in sufficient quantity so that it could entrap, bury, or otherwise injure and immobilize a person.

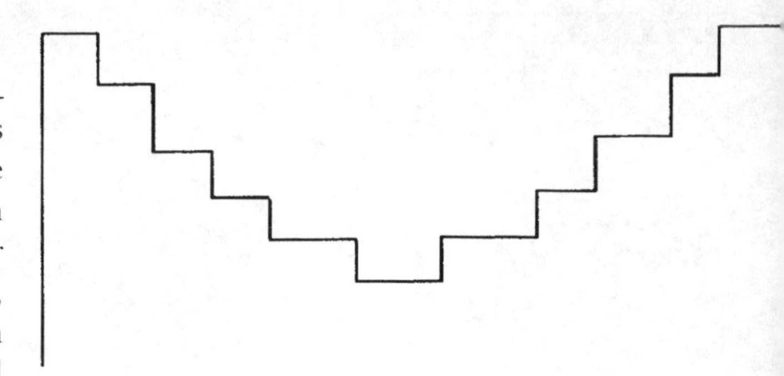

Figure 19-1 Example of multiple bench excavation Source: OSHA Technical Manaual, January 20. 1999.

- Competent person—one who is capable of identifying existing and predictable hazards in the surroundings, or recognizing working conditions that are unsanitary, hazardous, or dangerous to employees, and who has authorization to take prompt corrective measures to eliminate them.
- Excavation—any man-made cut, cavity, trench or depression in an earth surface formed by earth removal. This definition also includes trenches.
- Faces or sides—the vertical or inclined earth surfaces formed as a result of excavation work.
- Hazardous atmosphere—an atmosphere that is flammable, poisonous, corrosive, oxidizing, irritating, oxygen deficient, toxic, or otherwise harmful to the health or lives of workers.
- Protective system—a method of safe-guarding employees from cave-ins, from materials that could roll or fall from an excavation face into an excavation, or from the collapse of adjacent structures.
- Shield—a structure that is able to withstand the forces imposed on it by a cave-in and thereby protect workers within the structure (Figure 19-2).
- Shoring—a structure such as a hydraulic, mechanical, or timber system that supports the sides of an excavation and that is designed to prevent cave-ins.

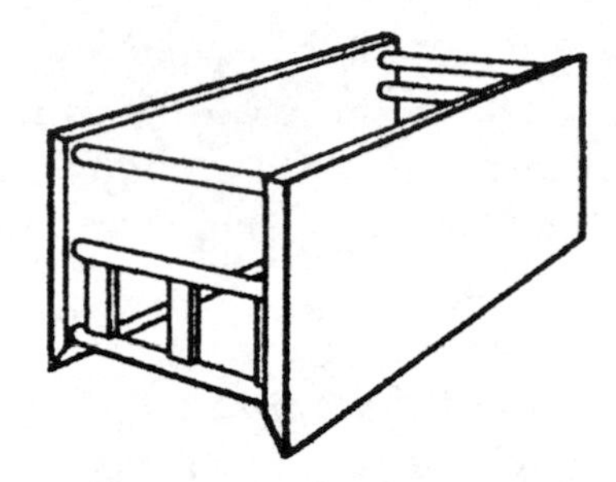

Figure 19-2 Trench shield

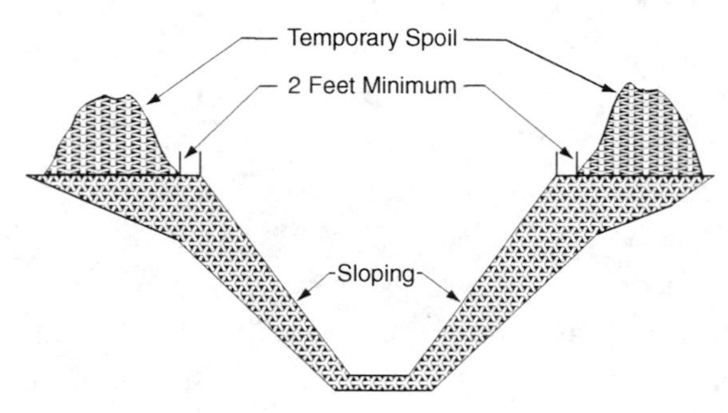

Figure 19-3 Generalized trench profile. From: OSHA Technical Manual.

- Sloping—a method of protecting employees by forming sides that are sloped away from the excavation to prevent cave-ins (Figure 19-3).
- Spoil—the material removed from an excavation and generally piled beyond the trench walls. Spoil can be classified as temporary (placed no closer than two feet from the edge of an excavation and set so it channels rainwater away from the trench) and permanent (located "at some distance from the excavation") (Figure 19-3).
- Stable rock—natural, solid mineral material that can be excavated with vertical sides and that will remain intact while exposed.
- Support system—a structure such as an underpinning, bracing, or shoring that provides support to an adjacent structure, underground installation, or the sides of an excavation.
- Trench—an excavation that:
  —has a narrow width in relation to its length;
  —is made beneath the surface of the ground;
  —in general, is deeper than it is wide; and
  —has a width of less than 15 feet when measured at the bottom.

19.2.2 THE COMPETENT PERSON

OSHA requires that a competent person inspect the excavation and adjacent areas at least once a day for possible cave-ins, failures of protective systems and equipment, and hazardous atmospheres or other hazardous conditions. This competent person has been trained in or has knowledge of the OSHA excavation standard. They should be able to demonstrate knowledge of soil analysis, should have the authorization to take prompt corrective measures to eliminate any hazards, and they may be responsible for the coordination and direction of emergency response procedures. In some cases, they may have to identify unsanitary conditions and point out physical hazards to avoid puncture wounds or exposure to biologic materials. Large operations should have a full-time safety official who serves as the competent person and makes recommendations to improve on-the-job safety. A smaller operation may have a part-time safety official who may also be a supervisor. In any case, this individual must survey the excavation site at least once a day to look for hazardous conditions.

19.3 Preparation and Excavation Safety

A number of on-the-job accidents can be avoided with adequate initial planning. Safety checklists should be developed prior to the start of a job to make certain that adequate information is available and that necessary paperwork and tools are on hand. Such lists should incorporate information necessary for safe operations and should include hazard identification and control measures for, but not necessarily limited to:

- buried utilities (electrical power, natural gas, and water lines) and overhead utilities;
- foot and vehicle traffic;
- hazardous atmospheres;
- the nearness of structures and their ability to prevent falling hazards;
- weather and its potential for change;
- access and egress;
- exposure to falling loads;
- warning systems for mobile equipment;
- heavy equipment operation;
- surface and ground water, including the water table;

- soil and rock stability;
- emergency rescue equipment; and
- inspections.

These and other conditions can be determined by job site studies, observations, test borings for soil types or conditions, geology reports, and consultations with local officials and utility companies.

19.3.1 UTILITIES

Utility companies must be contacted at least 24 hours before excavation begins. Companies will usually dispatch a technician to locate the utilities (within 3 to 5 feet) on the site where the excavation is to occur. Some areas have established a "digger's hotline" number so that one phone call may cover all potential installations. Once the underground utilities have been located, they must be protected from activity on the site. This could mean that they need to be deenergized, discharged, covered, supported, or removed in order to protect workers.

A further refinement of the site is removing or relocating surface items that may encumber the work or prove to be sources of injury for laborers. Such items include among other things logs, trees, boulders, garbage, concrete, sidewalks, overhead power lines, and phone lines.

19.3.2 TRAFFIC AND MOVING HAZARDS

When an excavation is near moving traffic, or can be affected by vibrations caused by traffic or heavy construction vehicles, workers face danger from collisions, falls, and cave-ins. Hazard controls include wearing warning vests of reflective or highly visible material. Vests are not useful as PPE, however, if workers do not wear them. The manager or competent person at the site must enforce their use. Warning signs or flag persons should also be used for public traffic. This should include foot traffic as well as vehicular traffic. Reflective tape and/or barricades add another level of warning. Finally, if a protective support system is in place in the excavation, it should be heavy enough to withstand the number and weight of passing vehicles.

For example, a plumbing company employee and an independent backhoe operator were making a sewer line connection in a 13-foot deep trench located adjacent to a busy, multi-laned highway. A portion of the trench wall caved in (probably from traffic vibrations, according to OSHA reports), burying the employee. The backhoe operator was buried up to his chest in the trench.

19.3.3 HAZARDOUS ATMOSPHERES

Because a depression is being created in the earth and because certain gases have a density greater than air and thus flow into low-lying areas, workers may suffer the deadly effects of suffocation, poisoning, or explosion unless the suspect air is tested and regularly monitored. Liquid petroleum gas and acetylene pose flammability and explosion threats from accumulation in the excavation, from cutting operations, or from other spark sources. Carbon monoxide can be a serious risk when the gas is introduced from the exhaust of internal combustion engines, including those used to pump accumulated water from the trench. Carbon monoxide, hydrogen sulfide, and other confined space hazards exist on a regular basis in E&T work including landfill areas or in areas where hazardous substances are stored nearby. Workers remediating hazardous waste sites may be exposed to even more virulent combinations of gases, liquids, and even solids. Because of this, monitoring of suspect environments should take place before and during trenching.

Proper breathing aids, such as supplied air, respirators, or site ventilation, must be used when necessary. Testing of the atmosphere must be complete before unprotected workers can be let into trenches greater than four feet deep. Employees must also not be exposed to an atmosphere containing more than 20 percent of the lower flammable limit of flammable gases. It may be necessary to further protect trenching workers from hazardous atmospheres by using lifelines and retrieval harnesses.

19.3.4 NEARBY STRUCTURES AS FALLING HAZARDS

In some instances the excavated trench may be "safe" as far as regulations are concerned, but the potential for collapse of adjacent buildings, walls, or sidewalks constitutes as a hazard to workers. Support systems such as shoring, bracing, or underpinning must be provided to ensure the stability of adjacent structures. In some cases, it may be more feasible to remove the structure. No digging should be done under any structure unless the excavation is in stable rock or unless a proper protection system is in place and a registered professional engineer approves of the job.

For example, an employee was in the process of locating an underground water line. A trench had been dug to a depth of approximately four feet along the side of a brick wall that was seven feet high and five feet long. The wall collapsed onto the victim who was standing in the trench. The injuries were fatal.

19.3.5 WEATHER

The presence of wind, rain, or other weather elements can change the conditions of an excavation enough to cause drowning, cave-ins, slips and falls, or the falling of heavy objects into a trench. Within a few hours, rain, frost, snow, or even extreme heating and drying of the soil can change site conditions and introduce new hazards. Workers at risk from any site safety change must be removed until corrective steps are taken. Some methods to decrease weather hazards include the installation of support systems strong enough to withstand weather conditions, and reinspection of the excavation site and trench supports following each weather change.

19.3.6 ACCESS AND EGRESS

It may be necessary to move people, material, and possible machinery from the excavation. No matter what is being moved, ramps or ladders for personnel, and structural ramps for equipment must be designated by a competent person. This means that they should ensure that ramps are stable, connected together (if necessary), of uniform thickness, and designed to prevent slipping or tripping. Excavations more than four feet deep should have a fixed means of getting out. Ladders and other means of exit must be located so that workers need not travel more than 25 feet to an egress route. Ladders must also be secured and extend a minimum of three feet above the landing. In all cases, metal ladders should be used with caution, particularly when electric utilities are present or when a potentially explosive atmosphere may be present and sparking is a hazard concern.

For example, an employee was working in a trench four feet wide and seven feet deep. About 30 feet away, a backhoe was straddling the trench. The backhoe operator noticed a large chunk of dirt falling from the side wall behind the worker in the trench. He called out a warning. Before the worker could climb out, six to eight feet of the trench wall had collapsed on him and covered his body up to his neck. He suffocated from the weight of the soil compressing his chest before the backhoe operator could dig him out. There were no exit ladders. No sloping or shoring had been used in the trench.

19.3.7 EXPOSURE TO FALLING LOADS

Employees must be protected from loads or objects falling from operations in or nearby the excavation. In order to fulfill this requirement, no worker is permitted underneath raised loads being handled by lifting or digging equipment. Employees are also required to stand away from equipment being loaded or unloaded in order to avoid being struck by any falling materials. Obeying these rules appears easy until the reality of loading or unloading materials from vehicles and moving those materials into a narrow trench with accurate placement actually takes place. The usual reaction is to help "manhandle" the material into the proper location. It is this sort of situation that calls for the presence of a competent person to prevent such hazardous actions.

19.3.8 WARNING SYSTEMS FOR MOBILE EQUIPMENT AND HEAVY EQUIPMENT OPERATION

Built-in safeguards that are supposed to help ensure safety at an excavation sometimes fail to work in the manner originally intended. Operators have been known to disconnect back-up alarms or fail to report to workers that they may be malfunctioning. If a malfunction does take place, hand or mechanical signals must be used. Back-up alarms from several machines may confuse workers or, after hearing such alarms for a period of time, workers may become desensitized to possible danger. Employees not paying attention account for many excavation injuries and fatalities.

For example, a contractor at an excavation site was operating a backhoe when an employee attempted to walk between the swinging superstructure of the backhoe and a concrete wall. As the employee approached the backhoe from the operator's blind side, the superstructure hit the victim, crushing him against the wall. Although this vehicle did not fall into the trench, the incident points out that safety is not just for those individuals working inside the excavation. Barricades are possible safety precautions that could have been put in place to restrict approach.

Heavy equipment used on many excavation sites provides its own hazards. For example, vibrations in and around the site may weaken side walls and cause cave-ins. Heavy equipment operators may also inadvertently dig into a hole containing people who may not be readily visible from the operator's position. Proper maintenance of the machines is also crucial. Timely repairs may prevent accidents from occuring.

For example, a three-man crew was digging a trench for a new sewer line using a backhoe that was very poorly maintained. Among other problems, its starter button did not work and the safety catch on the gear box was broken. In order to start the machine, the operator used a screwdriver to engage the starter. When the machine started, the gear shift engaged and the vehicle lurched foreword, running over the victim.

The use of equipment by underqualified individuals also contributes to safety failure. Even qualified operators may be asked to take their expertise to the limit when deadlines or extra costs are in question. Instead of ordering the proper equipment and perhaps paying higher rental cost or employee costs in salary, a contractor may try to extend the "reach" of the equipment by edging closer to the side wall of the trench or by attaching the backhoe to a truck or tree for greater stability. In any case, safety is compromised.

19.3.9 SURFACE AND GROUND WATER

Water can cause worker drowning or site cave-in. The OSHA standard prohibits employees from working in excavations where water has accumulated (or is accumulating) unless adequate protection is provided. The site must be covered by a tarp or shelter; surface water must be diverted with dikes and ditches; water must be drained by a process such as pumping; or a lifeline or retrieval harness must be used. If water removal equipment is used to control or prevent water from accumulating, the equipment and its operations must be monitored by a competent person in order to ensure proper use. While pumps can drain trench areas of water, care should be taken before entering the excavation because carbon monoxide gas from the generator or motor may settle into the hole, causing a hazardous atmosphere.

Any system meant to provide special support or shielding from standing water must be approved by a registered professional engineer. When it is raining, employees should be removed from excavations and not allowed to return until a competent person has inspected the excavation to determine its stability.

19.3.10 SOIL AND ROCK STABILITY

In E&T practices, "soil" is defined as any material removed from the ground to form a hole, trench, or cavity for the purpose of working below the earth's surface. The material is most often weathered rock and clays, but can also be gravel, sand, and solid rock. It is necessary to know the characteristics of the soil since each site presents its own unique situations. It is the condition of the soil, as well as external conditions (discussed previously) that will ultimately determine the type of mechanisms used to protect the workers. The classification of soils will be found in a later section, as will the shoring, shielding, and slope configurations associated with worker protection.

Protection for workers in excavation circumstances is necessary, due in large part to the physical characteristics of soil. Soil is an extremely heavy material and may weigh more than 1,000 pounds per cubic yard. This mass of soil, dropping from only a few feet onto a human can cause instant crushing, or a slow, squeezing action in which the skeletal mass collapses followed by the lungs and other vital organs. Furthermore, wet soil, rocky soil, or some types of solid rock may be even heavier.

To get an idea of the physical forces being exerted within the side walls of a trench, imagine a column of cubic foot-sized soil blocks, each weighing 100 pounds and stacked five high. This means the block at the bottom is supporting its own weight as well as the 400 pounds resting on it. The combined weight of this column is 500 pounds spread over a one foot square area. The column exerts not only a vertical force, but lateral forces as well, pushing outward in all directions. As the weight of the column increases, the soil blocks at the bottom have a tendency to compress and spread out in the direction of least resistance. If the column is contained within the earth, the pressure is generally equalized on all sides, and equilibrium is maintained.

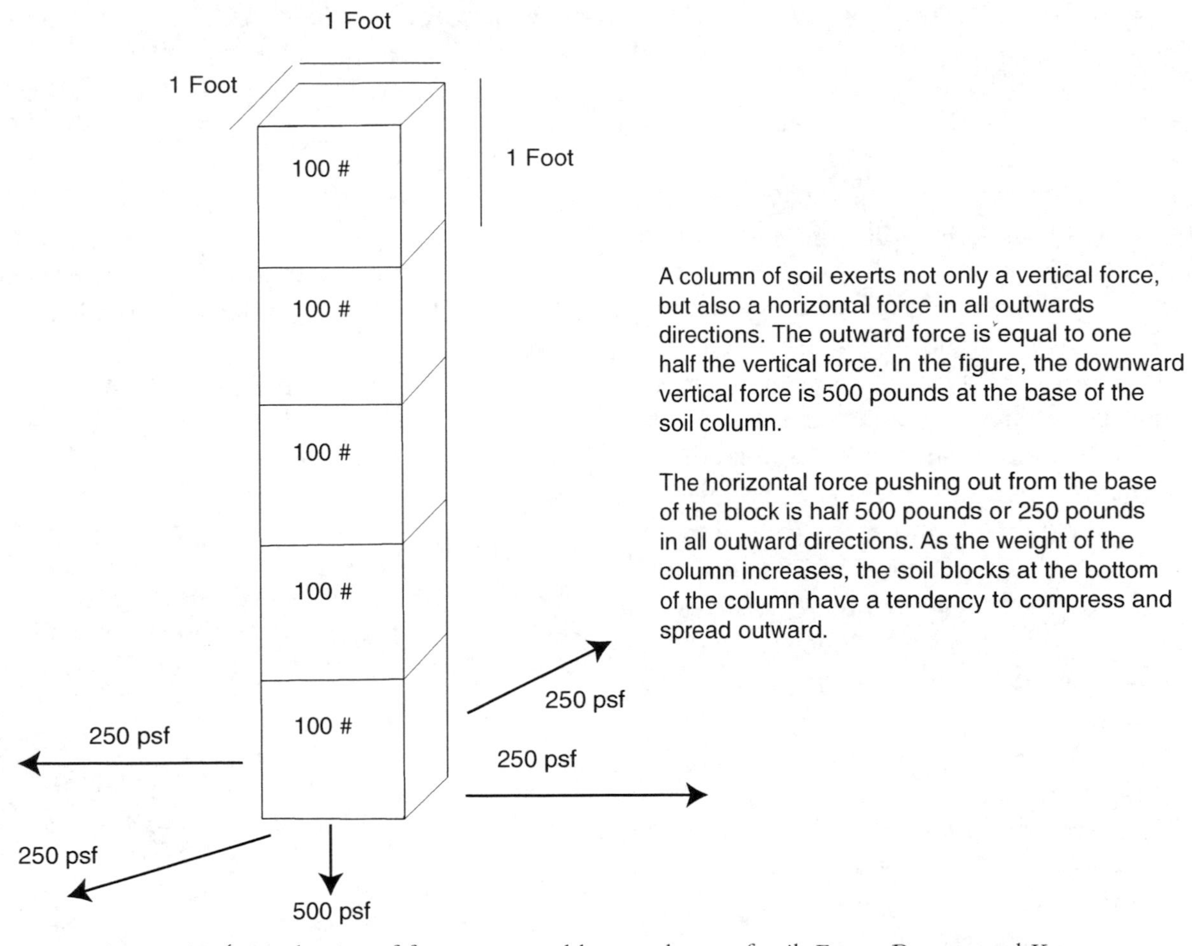

Figure 19-4 Mechanics of forces exerted by a column of soil. From: Brown and Kramer.

However, when excavation takes place and one side of this column is exposed, the compressive and lateral forces will work to "squeeze out" into the newly-opened direction (Figure 19-4).

The first failure of the side wall may come when the bottom of the wall moves into the trench leaving behind an undercut area. There is often a second movement in which more of the wall erodes. Finally, the erosion at the base of the trench leaves the upper part of the column poorly supported and more soil may fall into the excavation. Many rescue attempts are unsuccessful because rescuers tried to save victims before the second and third movements took place. This often traps the would-be rescuers along with the initial victims.

For example, employees were laying sewer pipe in a trench 15 feet deep. The sloped sides of the trench, four feet wide at the bottom and 15 feet wide at the top, were not shored or protected to prevent a cave-in. Soil in the lower portion of the trench was mostly sand and gravel and the upper portion was clay and loam. The trench was not protected from vibration caused by heavy vehicle traffic on the road nearby. To leave the trench, employees had to climb over the backfill. As they attempted to leave the trench, there was a small cave-in covering one employee to his ankles. When one employee went to his co-worker's aid, another cave-in occurred covering him to his waist. The first employee died of a rupture to the right ventricle of his heart at the scene of the cave-in. The other employee suffered a hip injury.

19.3.11 EMERGENCY RESCUE EQUIPMENT

Emergency rescue equipment, such as a breathing apparatus, a safety harness and line, or a basket stretcher, should be readily available for dealing with hazardous atmospheres (see section 19.3.3 in this unit). Harnesses may be used for trenches where egress is more difficult than usual or at the request of employees.

In addition, every contractor should have emergency response procedures in place in case an accident occurs. These include having a list of phone numbers for physicians, medics, 911 services, or the names and numbers of members of the on-site Emergency Response Team. In case of an accident, the Emergency Response Team or 911 should be contacted with vital information on the site situation. Knowing the exact location, number of victims, trench measurements, and any special hazards will aid in the rescue efforts. In the meantime, it is necessary to keep all life-support and dewatering systems operating. Also, clear workers away from the excavation, shut down any heavy equipment, and prepare to meet the rescue personnel. Generally, on-site personnel should not try to dig the victim out with heavy equipment. They should also not allow others into the trench and should keep passers-by and the media away from the scene, allowing faster access for the rescuers. If no on-site rescue team is available, the supervisor should know what local authorities will respond with an idea of the relative time necessary for travel to the site.

In order to keep an orderly flow of actions on the site, one or two individuals should be in charge of notifying authorities and rescue personnel. An individual is needed to meet, advise, and direct rescue personnel at the job site. Additionally, rescuers may want to know what emergency response equipment is available at the site, where is it kept and who is trained to use it.

As job duties, job site locations, and even daily work practices change, emergency response plans should be revised to accommodate those changes. This may require frequent review and reassignment of responsibilities. Training for emergencies is essential. Other training may be required such as for the entry of confined spaces, since some excavations may also be classified as a confined space.

19.3.12 INSPECTIONS

Since construction excavation projects are inherently customized projects, with each having a unique set of parameters, inspections become an important part of the site process. Inspections shall be made by a competent person and should be documented. The E&T standard establishes the frequency of and reasons for inspections a follows:

- daily, before the beginning of each shift, and as needed throughout the shift;
- as dictated by the work being done in the trench;
- after every rainstorm;
- after other events that could increase hazards;
- when fissures, tension cracks, sloughing, undercutting, water seepage, bulging at the bottom, or other similar conditions occur;
- when there is a change in the size, location, or placement of the spoil pile; and
- when there is any indication of change or movement in adjacent structures.

Such inspections should focus on situations that could result in possible cave-ins, indications of failure of protective systems, hazardous atmospheres, and other hazardous conditions. When the competent person finds evidence of any of these situations, employees shall be removed from the hazardous area until the necessary precautions are taken to ensure their safety. In addition, walkways (with handrails, if required) shall be provided where employees or equipment are permitted to cross over the excavation.

Of the one dozen trenching accidents cited in OSHA's "Fatal Facts" publication in November 1999, only one site had inspections and only two had safety and health programs in place including the training of employees.

19.4 Protective Systems

The E&T standard states that each employee in an excavation shall be protected from cave-ins by an adequate protective system designed in accordance with section 1926.652. The means of protection include sloping, benching, shoring, and shielding, depending on the type of soil and the external conditions that may be present during the operation. Unless excavations are entirely in stable rock or will be less than five feet deep at a site that is thoroughly inspected and shows no potential for collapse, OSHA requires a protective support system. Design specifications are clearly outlined in the OSHA standard. Other approved designs, custom-made for the site by a registered professional engineer, must also be OSHA approved. The current excavation standard does not, however, specify a minimum length that an excavation must be before a protective system is required.

OSHA levied fines of $230,400 and $371,400, respectively, in cases involving potential life-threatening conditions and an actual fatality. In the former case, the company was reported by members of the community in which the work was being performed after they observed dangerous practices taking place. With such amounts of money involved, the use of proper protective systems, as well as inspections and safety systems seem to be minimal investments.

Another example occurred when four employees were boring a hole and pushing a 20-inch pipe casing under a road. The employees were in an excavation approximately nine feet wide, 32 feet long and seven feet deep. Steel plates eight feet by 15 feet by ¾-inch were being used for shoring and were placed vertically against the north and south walls of the excavation. They rested at an angle of about 30 degrees. There were no horizontal braces between the steel plates. The steel plate on the south wall tipped over, pinning an employee (who was killed) between the steel plate and the casing. At the time the plate tipped over, a backhoe was being operated adjacent to the excavation.

19.4.1 SLOPING AND BENCHING

Designs for sloping and benching as well as other specific requirements are presented in 29 CFR Subpart P, section 1926.652(b) and I Appendix B to Subpart P. To slope and bench a site means cutting the walls of the excavation back at an angle relative to its floor (see Figures 19-1 and 19-3). Sloping uses angled cuts on the sides, while benching consists of one or more terraces or steps sloping away from the deepest point of the excavation. These angled cuts are allowed by OSHA for excavations up to 20 feet deep. In solid rock, the angle of the wall may be 90 degrees, while in more unconsolidated material, the angles will be less. In addition, workers cannot work on the faces of sloped or benched excavations above other employees, unless the workers at the lower level are protected from falling material or equipment.

19.4.2 SHORING

Shoring is the provision of a support system for trench faces used to prevent the movement of soil, underground utilities, roadways, and foundations. It provides framework for soil emplacement and supports excavation walls. In theory, the closer the shores are placed together, the greater the support that will be offered. Shoring is used when the location or depth of the cut makes sloping back the maximum allowable slope impractical, such as when excavating near houses, streets, major roads, or transportation lines. Shoring systems consist of posts, struts, sheeting, and wales. The two basic types of shoring are timber frame and aluminum hydraulic (Figure 19-5).

The use of any pre-manufactured materials, such as hydraulic shoring, trench jacks, or shields (among others), must follow the manufacturer's installation specifications. Any deviations must be approved in writing by the manufacturer. Shoring should be put in place from the bottom of the excavation, working up to the top. Sheeting covering the sides of the trench should be taken out from above with the excavation being back-filled as soon as the support material is removed.

Prefabricated strut and wale systems represent correct technology. Made of aluminum and steel, hydraulic shoring provides a safety advantage over timber shoring since workers do not have to enter the trench to install or remove the materials. Side sheets and bracing can be pre-spaced before being lowered into the trench by machine. Aluminum materials are light enough to be installed by one worker, pressures are regulated along the length of the trench and the sections can be adapted to different trench depths and widths. In order to maintain maximum function of this system, hydraulic shoring should be inspected at least once per shift for leaking hoses and/or cylinders, broken connections or any other cracked or defective parts.

Pneumatic shoring is similar to that of hydraulic shoring. Rather than hydraulic pressure, air pressure is used. The major disadvantage of pneumatic shoring is that an air compressor must be on-site. Screw jacks are used to stabilize adjacent structures and to secure side wall shoring. Underpinnings, the material and foundation used to support a structure, may be used to stabilize adjacent structures.

19.4.3 SHIELDS

Shielding is a structure providing both sheeting and shoring in one package. According to OSHA, shielding

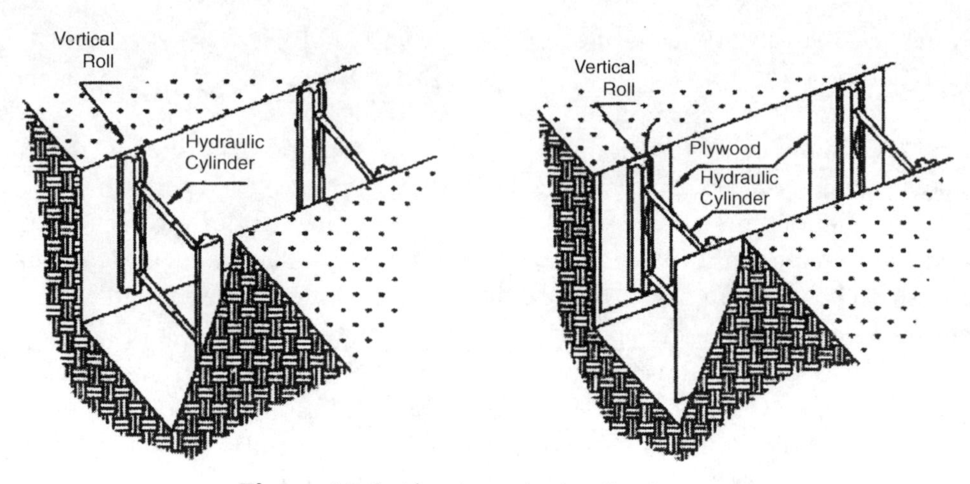

Figure 19-5 Aluminum hydraulic shoring

means a structure that can withstand the forces encountered from a cave-in and still protect workers within the structure. This differs from shoring that is used to support the excavation walls from moving. Shields can be permanent structures or can be portable and transported from site to site. Whether pre-manufactured or built on the job site, they must be in accordance with the OSHA standards. Shields that are used in trenches are called trench boxes or trench shields, but still serve the same function. The boxes are placed into and removed from the excavations by heavy equipment. Since they generally do not have to fit tightly in the excavation, the same heavy equipment might be used to backfill between the box and the trench wall to prevent lateral movement.

In all soil systems, the support or shield system must extend at least 18 inches above the top of the vertical trench side. Excavation is only permitted to a depth of two feet below the bottom of the shield, and then only if the shield is rated to withstand the forces calculated for the full depth of the trench.

Sometimes trench boxes are used in combination with sloping or benching. This may be a viable alternative when there is a difference in soil types within the excavation. Again, the top of the box must extend above the start of the slope or the lowest bench to prevent materials from falling into the trench.

When shielding, specific safety issues must be considered. First, for excavations where employees may be exposed to unstable ground, qualified personnel using practices that are compatible with set standards must design support systems. These systems must meet all professional engineering requirements to contain the walls. Next, protective systems must meet professional engineering requirements for all sides, slopes, and faces of the excavation. Never excavate below the base of a footing or a retaining wall unless appropriate precautions have been taken such as underpinnings, which have been certified by a professional engineer. Finally, if it is necessary to operate heavy equipment on a level above and adjacent to an excavation, the side of the excavation must be sheet-piled, shored, braced, or sloped to resist the additional pressure from this equipment.

19.5 Soil Classification

Classifying a soil's stability is an important part of evaluating the site. This must be done by a competent person, and must be based on the environmental conditions in the field immediately before work begins. Soil categories are determined by an analysis of the properties and performance characteristics of the deposits as well as the environmental conditions under which they are exposed. A proper soil classification includes a visual and a physical analysis. The competent person may also take a core sample which may reveal the presence of two or more soil types, all with their inherent characteristics and potential problems.

19.5.1 VISUAL TEST

The visual test is a qualitative evaluation of conditions around the site. The entire work site is inspected including the soil adjacent to the site as well as the soil being excavated. The competent person looks for indicators such as those shown in Figures 19-6 and 19-7. This may indicate a type of soil that can hold a great deal of water, so that when it dries, it shrinks and separates. It may also be an indicator of earthquake activity or tension cracks. Tension cracks usually form at a horizontal distance of 0.5 to 0.75 times the depth of the trench, measured from the top of the vertical face of the trench (19-7). Another visual record is layered soils, which may be the "normal" soil horizons in a specific section of the country. However, they may also indicate in-fill, dumping (where soils are disturbed), utility burial, or former earth usage (gravel quarries or agriculture). In such cases, it is a safe precaution to search records for former land use and determine how that use might have impacted the soil. Seepage signs or staining may indicate the presence of unknown buried substances or of surface dumping. Unknown stains should be evaluated to be certain that they will not pose a threat to the atmosphere inside the trench when excavation begins. Poor drainage, standing water, or very fast draining soil also denote a variety of conditions that would influence the type of protective system to be used.

The competent person should also look for signs of bulging or subsidence, which, in an unsupported excavation, creates unbalanced stress in the soil, causing the potential for face failure in the trench. In a trench that has already been opened and is inspected regularly, heaving or an up-thrust of soil can sometimes be seen in the bottom of the feature. This indicates that the pressure of the adjacent soil is forcing the bottom material upward. This may happen even when protective shielding is in place. Boiling of the soil takes place by the upward flow of ground water forcing soil with it into the cut. This happens when the water table is high or close to the bottom of the trench (Figures 19-6 and 19-7).

19.5.2 MANUAL TEST

The E&T standard allows for several types of physical tests for soil conditions. The characteristics of the soil that should be evaluated include its plasticity, dry strength, thumb penetration, and tests using tools such as a pocket penetrometer and a hand-operated shear vane.

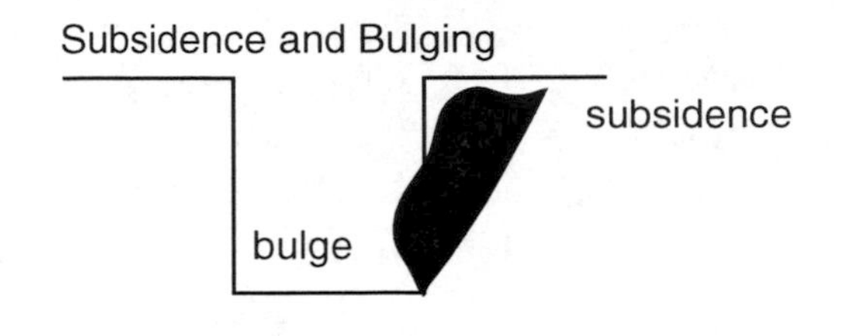

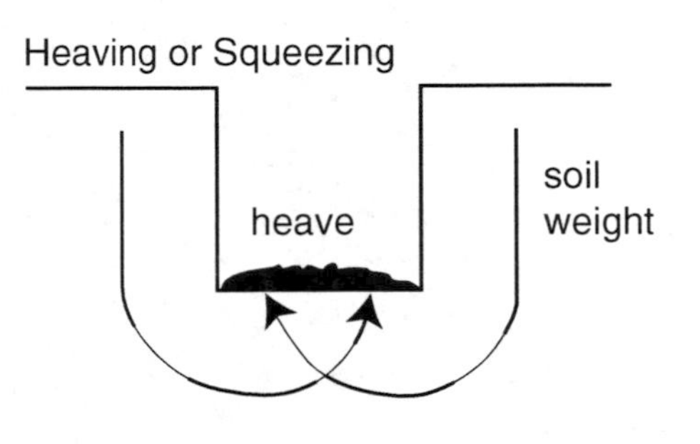

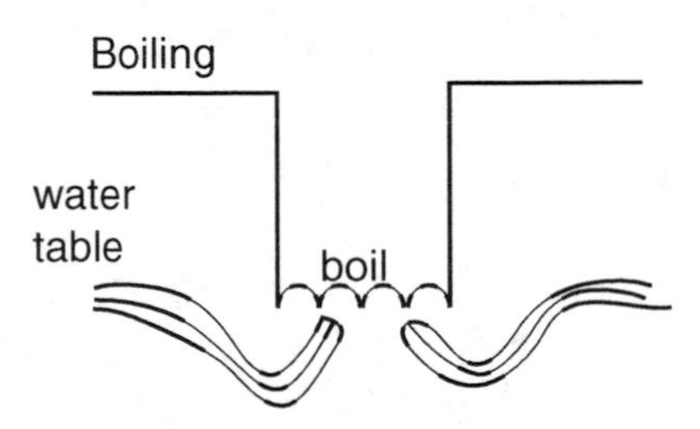

Figure 19-6 Examples of trench failure. From: OSHA Technical Manual.

19.5.2.1 Pocket Penetrometer

Penetrometers are direct-reading, spring-operated instruments used to determine the unconfined compressive strength of saturated cohesive soils. It is pushed into the soil while an indicator displays the reading. The instrument is calibrated in either tons per square foot or kilograms per square centimeter. It should be noted that these instruments have error rates of plus or minus 20 to 40 percent.

19.5.2.2 Shearvane

Also known as a Torvane, this instrument is used to determine the unconfined compressive strength of the soil. A series of vanes are pressed into undisturbed soil and a torsional knob is slowly turned until soil failure takes place. The direct instrument reading is

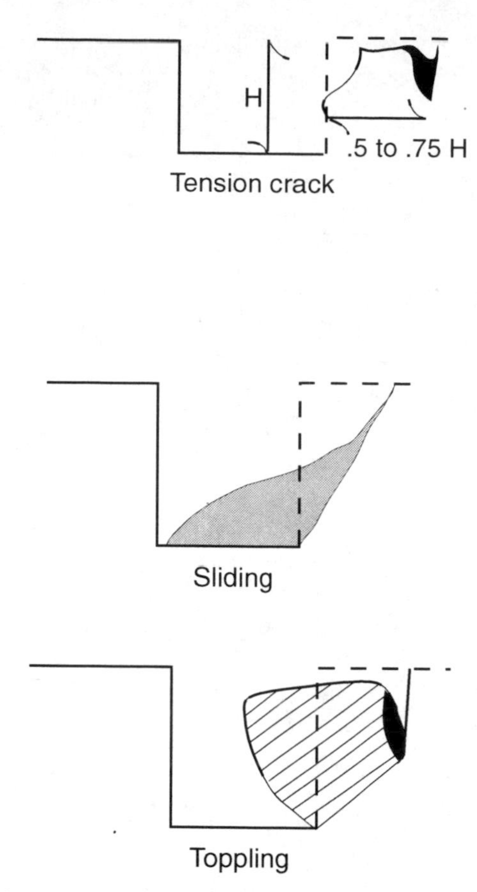

Figure 19-7 Examples of face failure. From: OSHA Technical Manual.

then multiplied by two to provide results in tons per square foot or kilograms per square centimeter.

19.5.2.3 Thumb Penetration

Thumb penetration is a qualitative procedure that involves an attempt by the evaluator to press their thumb into the soil being analyzed. If the thumb makes an indentation in the soil, but only with great difficulty, the soil is possibly Type A. If the thumb penetrates no further than the length of the nail, it is probably Type B. If the thumb penetrates the full length, it is Type C soil. The thumb test is subjective and thought to be the least accurate of the methods. However, soil scientists do find it a relatively accurate first-look evaluation technique.

19.5.2.4 Dry Strength Test

Dry soil that freely crumbles into individual grains (or with moderate pressure) is considered to be granular. Sand is such a material. Dry soil that falls into clumps and then breaks into smaller clumps (which can then be broken only with difficulty) is probably clay in combination with gravel, sand, or silt. If the soil breaks into clumps, but does not break into smaller clumps (and can only be broken with difficulty), the soil is considered to be unfissured and is probably of a high clay content.

19.5.2.5 Plasticity or Wet Thread Test

In this test, a moist sample of soil is molded into a ball. An attempt is then made to roll it into a thin thread. Once rolled to approximately ⅛ inch in diameter and two inches in length, the sample is then held by one end. If the sample does not break or tear, the soil is considered to be cohesive. This generally indicates clay content if it holds, and sand or silt content if it breaks.

19.5.3 SOIL TYPES

Classifying a soil's stability is an important part of evaluating the site. In general, soil is divided into four classes, from most stable to least stable. These classes are (1) Stable Rock, (2) Type A, (3) Type B, and (4) Type C. The categories are determined based on an analysis of the properties and performance characteristics of the deposits as well as the environmental conditions of exposure. A full classification of soils is found in 29 CFR, Subpart P, Appendix A.

19.5.3.1 Stable Rock

Stable rock is natural, solid mineral material that can be excavated with vertical sides and that will remain intact while exposed. It is usually identified by a rock name such a granite or sandstone. The presence of a solid–appearing material does not always indicate safe digging. Hidden cracks and fissures within the material may be present and cause subsequent buckling and toppling. Some rock is under just enough pressure in the earth to keep it solid. When an excavation is made, this stability is challenged, creating failure of the rock mass and thereby compromising the trenching procedure.

19.5.3.2 Type A

Type A soils are cohesive (stick together) such as clay, silty clay, or a cemented and compacted soil layer that is impenetrable to roots, called hardpan. They have

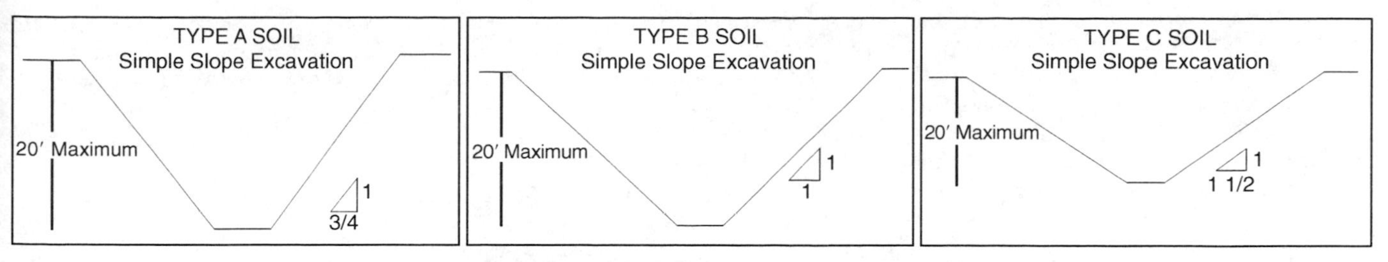

Figure 19-8 Soil classifications for trenching

an unconfined compressive strength of 1.5 tons per square foot or more, indicating some degree of stability. A soil is not classified as Type A if it is fissured (has a tendency to break along fracture planes with little resistance or exhibits open cracks in an exposed surface); is subject to vibration of any type such as from heavy traffic or pile driving; has previously been disturbed; is part of a steeply-sloped layered system where the layers dip into the excavation on a slope of 4 horizontal to 1 vertical or greater; or has seeping water. A layered system means two or more distinctly different soil or rock types arranged in various strata, one over the other (Figure 19-8).

19.5.3.3 Type B

Type B soils are granular soils such as silt, sandy loam, unstable rock or any unstable or fissured Type A soil. They have an unconfined compressive strength greater than half a ton per square foot, but less than 1.5 tons per square foot. They do not exhibit cohesive tendencies but rather crumble and break up easily when dry. Other examples are angular gravel, previously disturbed soils unless otherwise classified as Type C, soils that meet the compressive requirements of Type A but are fissured or subject to vibration, and dry and unstable rock. It also includes layered systems sloping into the trench at a slope less than 4 horizontal to 1 vertical (Figure 19-8).

19.5.3.4 Type C

Type C soils have an unconfined compressive strength of half a ton per square foot or less. They include granular soils such as gravel, sand and loamy sand, submerged soil, soil from which water is freely seeping, and submerged rock that is not stable. Any material in a sloped, layered system where the layers dip into the excavation or have a slope of 4 horizontal to 1 vertical or greater is also considered to be Type C (Figure 19-8).

19.5.3.5 Layered Geological Strata

Where soils are configured in layers or where a geological boundary exists, the soil must be defined on the basis of the soil classification for the weakest layer. Each layer may be classified individually if a more stable layer lies below a less stable layer, but protective measures must be taken based on the weakest section.

Keep in mind that outside disturbances during excavation may change even the best soil classification. Therefore, the soil should be inspected after any condition change.

19.6 Sources of Information

Brown, Larry C., K.L. Kramer, T.L. Bean, and T.J. Lawrence: "Trenching and Excavation: Safety Principles." Columbus, OH: Agricultural Engineering Department, Ohio State University Extension, August 1992.

Bureau of Labor Statistics, *Fatal Occupational Injuries By Event or Exposure and Age, 2000,* www.BLS.gov, August, 2001.

Coastal Video Communications: *Trenching and Shoring*. Coastal Video Communications Corp., 1990.

J.J. Keller and Associates, Inc.: *Keller's Construction Safety Series: Trenching and Shoring*. Neenah, WI: J.J. Keller and Associates, 1992.

Johnson, Linda F.: Know the Risks before the Digging Starts. *Occupational Health and Safety 65, Number 10*:192–193 (October 1996).

Nichols, Herbert L., D.A. Day, and D.H. Day: *Moving the Earth: The Workbook of Excavation*. Berkshire, England: McGraw-Hill Europe, 1998.

NIOSH ALERT: DHHS (NIOSH): *Preventing Deaths and Injuries From Excavation Cave-ins* by Millar, J. Donald, M.D., D.T.P.H. (Publication No. 85–110). Cincinnati, OH: U.S. Department of Health and Human Services, 1995.

Occupational Safety and Health Administration: *Excavations (revised)* by the U.S. Department of Labor (OSHA Publication 2226). Washington, DC: U.S. Government Printing Office, 1995.

Occupational Safety and Health Administration: *OSHA Fatal Facts, Numbers 9, 13, 22, 31, 41* developed by the U.S. Department of Labor. Washington, DC: U.S. Government Printing Office, 1999.

Occupational Safety and Health Administration: *National News Release: Community Involvement Makes Impact in Pennsylvania—Trenching Violations Bring Contractor $230,400 In Proposed Penalties*. Washington, DC: U.S. Government Printing Office, 1997.

"Occupational Safety and Health Act," (Public Law 91–596, 84 Stat. 1590) *U.S. Code*, Title 29, Section 1926 et seq. 1970 ed., updated June 2000.

Occupational Safety and Health Administration: *National News Release: OSHA Levies $371,400 Fine Against Danboro, PA, Contractor For Egregious Serious Violations In New Jersey and Pennsylvania*. Washington, DC: U.S. Government Printing Office, 1996.

Occupational Safety and Health Administration: *Technical Manual, Section V, Chapter 2, Excavations: Hazard Recognition In Trenching and Shoring*. Washington, DC: U.S. Government Printing Office, 1999.

Tyson, Patrick R.: *OSHA Directives, CPL 2.69–Special Emphasis Trenching and Excavation*. Washington, DC: U.S. Government Printing Office, 1985.

TWENTY

Resources

20 | CONTENTS

The importance of knowing basic chemical properties and toxicological characteristics of hazardous substances has been stressed in the previous units. This unit explains how to obtain much of this information through the use of printed, electronic, and other sources.

20.1 Published Sources of Hazardous Substances Information

The following is a review of information obtained from books and manuals dealing with the basic properties of chemicals. Approximately 150 books are currently in print that index and describe chemical properties. The books described in this section were selected because the material is presented in a useful manner for evaluating hazardous substances. Detailed reviews for many of the following texts are located on the World Wide Web (www). Some of the materials are accessible on the Internet. For example, MSDSs can be purchased in book form or CD format or obtained directly from a variety of web sites and manufacturers.

20.1.1 BUDAVARI, SUSAN (ED.), ET AL: THE MERCK INDEX, AN ENCYCLOPEDIA OF DRUGS, CHEMICALS, AND BIOLOGICALS. 12TH ED. WEST POINT, PA: MERCK & CO., INC., 1996.

This is an encyclopedia of chemicals, drugs, and biological substances. A typical entry contains the following:

- Chemical Name
- Compound Name
- History
- Chemical Structure
- Chemical and Physical Properties
- Human Toxicity (Acute and Chronic)
- CAS Number
- Company of Manufacture
- Use

The Merck Index is user-friendly in that it employs few abbreviations and some of the information is in sentence form. It is generally considered to be a good starting point when referencing special chemicals. It can be purchased in book form or on CD ROM.

20.1.2 LEWIS, RICHARD J. AND N. IRVING: SAX'S DANGEROUS PROPERTIES OF INDUSTRIAL MATERIALS. 10TH ED. 3 VOLS. NEW YORK: JOHN WILEY & SONS, 1999.

This three-volume reference set is a vast source of information on health and safety data, regulatory standards, toxicity, carcinogenicity, reproductive effects, and physical properties for over 20,000 chemical substances. It also offers new safety profiles for handling spills, accidental exposures, and fires. The Ninth Edition is also available on CD ROM and provides instant retrieval of some 22,380 chemicals according to listing name, hazard rating, chemical name, molecular formula, molecular weight, chemical and physical properties, synonyms, toxicity data with references, and safety profiles. While rather expensive for private libraries, it could form the core of a good hazardous materials reference collection.

20.1.3 MEYER, EUGENE AND B. GAMES: CHEMISTRY OF HAZARDOUS MATERIALS. INDIANAPOLIS, IN: 1997.

This text is primarily fire science but still a useful reference for individuals working with hazardous materials on a daily basis. It explains Department of Transportation (DOT) regulations and how they are related to the chemistry of the material. It details proper transport, handling, and storage of most major commodity chemicals and offers guidelines for emergency planning and response.

20.1.4 LEWIS, RICHARD J. SR.: HAZARDOUS CHEMICALS DESK REFERENCE. 4TH ED. NEW YORK: JOHN WILEY & SONS, 1996.

This edition has been updated with data available on over 6,000 industrial chemicals. It is considered the standard quick-access guide to key information on the hazardous properties of materials commonly encountered in industry, the laboratory, and the environment. Each entry provides the following information:

- Safety Profile
- CAS, NIOSH, and DOT Numbers
- OSHA Standards
- ACGIH, DFG/MAK Recommendations
- Carcinogenic Data

This reference is user-friendly in that it provides two cross-indexes to allow the reader to locate a chemical by CAS number or by synonym.

20.1.5 LENGA, ROBERT E.: THE SIGMA-ALDRICH LIBRARY OF CHEMICAL SAFETY DATA. 2ND ED. 2 VOLS. ST. LOUIS, MO: SIGMA ALDRICH CORPORATION, 1988.

This reference provides the following information fo a specific chemical:

- Appearance
- Toxicity Data
- Health Hazards
- First Aid
- Incompatibility

Although the above information can be found ir other references, the Sigma-Aldrich Library of Chemi cal Safety Data also provides information on decompo sition products (which are sometimes more lethal thar the parent materials), handling and storage, waste dis posal, procedures for spills and leaks, extinguishing media, and a diagram of the chemical structure. These books use abbreviations which must be understood before a meaning becomes clear.

20.1.6 ELANSKY, KENNETH B.: SUSPECT CHEMICALS SOURCEBOOK: A GUIDE TO INDUSTRIAL CHEMICALS COVERED UNDER MAJOR REGULATORY AND ADVISORY PROGRAMS. 7TH ED. ROYTECH PUBLISHERS, 1988.

This guide is user-friendly in that the user can reference a chemical by the chemical name, CAS number, SEQ number, or the PNN number. Once the chemical is located, the guide gives code reference numbers. These code numbers reveal extensive information on all the TSCA, CWA, SDWA, NIOSH, RCRA, CERCLA OSHA, and SARA regulations pertaining to the chemical, including agencies that are responsible for regulation of certain chemicals.

20.1.7 U.S. DEPARTMENT OF HEALTH AND HUMAN SERVICES: NIOSH POCKET GUIDE TO CHEMICAL HAZARDS. WASHINGTON, DC: U.S. GOVERNMENT PRINTING OFFICE, 1997.

This guide is user friendly in that a chemical can be found by either the chemical name, CAS number

RTECS number, or DOT identification and guide numbers. After locating the chemical, the following information is presented:

- Synonym, Trade Names
- Exposure Limits, time-weighted average (TWA)
- IDLH
- Physical Hazards
- Chemical and Physical Properties
- Incompatibilities and Reactivities
- Personal Protection
- Respirator Selection
- Health Hazards (Symptoms, First Aid, Target Organs)

Although there are abbreviations, most can be deciphered or quickly referenced at the beginning of the guide.

NIOSH can be accessed at www.cdc.gov/niosh/homepage.html. This site provides free publications available for downloading and order forms for purchasing other publications. This web site also provides an excellent starting point for links to a number of relevant sites.

20.1.8 U.S. DEPARTMENT OF TRANSPORTATION, RESEARCH, AND SPECIAL PROGRAMS ADMINISTRATION: 2000 NORTH AMERICAN EMERGENCY RESPONSE GUIDEBOOK. WASHINGTON, DC: U.S. GOVERNMENT PRINTING OFFICE, 1996.

This guide was developed as a joint effort by the transportation agencies of the United States, Canada, and Mexico in response to the North American Free Trade Act (NAFTA). The publication helps facilitate the transport of hazardous materials throughout North America and increases public safety by providing consistent emergency response procedures.

This guide primarily provides first responders (law enforcement officials and firefighters) with technical information and advice on emergencies connected with the transportation of hazardous materials. It is updated every three years to accommodate new products and technology. Response issues include quickly identifying and classifying the materials of the incident and protecting both the responders and the general public during the initial response phase of the incident. This guide book also references chemicals by name or DOT identification number. Specific information includes:

- Potential Hazards
  - —Health
  - —Fire or Explosion
- Recommended Emergency Action
- Extinguishing Media
- Spill Containment
- First Aid
- Initial Isolation and Protective Action Distances for Small or Large Spills

You can renew other publications online at www.dot.gov.

20.1.9 AMERICAN INDUSTRIAL HYGIENE ASSOCIATION: THE AIHA 2001 EMERGENCY RESPONSE PLANNING GUIDELINES AND WORKPLACE ENVIRONMENTAL EXPOSURE LEVEL GUIDES HANDBOOK. FAIRFAX, VA: AMERICAN INDUSTRIAL HYGIENE ASSOCIATION, 2001.

This pocket-sized reference guide presents a clear overview of two well-documented sets of exposure limits: the AIHA Emergency Response Planning Guideline Series (ERPGS) and Workplace Environmental Exposure Levels (WEELs). Recommended values for each series are contained in this booklet along with other important information. The format is valuable for quick reference, but the exposure limits should always be used in conjunction with the documentation provided in each full set of ERPGs and WEELs. Many other publications can be found at www.aiha.org.

20.1.10 ASH, MICHAEL AND I. ASH: THE SPECIALTY CHEMICALS SOURCEBOOK. VOLS. I AND II. ENDICOTT, NY: SYNAPSE INFORMATION, 2000. (ALSO AVAILABLE ON CD AND DISKETTE.)

This reference covers a broad spectrum of common industrial chemicals and detailed information on more than 7,400 specialty chemicals. Multiple cross-references include CAS numbers, UN/DOT reference numbers, classifications, trade names, and generic chemical synonyms. Chemicals may also be located based on their uses and industrial application areas through functional key word searches. Included is a *Manufacturer's Directory* containing locations and contact information for more than 4,000 trade name products and generic chemicals that are referenced in the book. Specific chemical monographs include:

- CAS numbers
- UN/DOT numbers
- Definition
- Toxicology
- Precautions
- Regulatory data
- Chemical synonyms
- Hazardous decomposition products
- FEMA numbers
- Classification
- Properties
- Storage
- Trade name synonyms
- Manufacturers
- Uses

20.1.11 AMERICAN CONFERENCE OF GOVERNMENTAL INDUSTRIAL HYGIENISTS: 2001 TLVs® AND BEIs®: THRESHOLD LIMIT VALUES FOR CHEMICAL SUBSTANCES AND PHYSICAL AGENTS, BIOLOGICAL EXPOSURE INDICES. CINCINNATI, OH: AMERICAN CONFERENCE OF GOVERNMENTAL INDUSTRIAL HYGIENISTS, 2001.

This reference is a small handbook that presents TWA and STEL values for more than 700 chemicals. It also gives information on physical agents such as noise levels, cold and heat stress, ionizing and nonionizing radiation, and ergonomics. There are nearly 40 listings for biological exposure indices (BEIs), which cover the biological monitoring of workers who may be exposed to a select list of chemicals. Other pertinent information is also presented and includes:

- Carcinogens
- Mixture TLVs
- Airborne particulate matter
- Chemical and physical substances and agents under study
- Heat stress
- Cold stress
- Hand-arm vibration
- Noise
- Radiation, including:
  —upper sonic and ultrasonic acoustic
  —ionizing
  —light and near-infrared
  —lasers
  —radio frequency and microwaves
  —ultraviolet

In a separate publication, the basis for each TLV is documented in detail. This is called *The Documentation of the Threshold Limit Values for Chemical and Physical Agents,* 7th. Ed, 2001. This resource provides the scientific basis and rationale for TLVs and is typically used by a professional industrial hygienist when evaluating or establishing an occupational health program.

20.1.12 AMERICAN CONFERENCE OF GOVERNMENTAL INDUSTRIAL HYGIENISTS: QUICK SELECTION GUIDE TO CHEMICAL PROTECTIVE CLOTHING BY D. FORSBERG AND S.Z. MANSDORF. 3RD ED. CINCINNATI, OH: AMERICAN CONFERENCE of GOVERNMENTAL INDUSTRIAL HYGIENISTS, 1997.

This reference provides recommendations for the selection of gloves, boots, suits, and other protective clothing based on permeation rates of 600 chemicals. Also included is a section with names, addresses and phone numbers of the suppliers and manufacturers of this clothing.

A number of other publications are available from ACGIH on the topic of PPE, including:

- *Personal Protective Equipment Pocket Guide;*
- *Personal Protective Equipment for Hazardous Materials Incidents: A Selection Guide;*
- *Guidelines for Selection of Chemical Protective Clothing, 3rd ed.; and*
- *PPE Made Easy*

For the entire selection of ACGIH publications, access their web site at www.acgih.org.

20.2 Computer Databases of Hazardous Substances Information

The following is a partial list of computer references and a summary of the information they contain. Most can be leased or purchased for company computers. Some can be accessed, reviewed, and ordered via the Internet. This data can be accessed by computers in emergency or rescue vehicles and at hazardous waste sites, schools, laboratories, etc. The availability of such data is proving invaluable at emergency response sites.

20.2.1 TOMES®—TOXICOLOGY, OCCUPATIONAL MEDICINE, AND ENVIRONMENTAL SERIES

The TOMES database was designed to help health care professionals evaluate and treat patients exposed to potentially toxic chemicals. It also provides rapid, easy access to medical and hazard information needed for managing chemicals safely in the workplace, evaluating exposures, responding quickly to emergency situations, and complying with regulations. The TOMES database contains four sections:

- MEDITEXT—Medical Management
- HAZARDTEXT—Hazard Management
- INFOTEXT—Regulations, Standards and General Information Documents
- 2000 North American Emergency Response Guidebook

20.2.2 TOMES PLUS®—TOXICOLOGY, OCCUPATIONAL MEDICINE, AND ENVIRONMENTAL SERIES PLUS

This system offers fast, easy access to medical and hazard information for the safe management of chemicals in the workplace and in the environment. Its references include government as well as proprietary databases available only from MICROMEDEX. An integrated simultaneous search of more than 15 databases with 2 million synonyms is part of the system.

The TOMES Plus database is an expansion of the TOMES system. In addition to Meditext, Hazardtext, Infotext, and the NAERG, 10 databases are included to assist the environmental and safety professional. The following is a list of the databases and the information that can be referenced within them. (Many of these are available at no cost from the National Library of Medicine and the NIOSH web pages.)

- HSDB—Hazardous Substance Data Bank from the U.S. National Library of Medicine
- CHRIS®—Chemical Hazard Response Information System from the U.S. Coast Guard
- OHM/TADS—Oil and Hazardous Materials/Technical Assistance Data System from the EPA
- IRIS—Integrated Risk Information System from the EPA
- RTECS—Registry of Toxic Effects of Chemical Substances from NIOSH
- REPRORISK—Reproductive Risk Information System (includes 4 databases)
- New Jersey Hazardous Substance Fact Sheet
- NIOSH Pocket Guide to Chemical Hazards

20.2.3 CAMEO®—COMPUTER-AIDED MANAGEMENT OF EMERGENCY OPERATIONS.

The CAMEO database program is available through the National Safety Council (NSC). Due to its mapping and graphics capabilities, this program can help in preplanning activities. An air mapping program can help predict the dispersion of a gas or vapor. From the associated web site (www.nsc.org), a number of chemical links are available, including:

- The EPA Chemical Emergency Preparedness and Prevention Office (CEPPO)
- Chem Finder
- MSDS keyword searches
- EPA Chemical Facts Sheets
- Agency of Toxic Substances and Disease Registry (ATSDR)
- Tox Frequently Asked Questions (FAQs)
- Chemical Backgrounds—chemical fact sheets developed by the Environmental Health Center of the National Safety Council
- Extension TOXicology NETwork—toxicology information including news releases, resources, mailing lists, and technical information

- Toxic Release Inventory System (TRIS)—information from the EPA about the transfer and release of toxic chemicals into the environment

Another part of the CAMEO®system is the Chemical Emergency Management Crossroads (www.nsc.org/xroads/about.htm). This web site is designed to supplement the CAMEO program with a library of links related to emergency response and planning.

20.2.4 HMIX—HAZARDOUS MATERIALS INFORMATION EXCHANGE

The Hazardous Materials Information Exchange (HMIX) is a computerized bulletin board designed especially for the distribution and exchange of hazardous materials information. The HMIX provides a centralized database for sharing information pertaining to hazardous materials emergency management, training, resources, technical assistance, and regulations. Access the HMIX at www.hmix.dis.anl.gov or through the EPA's website at www.epa.gov.

20.2.5 HASP—HEALTH AND SAFETY PLAN GUIDELINES BY THE U.S. EPA, DEPARTMENT OF COMMERCE, NATIONAL TECHNICAL INFORMATION SERVICE. (USERS GUIDE VERSION 3.0/4.0; DISK)

The HASP will assist in preparing a site-specific safety plan using an automated decision-making process which decreases the time normally required to retrieve and integrate data. HASP will recommend the level of protection required based on the tasks being performed and the chemicals that will be encountered. At critical decision-making points in HASP, the user will be able to either accept the recommendations made by HASP, or change or modify the required levels of protection. HASP includes standard protocols for "typical" waste remediation and assessment activities, and the health and safety plan format is consistent with the OSHA's 29 CFR 1910.120 requirements. Access the HASP Guidelines through the EPA's website at www.epa.gov.

20.2.6 ARCHIE—AUTOMATED RESOURCE FOR CHEMICAL HAZARD INCIDENT EVALUATION

The purpose of ARCHIE is to provide planning personnel with integrated planning methods for use in assessing the vapor dispersion, fire, and explosion impacts related to discharges of hazardous materials to the environment. This program enhances understanding of the chemical nature, sequence of events, and hazards associated with potential accidents, and also provides a basis for emergency planning. ARCHIE often requires additional data and evaluation for modification to the "typical" site model in order to account for actual site conditions and activities. The user must provide chemical data. This system does not include a chemical database.

20.2.7 FORSBERF, KRISTER AND L.H. KEITH: INSTANT GLOVES AND CPC DATABASE. 9TH ED. BLACKSBURG, VA: INSTANT REFERENCE SOURCES, INC., 1999.

Developed to provide technical help in selecting chemical protective clothing (CPC), this is a large industrial database of chemical permeation and degradation data. Over 10,500 permeation tests are reported with 350 models of CPC and 860 chemicals and mixtures. More than 3,000 chemical degradation tests plus 50,000 associated pieces of data on garment types, thickness, and CAS numbers are part of the database. Access is at www.instantref.com/cpc.htm.

20.3 Other Sources of Hazardous Substances Information

The Federal Government, private industry, communities, educational institutions and other associations and consortia provide chemical data, publications, free materials, real-time information for emergencies, and toll-free phone numbers for the emergency reporting of spills. Space precludes citing all such sources since more are continuously being added, while phone numbers and facility locations may change.

The following lists include examples of emergency agencies, informational sites, and miscellaneous references currently available that can provide useful information when working with hazardous substances. Individuals with Internet access may find a perusal of search engines to be a viable first step in information gathering. Many of the following sites can also provide valuable links to related material. Phone numbers and web site addresses, when included, were accurate as of August 2001, but may be subject to change without notice.

20.3.1 INFORMATION ON HAZARDOUS SUBSTANCES

20.3.1.1 AIHA—American Industrial Hygiene Association, 1-703-849-8888, www.aiha.org

The AIHA supports industrial hygienists and other occupational health, safety, and environmental professionals in improving the health and well-being of workers, the community, and the environment. AIHA is the largest association of occupational health professionals in the world. Within the AIHA are many technical committees that address indoor air quality, exposure and risk assessment, hazardous waste, ergonomics and cumulative trauma disorders, reproductive health hazards in the workplace, nonionizing radiation, noise hazards, construction health and safety, exposure limits, protective clothing, and respirators. The AIHA is deeply involved in government affairs at both Federal and state levels. Associated with the AIHA are over 75 local sections in many states and other countries.

20.3.1.2 ASSE—American Society of Safety Engineers, www.asse.org

The mission of this group is to advance the safety profession and enhance the knowledge and capabilities of safety engineers. To this end, the Society has a commitment to be proactive in governmental affairs relating to health and safety. It works closely with other professional societies, maintains contact with government agencies including OSHA, the Mine Safety and Health Administration, the EPA, the Consumer Product Safety Commission, the National Highway Traffic Safety Administration, and NIOSH.

20.3.1.3 TOXLINE®—Toxicological Literature On-Line, 1-301-496-1131 or 1-888-346-3656

TOXLINE® is a bibliographic database produced by the U.S. National Library of Medicine (NLM). TOXLINE® and TOXLINE 65® jointly contain more than 2.5 million bibliographic citations with abstracts and/or indexing terms as well as CAS registry numbers. Data covers the pharmacological, biochemical, physiological, and toxicological effects of drugs, pesticides, herbicides, environmental pollutants, animal venom, chemically induced diseases, occupational hazards, waste disposal, and mutagens and teratogens on humans or on control animal groups. TOXLINE® is composed of sixteen subfiles which contain publications and data from 1940 to the present.

20.3.1.4 ATSDR—Agency of Toxic Substances and Disease Registry, 1-404-639-0500, www.atsdr.cdc.gov/

The purpose of ATSDR, as part of the U.S. Department of Health and Human Services, is to perform public health assessments of waste sites, provide health consultations for specific hazardous substances, and respond to emergency releases of hazardous substances. It also performs research, conducts and develops information, disseminates such information, and provides education and training concerning hazardous substances. Its web site includes numerous links to chemical information sites as well as TOXFAQs, a series of summaries about specific hazardous substances. The HAZDAT database provides access to information on the release of hazardous substances from Superfund sites or from emergency events as well as the effects of hazardous substances on human health.

20.3.1.5 NASD—National Agriculture Safety Database, www.cdc.gov/

This database is produced by NIOSH's Education and Information Division. It is compiled of information from safety professionals and organizations in order to promote safety in agriculture.

20.3.1.6 TRI-US—Toxic Release Inventory User Support Service, 1-202-260-1545, www.epa.gov/

TRI-US originates from the EPA Office of Prevention, Pesticides, and Toxic Substances, Office of Pollution Prevention and Toxics, and the Information Management Division. This service offers specialized assistance to individuals needing Toxic Release Inventory Data collected by the EPA under EPCRA. It provides comprehensive search assistance for TRI on-line and CD-ROM databases.

20.3.1.7 IARC—International Agency for Research and Cancer—(World Health Organization) Lyon, France, 1-+33(0)4 72 73 84 85, www.iarc.fr/pageroot/top.html

The International Agency for Research on Cancer is part of the World Health Organization. The mission of this agency is to coordinate and conduct research on the causes of human cancer, the mechanisms of carcinogenesis, and the development of scientific strategies for cancer control. The agency is involved with both epidemiological and laboratory research, and disseminates scientific information through publications and meetings.

20.3.1.8 TSCA—Toxic Substance Control Act, 1-202-554-1404, www.epa.gov/

The TSCA Assistance Information Service provides informational assistance to chemical manufacturers, processors, users, storers, disposers, importers, and exporters concerning regulations under TSCA. It also operates as an information center for labor associations, government agencies, and private citizens.

20.3.1.9 The Chemical Industry Home Page, www.neis.com

This home page provides of list of chemical industry associates, management resources, chemical and process engineering reviews, environmental resources, and analytical chemical resources.

20.3.1.10 National Pesticide Telecommunications Network, 1-800-858-7378, www.epa.gov/

This resource provides assistance with spill control, handling, disposal, cleanup, and health effects. Available information includes the recognition and management of pesticide poisoning, toxicology, and environmental chemistry, as well as published information on more than 600 active ingredients incorporated into over 50,000 pesticide products registered for use in the U.S. since 1947.

20.3.1.11 Pesticide and Toxic Substance Hotline

—Hazardous Materials and Oil Spills (EPA)—1-800-424-8802; National resource center in the event of hazardous materials spills

—National Pesticide Telecommunications Hotline—1-800-858-7378; Information regarding all aspects of pesticide handling

—Poison Control Center, Washington, DC, 1-202-626-3333; Provides information on exposure to chemicals, poisons, or drugs

20.3.1.12 EPA RCRA, 1-800-424-9346, www.epa.gov/

This source provides information on Superfund and RCRA regulations, and responds to requests for certain documents in the *Federal Register*, or to requests for specific information from regulated communities and Federal, state, and local governments.

20.3.1.13 National Animal Poison Control Center of the University of Illinois, 1-888-426-4435, www.napcc.aspca.org/

This resource provides 24-hour consultation on the diagnosis and treatment of suspected animal poisonings or chemical contamination that affects animals. Licensed veterinarians and board-certified veterinary toxicologists are available to advise animal owners.

20.3.1.14 Chemical Information System, www.oxmol.co.uk/prod/cis/

This is a comprehensive collection of on-line chemical and hazardous waste site information. It provides access to more than 30 databases, TSCA, test data, and several cross-reference information systems such as CAS numbers. Downloading and printing is possible.

20.3.1.15 The United States Department of Labor, Occupational Safety and Health Administration, Safety and Health Internet Sites, www.osha.gov/safelinks.html

Provided by OSHA, this site is continuously updated with helpful and informative web sites, both in government and public/private sectors.

20.3.2 CHEMICAL REACTIVITY AND INCOMPATIBILITY

20.3.2.1 Chemical Reactivity Worksheet www.noaa.gov/

Reactivity, or the tendency of substances to undergo chemical change, is an ever-increasing problem with the proliferation of hazardous and non-hazardous substances. The Chemical Reactivity Worksheet (CRW) is a free program dealing with the reactivity of substances or mixtures of substances. It consists of an on-going database of more than 4,000 common hazardous materials, including information about the special hazard of each chemical and whether a chemical reacts with air, water, or other materials. It also presents a modeling program for the user to "mix" chemicals on-line to find out what, if any, reaction may occur and how hazardous the byproducts might be.

20.3.2.2 Chemical Reaction Hazards Forum, www.chemcall.demon.co.uk/index.html

Formed in 1995 in the United Kingdom, this forum allows interested parties to discuss chemical reactions and process hazards. Discussion includes reaction hazard assessments and the relative merits of new technologies in hazard control. Users may also exchange information and assist each other in providing quality service to industry and to the public.

20.3.3 EMERGENCY RESPONSE

20.3.3.1 OHMTADS—Oil and Hazardous Material Technical Assistance Data System, 1-410-321-8440

OHMTADS was developed to aid the EPA response division in responding to oil and hazardous materials emergencies. This source contains data comprised from published literature on over 1,400 materials that have been designated as oil or hazardous materials. It also contains information on emergency response, government regulations, and health and safety issues, among other topics. Of importance is data pertinent to clean-up efforts such as toxicity to humans, flora, fauna, reactions with other substances, necessary protective equipment for cleanup operations, transportation information, and methods of evacuation.

20.3.3.2 CHRIS—Chemical Hazard Response Information System, 1-800-424-8802 (emergency reporting), www.uscg.mil/

CHRIS provides information from the Coast Guard to assist with emergencies that occur during water transportation of hazardous chemicals. However, the information is not limited to water emergencies. The written reference includes first responder information, a chemical data base, modeling calculations for the determination of spill dispersal, and response method recommendations. The electronic CHRIS is interactive, producing model studies and response recommendations. Besides event modeling, each reference resembles an extensive MSDS and NIOSH collection of chemical data, reactivities, TLVs, etc.

20.3.3.3 Napoleon Volunteer Fire Department, Napoleon, OH, www.seidata.com

Representative of smaller fire departments and what items and procedures are available to them, this site provides examples of equipment, apparatus, turnout gear, and SOPs for emergency response.

20.3.4 TRANSPORTATION

20.3.4.1 CHEMTREC®—Chemical Transportation Center, 1-800-262-8200, www.cmahq.com

Supported by resources from the Chemical Manufacturer's Association, CHEMTREC® supplies shippers and carriers of hazardous materials with information to comply with Federal DOT regulations. These services can also be used to assist companies needing an emergency contact for in-house operations, a safety plan (HASP) and MSDS development. Information and help require membership with CHEMTREC®, which offers a 24-hour call center, chemical product safety specialists, and 2.8 million MSDSs.

20.3.4.2 NRC—National Response Center, 1-800-424-8802, www.nrc.uscg.mil/

The NRC is the only Federal point of contact for reporting oil and chemical spills. By either calling the toll-free number or by checking the web site, information can be located for reporting requirements and procedures. The Freedom of Information Act requires that all Federal agencies make data that will serve the public interest available in electronic form. The NRC has implemented an on-line query system that makes all oil and chemical spill data collected since 1990 available via the Internet. Data from the National Railroad Hotline (1-800-525-0210) is also available, as are reports taken during drills or spill exercises. Searches can be done based on spill, location, material involved, state, county, etc.

20.3.4.3 DOT Hotline Emergency 1-800-424-8802, Non-emergency 1-800-467-4922, www.dot.gov/

Information/assistance is provided for the federal regulations concerning the transportation of hazardous materials.

20.3.4.4 Dangerous Goods Information Online, www.iata.org

Compiled and updated by the International Air Transport Association, this site provides basic information on the transportation of dangerous goods by air and other means. Includes IATA's Dangerous Goods programs, regulations, FAQs, related links, training schools, and products and services.

20.3.4.5 TRIS—Transportation Research Information Services, www.nas.edu/trb/about/tris.html

TRIS is the Transportation Research Board's database of transportation research. It consists of document abstracts describing the published literature of highway, railroad, maritime, and air transport research. The online version includes abstracts of unpublished research in progress. In order to use the entire database, one must be registered. However, research reports are free.

20.3.5 MSDS INFORMATION, www.pp.okstate.edu/ehs/links/msds.html

A large number of MSDS web sites exist. Since many of these update in a timely fashion and can be copied for inclusion in safety manuals, they may make purchased, hard-copy texts obsolete. A very useful site is maintained by Oklahoma State University Environmental Health and Safety page. This site contains many useful links about chemical safety. Additionally, many chemical vendors have posted links to MSDS information.

Other starting points for searches include:

—Vermont SIRI (Safety Information Resources on the Internet), www.hazard.com/msds
—Independent Chemical Information Services, www.icislor.com/
—Agency for Toxic Substances and Disease Registry, www.epa.gov/
—Chemfinder Database searches, www.2.thomasregister.com, Chemical company phone numbers and products;
—www.chem.uky.edu From the welcome page, select Department Resources. Go to the MSDS section.

20.3.6 PERSONAL PROTECTIVE CLOTHING AND EQUIPMENT

Personal protective clothing and equipment is covered in sources like NIOSH publications and OSHA regulations. A number of manufacturers, educational institutions, private and public organizations, and individual fire protection districts have also compiled lists of equipment, special chemicals, and protective materials. Just a sample of the information available in text form and/or online follows. With any source, the user should get additional advice to verify data when the information is questionable or incomplete.

—OSHA Technical Manual: *Chemical Protective Clothing*, 1987. Free online at www.osha.gov/, then click on "technical links."
—OSHA Federal Register: Personal Protective Equipment for General Industry—59:16334-16364. 04/06/94. Available online at www.osha.gov/.
—*TDI User Guidelines for Chemical Protective Clothing Selection*, December, 1997. Available online at www.osha.gov/.
—*Recommendations for Chemical Protective Clothing: A Companion to the NIOSH Pocket Guide to Chemical Hazards*, online at www.dcd.gov/niosh/ncpcl.html Search of the NIOSH web site for "chemical protective clothing" produced 70 documents including a Chemical Protective Clothing Database;
—Forsberg, Krister, Mansdorf, S.Z., *Quick Selection Guide to Chemical Protective Clothing*, Van Nostrand Reinhold, New York, NY, 1997.
—National Farm Medicine Center, www.marshmed.org; provides a list of mail order supplies for farmers and ranchers.
—*Virginia Occupational Safety Program*, www.virginia.edu; a large university plan for

PPE and safety requirements, FAQs, hazard assessment summaries PPE for specific activities; very good for setting up such a program.

—Crossroads on Personal Protective Equipment: National Safety Council. Provides links for detailed PPE information. This web site has been set up for emergency responders and planners to help them determine the correct classification and equipment effectiveness for a specific chemical;

—Ness, Shirley A., *NIOSH Case Studies in Personal Protective Equipment*, NIOSH, Cincinnati, OH, 1996.

—Stull, Jeffery O., *PPE Made Easy: A Comprehensive Checklist Approach to Selecting and Issuing Personal Protective Equipment*, Government Institutes, Rockville, MD, 1998.

—Stull, Jeffery, A.D. Schwope (eds.), *Performance of Protective Clothing, Volume 6*, American Society for Testing and Materials, West Conshohocken, PA, 1997.

20.3.6.1 Skin Protection and Risk Analysis Research Group, www.tno.nl/instit/pml/organ/hr.htm

This group conducts research on the properties of protective clothing materials and performs risk analysis studies on exposure levels, hazard areas, and atmospheric dispersion from released toxic substances. The main focus is on chemical warfare agents with accrued knowledge adapted for civil projects. The group is working toward an accreditation for testing protective clothing.

20.3.6.2 CCOHS—Chemical Protective Clothing; Canadian Center for Occupational Health and Safety, www.ccohs.cs/oshanswers.html

This is a source for user-friendly information, CPC, PPE, gloves, etc. from the Canadian equivalent of OSHA. It provides links to other health and safety sites.

20.3.7 PLANNING AND MANAGEMENT

20.3.7.1 OESH—Office of Environment, Safety, and Health—Department of Energy, www.doe.gov

This division of the DOE has the mission of protecting the environment, workers, and the public from hazards posed by DOE facilities and operations. As such, office functions include the development and establishment of environmental, occupational safety and health, and medical policies and rules for operating DOE facilities; conducting oversight activities; and providing technical assistance to the department's programs.

20.3.7.2 EPA Small Business Hotline—1-800-368-5888, www.epa.gov/smallbusiness/help.htm or www.epa.gov

Advice and information on problems of small-quantity generators of hazardous waste can be obtained from this source. Also available are state environmental assistance programs, environmental software, training programs, chemical accident programs, pesticide programs, pollution prevention programs, and EPA Hotlines.

20.3.7.3 Environment, Safety, and Health Information Portal, www.tis.eh.doe.gov/portal/

This is an excellent source of information on environmental safety and health across a complex array of interconnected web sites. It contains feature stories of new regulations, hearings, community concerns, and management tools, as well as archives.

20.3.7.4 Environmental Chemistry.com www.environmentalchemistry.com

This site provides chemistry, environmental, and hazardous material educational resources including:

—a detailed periodic table of elements;
—articles on chemistry;
—environmental and hazardous materials issues;
—hazardous materials transportation; and
—reference materials.

These are only a few of the available sources, especially on the World Wide Web. The user should carefully consider the sources and get additional advice for verification when data sources or information are questionable or incomplete.

APPENDIX 1

Acronym List

ACGIH	American Conference of Governmental Industrial Hygienists
ACP	Access Control Point
AIDS	Acquired Immunodeficiency Syndrome
AIHA	American Industrial Hygiene Association
ALARA	As Low as Reasonably Achievable
APF	Assigned Protection Factors
APR	Air-purifying Respirator
ARAR	Applicable or Relevant and Appropriate Requirements
ARCHIE	Automated Resource for Chemical Hazard Incident Evaluation
ASR	Atmosphere Supplying Respirator
ASSE	American Society of Safety Engineers
ASTM	American Society for Testing and Materials
ATSDR	Agency of Toxic Substances and Disease Registry
BAT	Best Available Technology
BLEVE	Boiling Liquid Expanding Vapor Explosion
BLS	Bureau of Labor Statistics
BP	Boiling Point
BTT	Breakthrough Time
°C	Degree Centigrade = (°F – 32) × 5/9
Ca	Carcinogen
CAA	Clean Air Act
CAS	Chemical Abstract Services
CASRN	Chemical Abstract Services Registry Number
CERCLA	Comprehensive Environmental Response Compensation, and Liability Act
CFR	Code of Federal Regulations
CGI	Combustible Gas Indicator
CHEMTREC	Chemical Transportation Center
CHRIS	Chemical Hazard Response Information System
CIH	Certified Industrial Hygienist
CLIA	Clinical Laboratory Improvement Act
CPC	Chemical Protective Clothing
CPR	Cardiopulmonary Resuscitation
CESQG	Conditionally Exempt Small Quantity Generator
CPSA	Consumer Product Safety Act
CRC	Contamination Reduction Corridor
CRZ	Contamination Reduction Zone
CSP	Certified Safety Professional
CTD	Cumulative Trauma Disorders
CWA	Clean Water Act
DFG	German Research Society
DNA	Deoxyribonucleic Acid
DOE	Department of Energy
DOL	Department of Labor
DOT	Department of Transportation
DQO	Data Quality Objectives
E&T	Excavation and Trenching
EHS	Extremely Hazardous Substance
EKG	Electrocardiogram
EMT	Emergency Medical Technician
EPA	Environmental Protection Agency
EPCRA	Emergency Planning and Community Right-To-Know Act
ER	Emergency Response

ERPGs Emergency Response Planning Guidelines
ESA Endangered Species Act
ESCBA Escape Self-Contained Breathing Apparatus
eV Electron Volt
EZ Exclusion Zone
°F Degree Fahrenheit = (9/5 × °C) + 32
FAQ Frequently asked questions
FFA Flammable Fabrics Act
FFDCA Federal Food, Drug, and Cosmetic Act
FHA Federal Highway Administration
FHSA Federal Hazardous Substance Act
FID Flame Ionization Detector
FIFRA Federal Insecticide Fungicide, and Rodenticide Act
FRA Federal Railroad Administration
GHS Globally Harmonized System (labeling)
GI Gastrointestinal
HAP Hazardous Air Pollutant
HASP Health and Safety Plan
HASPG Health and Safety Plans Guidelines
HAVS Hand-Arm Vibration Syndrome
HAZCOM Hazardous Communications Standard
HAZMAT Hazardous Materials
HAZWOPER Hazardous Waste Operations and Emergency Response
HCS Hazard Communication Standard
HEPA High Efficiency Particulate Air
HIV Human Immunodeficiency Syndrome
HMIX Hazardous Materials Information Exchange
HMTA Hazardous Materials Transportation Act
HMTUSA Hazardous Materials Transportation Uniform Safety Act
HSDB Hazardous Substance Data Bank
E&T Excavating and Trenching
ICN International Classification Number (placards)
IDLH Immediately Dangerous to Life or Health
IP Ionization Potential
IRIS Integrated Risk Information System
LANL Los Alamos National Laboratories
LC50 Lethal Concentration (Airborne)—50%
LD50 Lethal Dose (Absorption or Ingestion)—50%
LEL Lower Explosive Limit
LEPC Local Emergency Planning Committee
LFL Lower Flammable Limit (same as LEL)
LUST Leaking Underground Storage Tank
mm millimeter = 0.039 inches
MAK Maximum Allowable Concentration (German)
MSDS Material Safety Data Sheet
MSHA Mine Safety & Health Administration
N *N*ot resistant to oil; respirator filter classification
NAERG North American Emergency Response Guidebook
NAFTA North American Free Trade Act
NCP National Contingency Plan
NEPA National Environmental Policy Act
NFPA National Fire Protection Association
NIHL Noise Induced Hearing Loss
NIOSH National Institute for Occupational Safety and Health
NOEL No Observed Effects Level
NPL National Priority List
NRC National Response Center
NRC Nuclear Regulatory Commission
OCSL Outer Continental Shelf Lands
OESH Office of Environmental Safety and Health (DOE)
OFAP Office of Federal Agency Programs
OHMTADS Oil & Hazardous Materials Technical Assistance Data System
OMB Office of Management and Budget
ORM Other Regulator Materials
ORMD Other Related Materials (placards)
OSH Occupational Safety and Health
OSHA Occupational Safety and Health Administration
OSHRC Occupational Safety and Health Review Commission
OVA Organic Vapor Analyzer
P Oil *P*roof respirator filter classification
PCB Polychlorinated Biphenols
PDS Personnel Decontamination Station
PEL Permissible Exposure Limit

PELC Permissible Exposure Limit—Ceiling
PELP Permissible Exposure Limit—Peak
PF Protection Factor
PID Photoionization Detector
PM-10 Particulate Matter—10 Microns or less in Diameter
PNN Premanufacture Notice Number
PNOC Particulate not otherwise classified
PPE Personnel Protective Equipment
PPPA Poison Prevention Packaging Act
PRCS Permit Required Confined Space
PSA Pipeline Safety Act
PTS Permanent Threshold Shift
PVC Polyvinyl Chloride
PWSA Ports and Waterways Safety Act
QA Quality Assurance
QC Quality Control
R *R*esistant to oil; respirator filter classification
RBE Relative Biological Effectiveness
RCRA Resource Conservation and Recovery Act
REL Recommended Exposure Limit
REL-C Recommended Exposure Limit—Ceiling
REM Roentgen Equivalent, Man
REPRORISK Reproductive Risk Information System
RQ Reportable Quantity
RSPA Research and Special Programs Administration
RTECS Registry of Toxic Effects of Chemical Substances
SAP Sampling and Analysis Plan
SAR Supplied Airline Respirator
SARA Superfund Amendment Reauthorization Act
SCBA Self-contained Breathing Apparatus
SDWA Safe Drinking Water Act
SEQ Sequence Number
SERC State Emergency Response Commission
SL Short Term Exposure Limit Designation Under OSHA
SOP Standard Operation Procedure
SSHO SHF Safety and Health Officer
STEL Short Term Exposure Limit
SWDA Solid Waste Disposal Act
SZ Support Zone
TCE Trichloroethylene or Trichloroethene
TERIS Teratogenic Risk Assessment of Drugs; database
TLD Thermoluminescent Dosimeter
TLV Threshold Limit Value
TLV-C Threshold Limit Value—Ceiling
TLV-STEL Threshold Limit Value—Short Term Exposure Limit
TLV-TWA Threshold Limit Value—Time-Weighted Average
TOMES Toxicology, Occupational Medicines, and Environmental Series
TOXLINE Toxic Information On-Line
TSCA Toxic Substance Control Act
TSDF Treatment, Storage, and Disposal Facility
TTS Temporary Threshold Shift
TWA Time Weighted Average
UEL Upper Exposure Limit
UFL Upper Flammable Limit (same as UEL)
UN United Nations
UV Ultraviolet
VOC Volatile Organic Compound
WEELs Workplace Environmental Exposure Level Guides
WPF Workplace Protection Factor
www World Wide Web (Internet)

APPENDIX 2

OSHA Regional Offices

Region 1:
JFK Federal Building, Room E340
Boston, Massachusetts 02203
Connecticut, Massachusetts, Maine, New Hampshire, Rhode Island, Vermont
(617) 565-9860

Region 2:
201 Varick Street, Room 670
New York, New York 10014
New Jersey, New York, Puerto Rico, Virgin Islands
(212) 337-2378

Region 3:
Gateway Building, Suite 2100
3535 Market Street
Philadelphia, Pennsylvania 19104
District of Columbia, Delaware, Maryland, Pennsylvania, Virginia, West Virginia
(215) 596-1201

Region 4:
61 Forsyth Street, SW
Atlanta, Georgia 30303
Alabama, Florida, Georgia, Kentucky, Mississippi, North Carolina, South Carolina, Tennessee
(404) 562-2300

Region 5:
230 South Dearborne Street, Room 3244
Chicago, Illinois 60604
Illinois, Indiana, Michigan, Minnesota, Ohio, Wisconsin
(312) 353-2220

Region 6:
525 Griffin Street, Room 602
Dallas, Texas 75202
Arkansas, Louisiana, New Mexico, Oklahoma, Texas
(214) 767-4731

Region 7:
City Center Square
1100 Main Street, Suite 800
Kansas City, Missouri 64105
Iowa, Kansas, Missouri, Nebraska
(816) 426-5861

Region 8:
1999 Broadway, Suite 1690
Denver, Colorado 80202-5716
Colorado, Montana, North Dakota, South Dakota, Utah, Wyoming
(303) 844-1600

Region 9:
71 Stevenson Street, Room 420
San Francisco, California 94105
Arizona, California, Guam, Hawaii, Nevada
(415) 975-4310
Technical Assistance (800) 475-4019
Publications (800) 475-4022

Region 10:
1111 Third Avenue, Suite 715
Seattle, Washington 98101-3212
Alaska, Idaho, Oregon, Washington
(206) 553-5930

Information as of March, 2000.
If these numbers are no longer active, refer to the telephone directory for listings under "Government, Department of Labor" and/or individual state listings for OSHA.

In case of emergency, call 1-800-321-OSHA